Phosphoric Acid

Purification, Uses, Technology, and Economics

Phosphoric Acid

Purification, Uses, Technology, and Economics

Rodney Gilmour

CRC Press
Taylor & Francis Group
Boca Raton London New York

CRC Press is an imprint of the
Taylor & Francis Group, an **informa** business

CRC Press
Taylor & Francis Group
6000 Broken Sound Parkway NW, Suite 300
Boca Raton, FL 33487-2742

First issued in paperback 2017

Version Date: 20140121

ISBN 13: 978-1-138-07741-6 (pbk)
ISBN 13: 978-1-4398-9510-8 (hbk)

Library of Congress Cataloging-in-Publication Data

Gilmour, Rodney.
Phosphoric acid : purification, uses, technology, and economics / author, Rodney Gilmour.
pages cm
Includes bibliographical references and index.
ISBN 978-1-4398-9510-8 (hardback)
1. Phosphoric acid. 2. Phosphoric acid--Industrial applications. I. Title.

TP217.P5G55 2013
661'.25--dc23 2013045711

Visit the Taylor & Francis Web site at
http://www.taylorandfrancis.com

and the CRC Press Web site at
http://www.crcpress.com

ad maiorem Dei gloriam

Contents

Preface

Within three months of joining Albright & Wilson (A&W) and talk of handover plans for the leadership of the corporate engineering department, I was asked to help with its dismantlement, along with corporate research, in a bid to cut company overheads. This was the beginning of a turbulent period, initially of cost saving within A&W and subsequently of rationalization of the combined assets of A&W and Rhodia. Although formal technical reports were secure in company libraries, much of the detailed technology know-how was lost as experienced employees left. Subsequently, business units were sold off and sometimes closed, with the further loss of corporate memory. In these circumstances, central libraries can become neglected or even disappear, and knowledge and understanding is lost. Other industrial phosphate companies were going through the same process in a giant chess game of global rationalization. Meanwhile, the pioneers of the technology, whose names appear on the patents, are now old or have passed away. Therefore, I have written this book partly as a review of the technology and its progress since the 1960s to signpost where it came from and where it has got to before all understanding was lost; I have felt at times like the Last Mohican.

Chapter 1 includes a brief historical review to place the current technology in context. As I began to write it, I suspected that over the centuries, a number of significant technological leaps would emerge, and this has proven to be the case. There is much learning to be gained from considering these developments, which have had a material effect on society. Some of those effects were observed in literature, and a few references are included to encourage the young engineer to develop, like Solzhenitsyn's "engineers of the twenties," an "agility and breadth of thought, the ease with which they shifted from one engineering field to another, and, for that matter, from technology to social concerns and art."

As part of my induction into the world of phosphates, I was directed to Van Wazer (Volumes I and II, 1958, 1961), Slack (1968), and Becker (1983) and for a more historical perspective, Waggaman (1960, first edition 1929). Of these, only Slack has three short essays on solvent extraction, which was in its nascent stage in 1968. This book builds on their foundation.

Much has happened in the industry since the 1960s as it rose, consolidated, and subsequently fell in line with market demands and environmental pressure (or possibly commercial pressures masquerading as environmental). The purification processes have been surrounded with great secrecy, which explains the dearth of literature, other than patents, since Slack. Now though, as all of the original patents have expired, and several purification plants have closed, it seemed a reasonable time to record the salient points of the development of the technology and its implementation and attempt to assess the relative merit of one process over another. Conscious of some sensitivity, I have attempted to strike a balance between uninformative blandness and genuine commercial interest. There are sufficient handholds for students undertaking design projects to arrive at meaningful designs.

Hopefully, sufficient theory is included, together with extensive references, to help explain various purification processes, especially solvent extraction. Judging what is sufficient is tough as one must stand beyond the point of discussion and look both backward and forward. Standing at the edge of the forest, pointing to the explorers in the bush, it is apparent that the thicket of the solvent extraction of crude phosphoric acid is dense and very difficult theoretical ground. This explains both why there are no robust solvent extraction models, other than the highly simplified, and why every new plant design and every new potential crude acid must undergo extensive testing and piloting. Carrying out and interpreting pilot trials of solvent extraction and the supporting processes, and converting this knowledge into a plant design, amounts to a high technical barrier to entry into this field.

The uses of purified phosphoric acid are wide and when taken with its commercial salts extensive and still growing. Perhaps the acid with the most unexplored application potential is not an acid but a mélange of acids, commonly referred to as "polyacid," or more formally polyphosphoric acid. The most common manufacturing processes, and product applications, for both polyacid and the principal phosphate salts are discussed.

There is general concern about the longevity of the world's natural resources, and phosphates are no exception. Any chemical process and its products and by-products should be assessed for its sustainability, and this is addressed, together with safety, health, and environmental considerations. There are both challenges and huge opportunities for the responsible stewardship of the global phosphorus cycle. The relatively small stream that is directed to industrial phosphates and on to products for human use or consumption is largely wasted (or more accurately, dispersed ultimately in the seas); the economic closure of this part of the phosphorus cycle is foreseeable within 20 years, which in turn may lead to smaller, more local production plants.

There is not an extensive body of literature on the commissioning of chemical production processes. The commissioning of a number of plants in this industry has been troublesome; therefore, it seemed appropriate to discuss this topic, which is done in the last chapter.

I hope the reader, whether a chemistry, chemical engineering, business, or industrial history student, or a new entrant to the industry, will find this book helpful and the more experienced, agreeable.

Rodney B. Gilmour

Acknowledgments

My induction into the world of industrial phosphates started when I joined Albright & Wilson at the beginning of 1998. The CEO at the time, Paul Rocheleau, suggested I get to know Alan Williams MBE, inventor on many A&W PWA patents and the inspiration behind many of the acid projects. Alan worked with Ivan Granger, head of phosphate process engineering, who had designed and commissioned the first two trains of the PWA plant at Aurora, NC; Ivan was able to identify Alan's best ideas and knock them into a process engineering scope. With chemist Dr. Mark Rose, the three druids thoroughly educated me in the very turbulent period 1998–2001. Alan in particular spent hours and hours with me, many of them in smoke-filled rooms around the world, explaining and debating; eventually, I was able to contribute to the discussion; I consider him both friend and mentor. My education continued outside A&W working with Ivan in consultancy until his sad and untimely passing early in 2007.

A chance meeting with Samuel R. Goodson on a flight from Christchurch, New Zealand, to Singapore led to many interesting projects and adventures in China. Sam was plant manager for A&W on the Aurora project and subsequently VP in Asia.

As well as many highly competent engineers and chemists in A&W, including Dr. John Godber, I have in consultancy had the privilege of meeting and working with individuals from other companies who have been granted patents in this field, including Dr. Richard Hall of FMC, Dr. David Gard of Monsanto, and Dr. Alex Maurer of Hoechst.

I feel I still know a lot less than many of these characters but thank them for their time and teaching and hope this text will at least qualify as an introduction to the subject.

Thanks are due to those who reviewed the original proposal and have subsequently commented.

I am grateful to two librarians for their help: Rupert Baker, library manager at The Royal Society, for papers relating to Sir Robert Boyle and Thomas Graham; and John Blunden-Ellis, librarian at the University of Manchester, for tracking down some elusive Japanese papers; also to Rachel Lambert-Jones, collections officer at WAVE Wolverhampton Art Gallery for the images of *The True History of the Invention of the Lucifer Match*.

I am also grateful to the following:

- Professor Alison Emslie Lewis at the University of Cape Town for comments and the graphs of sulfide and hydroxide precipitation
- Bob Tyler, now managing supervisor at Solvay, for permission to use a number of A&W images
- Hugh Podger, author, and Alan Brewin of Brewin Books Limited, publisher of *Albright & Wilson: The Last 50 Years*, for permission to use images in Chapter 2 and for producing such a useful and interesting book

- Ivan Batka, CEO of Fosfa akciová společnost, for permission to use the image of the thermal acid plant at Břeclav
- Ray McKeithan, manager, Public and Government Affairs at PotashCorp—PCS Phosphate Aurora, for the images of the Aurora plant and permission to use them
- The staff at Taylor & Francis Group: acquiring editor Allison Shatkin, Jill Jurgensen, Jennifer Stair, Kari Budyk, Arlene Kopeloff, and others behind the scenes, as well as Deepa Kalaichelvan and the team at SPi Global

Thanks are due to my children, Emily, David, and Michael: whether assisting me in kitchen chemistry making phosphoric acid crystals, calling the pitch of the food blender to estimate the rotational speed of the blades, helping with research and ideas, commenting on drafts, or just letting me get on with writing in my study.

Finally, I must thank my wife, Elizabeth, for her support, encouragement, patience, and forbearance (and the idea for the food blender).

Author

Rodney B. Gilmour is a chemical engineering and project management consultant with over 30 years experience in the industry. He has a special interest in the purification of phosphoric acid and the production and use of its phosphate derivatives. He has undertaken development work, carried out project assessments and designs, commissioned plants, and served as an expert witness in this technical field.

Rodney began his career with ICI Organics in the north of England after graduation in 1983 and worked in maintenance, commissioning, and project engineering roles. In 1989, he moved to ICI research and development to manage laboratory and pilot plant projects for the new chlorine-free refrigerants. From 1996, he was plant engineer and development manager of a vinyl chloride plant. In 1998, he was appointed senior project manager at Albright & Wilson, subsequently acquired by Rhodia, and worked in the industrial phosphates field. In 2003, he left Rhodia and became a director of Process Engineering Design on Line Limited, providing consultancy in chemical engineering and project management.

Rodney has an MA in engineering science from the University of Oxford and an MSc in chemical engineering and design from the University of Manchester.

Terminology and Units

There are many terms and abbreviations in the phosphates industry, and they are often used loosely. Those included in the following are used throughout the book; those more specific to a particular chapter are defined in that chapter.

Unless otherwise stated, SI units are used throughout. Tons denote metric tons.

P_2O_5 — Strictly, P_2O_5 represents phosphorus pentoxide, the product of burning phosphorus in dry air, and an item of commerce used for example as a desiccant or catalyst. Most commonly, P_2O_5 is the unit of currency in the phosphate industry. There are many products containing differing proportions of phosphate, and it is convenient, whether carrying out mass balances or evaluating profitability, to express different products in terms of their P_2O_5 content. The practice started with the fertilizer industry, which expressed the nitrogen, phosphorus, and potassium content of individual products (e.g., NPK fertilizers) in terms of oxides. Phosphate rock compositions are also expressed as oxides. Conversion between P, H_3PO_4, and P_2O_5 is easily deduced from the following stoichiometric equations:

$$P_2O_5 + 3H_2O \rightarrow 2H_3PO_4$$

$$P_4 + 5O_2 \rightarrow 2P_2O_5$$

Thus the standard purified phosphoric acid containing 85% H_3PO_4 is equivalent to 61.5% P_2O_5 or 26.9% P.

BPL — BPL or bone phosphate of lime is an antiquated expression for the tricalcium phosphate content of a phosphate rock or product. It is now only used to describe a phosphate rock, for example, 70–72 BPL Khourigba, which means a phosphate rock, from the Khourigba mine in Morocco that has been beneficiated to the extent that its bone phosphate of lime composition is 70%–72% $Ca_3(PO_4)_2$ by weight. BPL is converted to P_2O_5 by multiplying by 0.458.

WPA — Wet phosphoric acid, also referred to as WPPA (wet process phosphoric acid), green or black acid (because of its color due to impurities), or merchant grade acid (MGA). Loosely, all these terms cover a phosphoric acid made by acidulating phosphate rock with sulfuric or hydrochloric acid. More precisely, different names apply to different purities or concentrations of this acid.

PWA — Purified wet phosphoric acid, also called PPA (purified phosphoric acid), refers to a WPA that has undergone a series of purification steps including solvent extraction bringing it to a similar quality as thermal acid.

Thermal acid — Phosphoric acid made by burning phosphorus in air and condensing the phosphorus pentoxide vapor in water. Prior to solvent extraction, only thermal acid was sufficiently pure to be used in food grade applications.

1 An Introduction to the Industrial Phosphates Industry

1.1 HISTORY AND BACKGROUND

The development of the purification of phosphoric acid is intertwined with the history and development of both phosphorus and phosphatic fertilizers. The end of the twentieth century saw the consolidation of this industry with fewer corporations and larger plants taking their raw materials from fewer sources. It is quite possible that during the next 50 years that situation will reverse with small local plants utilizing locally recycled sources. Consequently, it may be that some of the lessons learned in the development of the industry as it grew globally could be useful should the industry move locally.

In his essay "Life's Bottleneck" [1], Isaac Asimov wrote "We may be able to substitute nuclear power for coal power, and plastics for wood, and yeast for meat, and friendliness for isolation—but for phosphorus there is neither substitution nor replacement." His thesis was that because phosphorus is a critical component of living matter and that this element was finding its way to the ocean via erosion, fertilizer runoff from the fields, and sewage, life would indeed come to a bottleneck as the resource ran out. In 2008, the Global Phosphorus Research Initiative was founded and the phrase *Peak Phosphorus* promulgated [2]. This topic has sparked intense debate and is discussed in Chapter 7.

Asimov's thesis as put forward in "Life's Bottleneck" is suspect; nevertheless, his quotation is spot on. Phosphorus (from the Greek Φωσφόρος phōsphoros, meaning *light bearing*) is a component of deoxyribonucleic acid (DNA), ribonucleic acid (RNA), and adenosine triphosphate (ATP). ATP is used to transport cellular energy and like DNA and RNA is fundamental to all living matter. Calcium phosphates are a major constituent of bones and teeth. An adequate supply of phosphorus is therefore essential for the healthy growth and maintenance of plant and animal life.

Phosphorus burns in the air and exists in nature as phosphates. It undergoes a natural, biogeochemical cycle over millions of years. Starting with plants, these absorb phosphates from water and soil; animals consume the plants and return some phosphates to the soil as waste. Both plants and animals die returning more phosphates to the soil. As well as being taken up by plants, phosphates are moved by water into streams and rivers and so into lakes and seas. Here, they settle to the bottom and in time become sedimentary rock. Geological movement may expose some phosphate-bearing rock to erosion. The natural processes of rock formation lead inevitably to quite significant differences in its chemical composition; in turn this does lead to important processing approaches when making phosphorus or phosphoric acid; it also has consequences for the purification of phosphoric acid. In the last 150 years, phosphate rock has been mined, processed to *available* phosphate in fertilizers, and put on the land to supplement natural phosphate. Figure 7.1 in Chapter 7, shows an indicative extent of the phosphorus cycle.

The ancients recognized the benefits of fertilizing the land although there is no record that they knew what the mechanisms were or any idea that phosphates played a key role. In the Bible, in the book of Isaiah, which was written about 700 BC, reference is made to straw being trampled into a dung heap. The book of Luke, written about 60 AD, quotes a man saying he will dig in manure around an unproductive fig tree. In his "Natural History," Pliny the Elder [3], writing in the first century AD, refers extensively to different manures, for different crops, including pigeon droppings and guano in general as well as the use of lime ash.

Arguably, the discovery and isolation of phosphorus, by Hennig Brandt in Hamburg, Germany, in 1669 marked the end of alchemy and the beginning of the science of chemistry. It also marked the beginning of the modern history of phosphorus [4,5]. Hennig Brandt trained as a glass maker, was a soldier, and his first wife came with a healthy dowry that allowed him to practice as an alchemist. He was in search of the philosopher's stone, making gold from base metals, and thought that its secret lay in the preparation of urine. Brandt's procedure was the high-temperature heating of vast quantities of evaporated, aged urine. His procedure was inefficient, and he threw away a high proportion of his product during his process. Nevertheless, he was left with a white, waxy substance that glowed in the dark. News of his discovery of this material, which he named *kalte feuer*, cold fire, spread throughout Europe (the absence of a Latin name perhaps supporting the later assertion that he was "an uncouth physician who knew not a word of Latin" [6]). Brandt is depicted in the evocative painting "The Alchymist in Search of the Philosopher's Stone, Discovers Phosphorus," painted by Joseph Wright A.R.A. of Derby (1734–1797), which is shown in Figure 1.1. The painting is part of a collection that also includes Wright's "A Philosopher Lecturing on the Orrery" and resides in the Derby Museum and Art Gallery.

Johann Kunckel (1630–1702) was the son of an alchemist in the court of the Duke of Holstein. He became a chemist, described the properties of phosphorus, and spent his last years in the service of King Charles XI of Sweden, who conferred on him the titles of Baron von Löwenstern and Counselor of Metals. Kunckel visited Brandt and wrote immediately to his friend Johann Daniel Krafft (1624–1697). Krafft, a commercial agent from Dresden, Saxony, went immediately to see Brandt and bought his secret recipe for 200 thalers (about $6000 today).

Krafft exhibited *das kalte feuer* at various European courts and was invited, for a fee of 1000 thalers, to show the phosphorus at the English court of King Charles II in 1677. When Krafft arrived in London, he was contacted by Robert Boyle (1627–1691) and asked to give a demonstration to the Royal Society at Ranelagh House in London.

The first demonstration by Krafft during September 1677 [7] of *gummous and liquid noctilucas* was not entirely successful so he returned a week later with a fresh piece of phosphorus the size of a pinhead. After the demonstration, Boyle elicited that the critical ingredient was *from man* and guessed this was either urine or feces. Boyle went on to develop his own process, with the help of his assistants, which he described in papers he lodged with the Royal Society in 1680, which were

FIGURE 1.1 "The Alchymist in Search of the Philosopher's Stone, Discovers Phosphorus," painted by Joseph Wright A.R.A. of Derby (1734–1797) (author's photograph).

sealed until his death [8]. In 1678, he had employed an alchemist Johann Becher and the 17-year-old Ambrose Godfrey Hanckwitz (1660–1741). Hanckwitz (born Ambrosius Gottfried Hanckwitz) was from Nienburg, Saale, Germany. They were set to work, boiling up urine and processing feces, and developed a process with a much better yield than Brandt and much higher purity. With the assistance of Becher and Hanckwitz, Boyle carried out extensive scientific studies into phosphorus, which he wrote down and submitted to the Royal Society [9]. Boyle's papers are remarkably clear and similar in style to modern scientific papers. The publication of these papers laid a firm foundation for others to build on and gives weight to the claim that Sir Robert Boyle is "the father of modern chemistry and the brother of the Earl of Cork."

Hanckwitz went on to produce phosphorus of such purity that it was sold throughout Europe. His business allowed him to purchase a laboratory in Southampton Street, London. In the early 1700s, the sales price was £3/oz equivalent today to about $1500, in other words $53 m/ton; he made, in his spare time, around 800 ounces per year.

The phosphorus process was based on urine until 1769, although Andreas Sigismund Marggraf (1709–1782) of Berlin improved the process through the use of red lead (Pb_3O_4) and charcoal.

The next big step came from Sweden where Carl Wilhelm Scheele (1742–1786) and Johan Gottlieb Gahn (1745–1818) corresponded with one another on the nature of bone. They tried dissolving bone ash in sulfuric acid so making phosphoric acid. The phosphoric acid was heated with charcoal releasing phosphorus. Here of course we see the beginning of the modern process for the production of phosphoric acid.

In the late eighteenth century, France became the center of phosphorus manufacture. Bernard Pelletier (1761–1797—"a chemist of considerable eminence" [10] who died of inhalation of chlorine gas) started to manufacture phosphorus on a large scale using the new process proposed by Gahn and Scheele. Total phosphorus production at this time was about 3200 ounces per year and reflected a small market for lights, theatrical uses, and flame proofing. Thus, the known applications at the time would not support a big enough market to support a phosphorus-based chemical industry.

The first matches are attributed to Chancel of Paris in 1805 [11], although with no phosphorus content. Dérosne and others prepared matches with phosphorus from about 1816, but they were both explosive and dangerous. Matches (known as *Lucifers* [12]) were invented in 1831. Their predecessors were known as Prometheans and required the user to dip a chemically prepared match into a bottle of fluid [13]. Safety matches prepared by Böttger in 1848 and patented by May (Bryant and May) in 1865 used red phosphorus. Before matches, starting fire for heating and cooking was quite a performance: a spark was struck with a flint, which ignited tinder (small pieces of dry cloth kept in a *tinderbox*); this made embers that were placed with kindling (small twigs and pieces of wood), blown on to create a flame that set fire to larger pieces of wood and coal, and so gradually a fire was made. That does not sound too bad in a cold but dry country in southern Europe but much more challenging in a cold, damp climate like England. As well as fires, candles and lamps were needed for light. In Chapter 2 of *A Tale of Two Cities*, Charles Dickens described how the guard of a horse-drawn coach would relight a coach lamp if it blew out: "he had only to shut himself up inside [the coach], keep the flint and steel sparks well off the straw, and get a light with tolerable safety and ease (if he were lucky) in 5 minutes." Clearly, the humble match was one of history's big products and it transformed the phosphorus industry. Figure 1.2 shows the cover and back of an account of the *True History of the Invention of the Lucifer Match* by John Walker. Now there was a substantial market for phosphorus and so phosphorus production moved up to industrial scale. M.M. Coignet, of Lyon, France, improved the Pelletier process of making bone ash, acidifying it with sulfuric acid to phosphoric acid and then converting this to phosphorus, as shown in the following equations:

$$Ca_3(PO_4)_2 + 3H_2SO_4 \rightarrow 3CaSO_4 + 2H_3PO_4 \tag{1.1}$$

$$H_3PO_4 \xrightarrow{\Delta} HPO_3 + H_2O \tag{1.2}$$

$$4HPO_3 + 12C \xrightarrow{\Delta} P_4 + 2H_2 + 12CO \tag{1.3}$$

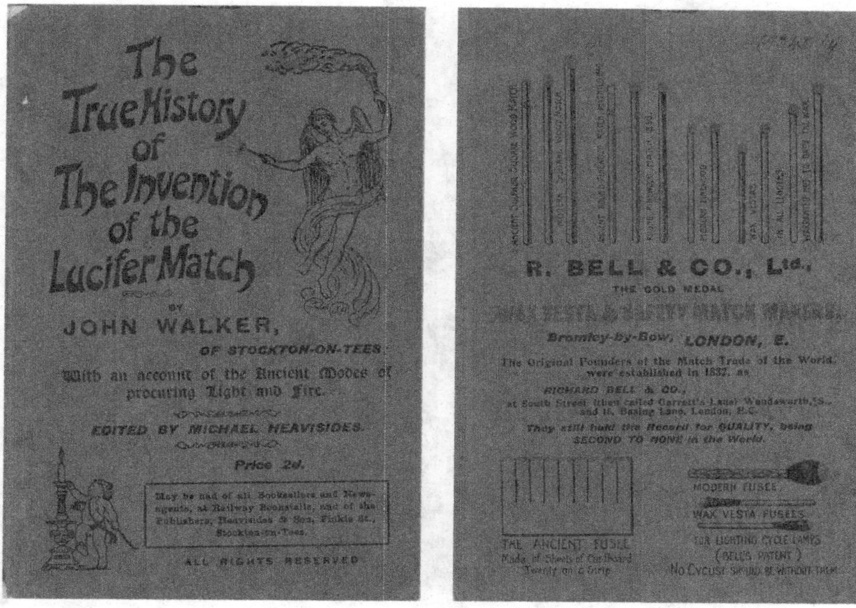

FIGURE 1.2 Front and back cover of *The True History of the Invention of the Lucifer Match* by John Walker, editor Michael Heavisides, publisher Heavisides & Son, Stockton-on-Tees, U.K., 1909. (Courtesy of WAVE Wolverhampton Art Gallery.)

All over France, kilns were built for the calcination of bones, an example of which is shown in Figure 1.3. The output boomed, in the 1840s export from France to England, was 4500 kg and the price had fallen to 21 shillings/lb about 2% of the price in the 1700s.

After serving an apprenticeship, Arthur Albright (1811–1900) moved to John & Edmund Sturge Ltd., a firm of manufacturing chemists in Birmingham, England [14]. He was taken into partnership in 1840 and in 1844 persuaded Sturge to start making phosphorus from bone ash from South America. Albright took samples to the Great Exhibition of London in 1851 and Paris in 1855. At that time the major producers were Coignet of Lyon, France; Zoeppritz of Freudenstadt, Germany; Riemann of Nuremberg, Germany; and smaller producers in Russia, the United States, and elsewhere. The phosphorus price had fallen to 3 shillings/lb (equivalent to \$300 k/ton today), but Albright and Sturge could still make a profit.

Albright left Sturge and moved to Oldbury, Birmingham; phosphorus derivative products are still processed there today. In 1855, he formed Albright & Wilson with John Wilson (1834–1907); the company seal is shown in Figure 1.4, the light-bearing cherub; phosphorus production was 26 tons per year. Still using what was basically the Pelletier process, Albright & Wilson produced 1 ton of phosphorus from 10 tons of bone ash—an impressive 80% yield. Albright & Wilson developed a purification process using potassium dichromate. In 1863, there were 27 furnaces and 648 retorts at the Oldbury works. Figure 1.5 shows the cross section of a single and double coal-fired phosphorus furnace. Figure 1.6 shows the phosphorus condensing arrangements

FIGURE 1.3 A French bone furnace (Four pour la calcination des os). (From Furne, L.F., *Les merveilles de l'industrie, ou Description des principales industries modernes*, vol. 4, Jouvet et Cie., Paris, France, 1860.) (Photograph by El Bibliomata, Creative Commons license.)

FIGURE 1.4 Albright & Wilson cherub company seal. (From Threlfall, R.E., *The Story of 100 Years of Phosphorus Making, 1851–1951*. Albright & Wilson, Oldbury, England, 1951. With permission of Solvay.)

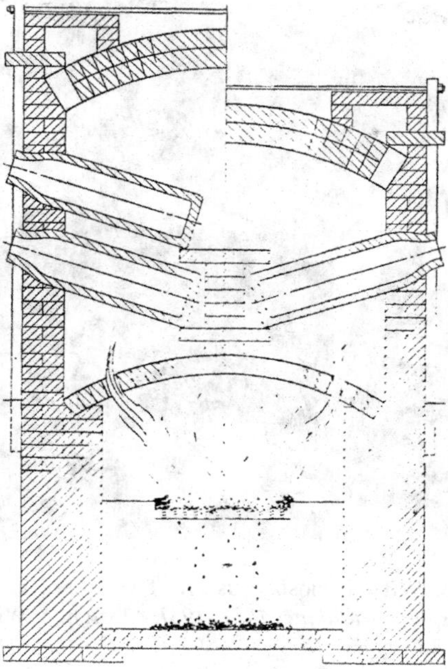

FIGURE 1.5 Albright & Wilson phosphorus retort. (From Threlfall, R.E., *The Story of 100 Years of Phosphorus Making, 1851–1951*. Albright & Wilson, Oldbury, England, 1951. With permission of Solvay.)

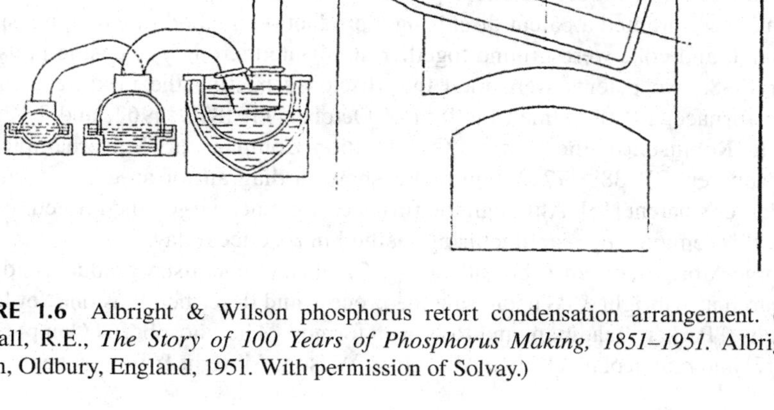

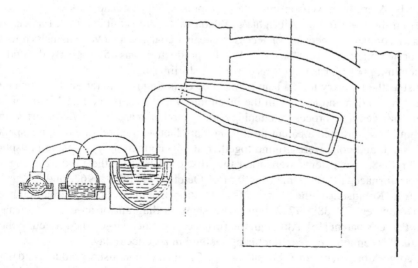

FIGURE 1.6 Albright & Wilson phosphorus retort condensation arrangement. (From Threlfall, R.E., *The Story of 100 Years of Phosphorus Making, 1851–1951*. Albright & Wilson, Oldbury, England, 1951. With permission of Solvay.)

FIGURE 1.7 Albright & Wilson phosphorus retort workers. (From Threlfall, R.E., *The Story of 100 Years of Phosphorus Making, 1851–1951*. Albright & Wilson, Oldbury, England, 1951. With permission of Solvay.)

and Figure 1.7 shows a group of workers with phosphorus retorts. The retorts were made from clay and the phosphorus was formed into 15 kg cheeses in iron receivers.

Bone became increasingly hard to source ("it is stated that the use of bones was so great in England during the eighties that the battlefields on the continent of Europe were turned to supply Great Britain's demand for phosphates" [15]). From 1870 onwards, Albright & Wilson imported sombrerite, a phosphate rock of guano origin from the West Indies. Albright & Wilson was more profitable than French and German competitors because of access to cheaper coal and a move to using synthesis gas for heating. In 1880, Albright & Wilson production was 450 tons. By the end of the century, phosphorus price was down to 1 shilling/lb.

Just as the industry took a leap forward by changing raw materials from urine to bone ash and demand leapt with the invention of matches, in the last decade of the nineteenth century, process technology took another leap. In 1867, Aubertin and Boblique were granted a patent disclosing a production method in which phosphate rock, sand, and coke were ground together and heated strongly releasing phosphorus. In 1888, two patents were filed that disclosed heating the mixture using an electric furnace: J.B. Readman on 18th of October, GB1888/14962, and T. Parker and A.E. Robinson of the Electric Construction Company of Wolverhampton on 5th December, GB1888/17719. Figure 1.8 shows a diagram of an electric furnace from Parker's patent [16]. Although the furnaces are much larger and refined in their design, this remains the manufacturing method in practice today.

George Albright, one of Albright's sons, a Cambridge chemistry graduate and head of research at Albright & Wilson, saw the process and described it as one "of brutal simplicity." Parker, Readman, and Robinson formed "The Phosphorus Company" in 1890 [17] and built a plant close to Albright & Wilson. Albright & Wilson neutralized

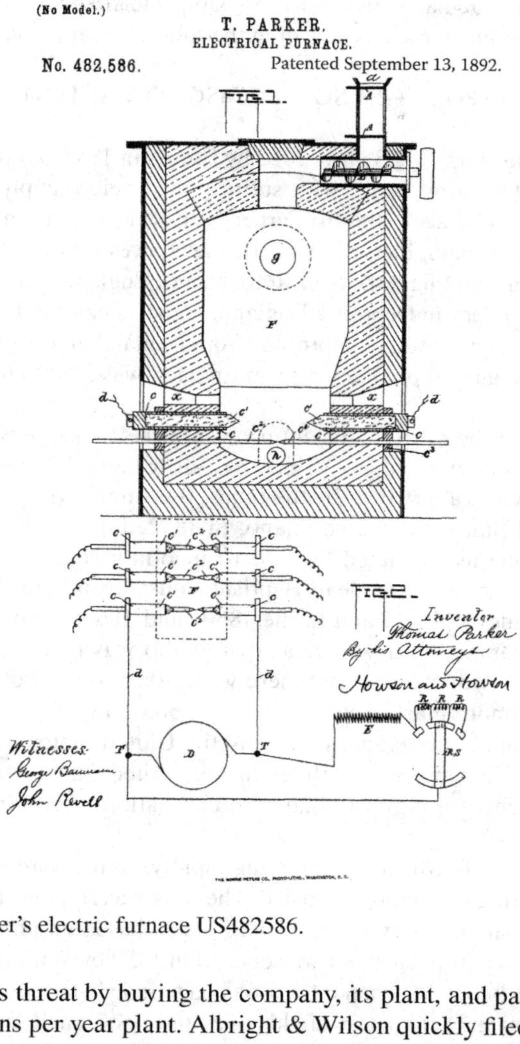

FIGURE 1.8 Parker's electric furnace US482586.

this grave business threat by buying the company, its plant, and patents and in 1893 built a new 200 tons per year plant. Albright & Wilson quickly filed patents in other countries and ventured into the North American market. A new subsidiary, Oldbury Electro-Chemical Company (OECCo), was established in 1896 with a large new phosphorus plant utilizing cheap hydroelectric power from Niagara Falls (production of 1 ton of phosphorus requires between 12 and 14 MWh electricity). In 1902, Albright & Wilson took over the Electric Reduction Company (ERCo), established on the Lievre River, Buckingham, Quebec, also for access to cheap hydroelectric power. By 1914, Albright & Wilson was at its zenith and the world's biggest phosphorus producer. New plants were built so that by 1918 the total capacity was 2700 tons per year; however, US companies had started to emerge to preeminence.

The emergence of the phosphorus and phosphate industry in the United States is linked less to phosphorus production and more to superphosphate, the fertilizer industry, and bakery leavening agents.

Superphosphate is the name given to a very simple fertilizer made by the action of sulfuric acid on phosphate rock as shown in the following equation:

$$Ca_3(PO_4)_2 + 2H_2SO_4 \rightarrow 2CaSO_4 + Ca(H_2PO_4)_2 \tag{1.4}$$

Justus von Liebig (1803–1873), preceded by Escher in 1835 and Ridgeway in 1839 [18], proposed that treating bones with sulfuric acid before applying them to soil increased their efficacy; according to Farber [19], Ridgeway used the word superphosphate in his 1839 pamphlet. Sir John Bennet Lawes (1814–1900) carried out a series of experiments and field trials in Rothamsted, England, submitted a patent in 1842, and set up a factory in Deptford, England, in the same year. Lawes started with bone charcoal but soon turned to coprolites from England and phosphate rock from Spain. By 1854, the annual production from England was 30,000 tons (compared to 26 tons of phosphorus).

The latter half of the nineteenth century would see a dash to secure guano as a phosphate source. Garcilaso de la Vega (1540–1616) noted in his history of Peru that guano was used as a fertilizer by the Incas. Baltimore, Maryland, was the first city in the United States to receive guano, in 1832. Junius Brutus Booth, father of the man who later assassinated President Lincoln, mixed the guano with bone dust he ground up on his farm. Jean Baptiste Boussingault (1802–1887) traveled to Peru to see guano deposits, and in the 1850s and 1860s, imports into England were 50,000 tons annually. It was clear that guano was better for the soil when treated by sulfuric acid and also that there were other sources of phosphate rock. By 1865, the Baltimore market was consuming 60,000 tons of guano at an average price of $50/ton. Total guano imports to the United States were 112,000 tons in 1861 [20]. The demand for fertilizer in the United States was driven by the depletion of nutrients through the intensive cultivation of cotton and tobacco in the Southern States.

As the demand for fertilizer was growing rapidly, so the searches for sources of guano were becoming ever more extensive. The time was right for the establishment of the superphosphate industry in the United States based on phosphate rock. The first American superphosphate plant was opened in 1850 by William T. Davison and T.S. Chapell of Baltimore, Maryland; Davison had founded Davison, Kettlewell & Co. in 1832 as "grinders and acidulators of old bones and oyster shells"; the company was the first in the United States to use a sulfuric acid chamber. Other early producers were based in Philadelphia. Initially, superphosphate was not as attractive as guano, because its manufacturing process also required sulfuric acid and at that stage its process technology was relatively undeveloped.

In the latter half of the nineteenth century, geologists discovered phosphate rock deposits in South Carolina (in 1867) and Florida (in 1888) [21]. In 1896, Judge S.O. Weatherly discovered phosphate rock at Mt. Pleasant, Tennessee: by the early 1900s, 15 phosphate fertilizer companies were operating in this area, including International Minerals and Chemical Corporation (which became IMC and most recently Mosaic, a global fertilizer player).

Initially, rock was recovered by hand, which was hard work; however, soon, mechanization was introduced, and together with sulfuric acid capacity, the phosphate fertilizer industry expanded rapidly in the United States. By 1900, total US fertilizer sales were 3.7 m tons, far exceeding the rest of the world.

By the end of the nineteenth century phosphatic fertilizers and phosphorus, both made from mined phosphate rock, were established. Two new product streams were beginning to develop at this stage; both remain important today and are still developing; both are based on pure phosphoric acid, which until the 1970s was only available on an industrial scale through the oxidation of phosphorus and the hydration of the resultant P_2O_5. The first was phosphoric acid as a metal treatment, and the second was food phosphates.

While on duty in the Indian subcontinent, William A. Ross, a captain in the Royal Regiment of Artillery, discovered that phosphoric acid applied to a clean surface of iron produced a coating that would "preserve the iron, steel, or metal so treated from the injurious action (in the shape of rust or other oxide) of water or damp air or perspiration" [15]. In 1869, Ross was granted a patent GB1869/03119. One application was the treatment of corset reinforcements. The application technology really started to move forward with the work of Coslett [22] and later Allen [23] (who suggested the *Parkerizing* process that is still in use today, particularly for hand guns) and Richards [24].

During the mid-nineteenth century, baking powders were developed in three countries: by Alfred Bird in England (best known for Bird's Custard powder, developed with cornflower because of his wife's allergies); in Germany by August Oetker (1862–1918), who sold his baking powders under the trade name *Backin* and formed the Oetker-Gruppe Company, which is now a major global brand still producing baking powders through its subsidiary Budenheim; and in the United States by Eben Norton Horsford (1818–1893). Horsford studied under Liebig in Germany and returned to develop baking powders; he formed the Rumford Chemical Works to manufacture his products. Baking powders provide carbon dioxide to aerate the dough making it rise both on mixing and on baking; this function has been carried out for thousands of years by yeast. In the early part of the nineteenth century, baking powders were based on sodium bicarbonate and potassium bitartrate (*cream of tartar*), a by-product of brewing or sour milk (containing lactic acid the active ingredient). Horsford reformulated the baking powders using monocalcium phosphate, producing a much more stable blend. He filed many patents, the first in 1856, US 14722, and continued development up to his last in 1880. Horsford's phosphate source was charred bones; the patent describes reacting the bones with sulfuric acid, filtering the calcium sulfate off, and concentrating the phosphoric acid in iron kettles. Bone ash was added to the acid making monocalcium phosphate, which was mixed with starch or flour, milled, and left to dry for several weeks.

In 1902, August Koch (1871–1960) founded Victor Chemicals in Chicago Heights, Illinois. He started producing monocalcium phosphate and selling baking powders; supposedly, Victor Chemicals relied on bones from the Chicago meat markets. As the company grew, it moved into sulfuric acid and ammonium phosphate production. Koch later founded Diversey in 1923. In the 1930s, like several others, Victor established phosphorus manufacture in Nashville, Tennessee, taking advantage of the local phosphate resource and cheap power. Victor Chemicals was acquired by Stauffer in 1960; Stauffer in turn was acquired by Imperial Chemical Industries (ICI), and the phosphate arm was sold on to Rhône–Poulenc Rorer. Rhône–Poulenc was the French chemical combine that brought together many French chemical companies, including, for example, Coignet, which had been part of the phosphorus story of the early nineteenth century. The phosphate arm of Rhône–Poulenc was spun off with other nonpharmaceutical divisions as Rhodia in the late 1990s. In turn, Rhodia bought Albright & Wilson in 2001. The North American phosphate operations were acquired by Bain Capital and established as Innophos in 2005. Today, Innophos still produces calcium phosphates in Chicago Heights and Nashville for the baking, dental, and pharmaceutical markets. Monocalcium phosphate is still produced and sold, as well as more sophisticated phosphate products for the food, beverage, and pharmaceutical markets.

In Germany in 1908, Ludwig Utz and Jean Hensel founded Budenheim in the village of that name, to make cream of tartar and, later, coffee granules. During the First World War, with raw materials in short supply, Budenheim commenced production of both calcium and sodium pyrophosphates. In 1923, Oetker became the majority shareholder, and Budenheim the principal supplier of phosphates to Oetker; this is still the case today.

Food blends are sold to the bakery industry and are formulated for different products such as biscuits, doughnuts, pizza bases, and cakes. The main components are a source of carbon dioxide, usually sodium bicarbonate, and an acid. The usual acid component is disodium dihydrogen pyrophosphate (also known as sodium acid pyrophosphate or SAPP), monocalcium phosphate, or a combination of these. The end use will govern the composition of the blend.

Both Bird in England and Horsford in the United States did good business with the military. At the beginning of the nineteenth century, in the run up to and during the First World War, military demands led to significantly increased production although without any great technical development. Toward the end of that war and into the 1920s, the process technology of the manufacture of phosphorus, wet acid, and thermal acid derived from burning phosphorus began to move forward. Over the next 40 years, these technologies were developed to a state very close to that still in existence today in the twenty-first century.

Compared to the plants of the nineteenth and early twentieth centuries, today, wet process phosphoric acid (WPA) plants are continuous; the largest plants produce over 2000 tons per day; the operating acid concentration is either 28%–30% or 45% P_2O_5 depending on the technology. Prior to 1915, WPA was made by a batch process with 16% sulfuric acid in 1–2 ton batches in wooden tanks, by hand. The resultant slurry was filtered in lead-lined, wooden filtering pans. The filtrate, containing approximately 8%–10% P_2O_5, was concentrated in rock or lead-lined pans. The annual production in the United States was around 2000 tons [25]. The WPA process basically comprises four steps: rock preparation, which usually amounts to milling on a phosphoric acid plant (rock beneficiation comprises a number of steps usually at the mine); reaction utilizing 98% sulfuric acid; filtration and washing, to separate the phosphoric acid and calcium sulfate; and concentration to the required level.

In 1915, the Dorr Company developed an air lift agitation system and countercurrent washing of the calcium sulfate in thickeners; this was a continuous process. By 1920, continuous filters developed by the Oliver Company had replaced the thickeners and allowed a 25% P_2O_5 product acid. One of the earliest plants was for Rumford Chemical in 1916. A further 30 plants were built between 1916 and 1929, including for Victor Chemicals in Chicago Heights, Illinois, Kuhlmann in Rieme, Belgium (much later to become part of Rhône–Poulenc and then Rhodia), and Badische in Germany (now BASF). The larger of these plants had a capacity of 25–40 tons per day. In 1932, Dorr built a three-train plant for Cominco at Trail, with a capacity of 40–50 tons per day. This plant incorporated calcium sulfate recycle to improve crystal growth and therefore filterability and countercurrent filter washing; it was state of the art at the time, producing a 30%–32% P_2O_5 acid at the filter exit, which leads to it being called the *strong acid process*.

Between 1927 and 1932, two groups, one Swiss with Dorr interests, Kunstdünger Patent Verwertungs A.G., KPV, and the other Swedish, Kemiska Patenter, headed by Mr. Sven Nordengren of the Swedish Superphosphate company with German company Lurgi, worked on higher P_2O_5 concentration processes. Their goal was to produce 40%–50% P_2O_5 acid directly at the filter. This was a very challenging task; it requires operating at a higher temperature than a 28%–30% P_2O_5 acid process with consequent corrosion problems and producing a different form of calcium sulfate (hemihydrate), which is much harder to handle than the dihydrate form produced at lower temperature. Nordengren patented this process [26] and developed a belt filter to handle the hemihydrate crystals known as the Landskrona or Lurgi filter. In the 1950s and 1960s, Japanese companies, Nissan, Nippon Kokan Kaisha (NKK), and Mitsubishi, developed hemihydrate–dihydrate recrystallization processes. Prayon of Belgium continuously developed the dihydrate process and over time has licensed the most plants.

To summarize, WPA process technology took big leaps forward with the development of the Dorr strong acid process; work commenced by Nordengren, KPV, and others laid the foundations for more complex but more efficient processes that were, in particular, developed by the Japanese. Since the 1970s, the developments in reactor, cooler, filter, and concentrator design have been incremental; the improved materials of construction, instrumentation, and control have made the difference; about one-half of WPA plants are Prayon (Belgium) dihydrate plants, one-sixth Rhône–Poulenc, and one-tenth Jacobs; just over a tenth are Japanese hemihydrate plants. The largest plant in the world is a Jacobs dihydrate plant in India, which was recently expanded to 3200 tpd.

The development of WPA process technology brought not only the ability to increase capacity output but also much lower processing costs. The drive to improve farm yields after 1945 lead to positive government encouragement to increase fertilizer production, which in turn gave the impetus to both WPA and sulfuric acid producers to put the technology to good use.

Just as WPA technology and capacity developed and grew, so too did phosphorus and thermal phosphoric acid (made directly from phosphorus) process technology and capacity, particularly between 1910 and the 1960s. The most significant developments of phosphorus production technology came from two countries, the United States and Germany. At the end of the nineteenth century and very early in the twentieth century, France contributed important developments, from Coignet and Billaudot, with the first commercial electric furnace for phosphorus and waterless condensing apparatus, respectively. In the United States in 1914, Hechenbleikner started the operation of a new 4000 kW furnace at Charlotte, North Carolina; this was much larger than the previous largest furnace of the Oldbury Electrical Company's 1500 kW installed in 1897.

The technology development was taken up by the US Department of Agriculture in 1915 as well as private corporations. In 1920, Federal Phosphorus established a 2000 kW furnace in Anniston, Alabama; this was the main source of phosphoric acid until 1924 and was taken on by the Swann Corporation and later Monsanto. In 1924, Victor Chemicals conducted pilot work on a blast furnace technology and by 1929 had established a large plant in Nashville, Tennessee; this was part of a rush to the Tennessee phosphate deposits by several companies; Victor's choice of technology was driven by difficulty accessing power in the area. In 1933, the Tennessee Valley Authority (TVA) was established to investigate and develop both phosphorus and phosphoric acid technology; TVA brought two 7500 kW units online in 1934/1935 and two more in 1937 and 1939; in due course, TVA produced a number of reports and detail designs for plants that were licensed for a nominal sum. Victor established a 12,000 kW unit in 1938 at Mt. Pleasant, Tennessee, and 24,000 kW in 1940. In 1949 and 1950, Westvaco (later FMC) built an 18 and a 20,000 kW furnace in Pocatello, Idaho; these were the first furnaces operating on the rock of the Western States; the only phosphorus production in the United States today is at Soda Springs, Idaho, operating on this rock. In the 1950s, many furnaces of increasing size were built. The largest to be built was for Monsanto, at Soda Springs in 1966 at 70,000 kW.

In Germany, three corporations developed technology in this area: Griesheim Elektron, Knapsack, and Piesteritz. In 1927, four phosphorus furnaces were commissioned at Piesteritz, developed from Griesheim. In 1953, phosphorus operations were established at Knapsack with a 10,000 kW unit; a 50,000 kW unit was added in 1956. At the time, this was the largest phosphorus furnace in the world.

In 1966, a new 60,000 kW plant was built, with Knapsack know-how, in Vlissingen, the Netherlands. This plant was the only phosphorus plant in Europe for several years and was operated by the company Thermphos (spun out of Hoechst, which was changed from Knapsack in 1974) until early 2013. German engineering and construction company, Uhde, builders of the plant at Vlissingen and part of the Hoechst group, also built plants in Tschimkent in Kazakhstan as well as in Kunming, China, using the same technology.

The zenith of the construction of phosphorus plants also proved commercially fatal to its owner. At the beginning of the twentieth century, Albright & Wilson had been the world leader in phosphorus manufacture but was soon overtaken by US corporations. In the mid-1960s, Albright & Wilson considered concentrating all its phosphorus capacity in a new, large, economic plant to meet increasing demands for phosphates, principally the detergent builder, sodium tripolyphosphate (STPP). Thinking that sulfur prices would prove unfavorable and attracted by the prospect of cheap electricity in Newfoundland, Albright & Wilson commenced a project for two 60,000 kW furnaces. In 1968, Albright & Wilson commenced commissioning the new plant at Long Harbour, Placentia Bay, Newfoundland. Albright & Wilson had deemed a quotation for a plant from Uhde too expensive so combined in-house experience on a much smaller plant, together with published designs including those from TVA to come up with an engineering scope. Unwise economies were then made to the scope. A further seed of disaster was sown when in good faith environmental aspects of the design were based on an assessment that the surrounding sea and air could handle quite significant discharges from the process. Following a very difficult and extended commissioning period, the plant underwent major modification to address the environmental, operational, and reliability issues. Eventually, the plant achieved a relatively reliable output. The plant was closed down in 1989; the principal cause was commercial failure aggravated by operational and environmental challenges; the commercial failure lead to Albright & Wilson losing its independence and becoming part of the Tenneco conglomerate a position it held until 1995 [17].

A key cause of the closure of most phosphorus plants in Europe and the United States was the rise of purified phosphoric acid (PWA) plants. Even the early PWA plants were able to produce acid of sufficient quality to displace all but the highest purity acid demands that were satisfied by thermal acid plants. By the 1990s, most of the phosphorus manufacture, outside China, was directed to phosphorus derivative products.

The great expansion of phosphorus capacity in the 20 years from the late 1940s was driven primarily by STPP demand. Its prime use was as a detergent builder; its development was as significant as the match in the early nineteenth century. In 1946, Procter & Gamble launched a new synthetic detergent Tide®. This new detergent, one part alkyl sulfate, three parts STPP, was effective in cleaning heavily soiled clothes, in hard water, without leaving a soapy scum; it was far superior to any other cleaning product. The two other big *soapers*, Colgate and Unilever, scrambled to catch up. STPP is made by reacting phosphoric acid with either sodium bicarbonate or sodium hydroxide; the resultant sodium phosphate liquor is dried and heated to the desired transformation temperature. The quality of the acid, its purity, is important; too high a level of some metals can affect the color or properties of the final detergent blend. On the other hand, there is no need for acid of, say, food grade quality, which, once arsenic has been attended to, is the standard purity of thermal acid derived from phosphorus. Levels of sodium, for example, can be very high as the next step in the process is to react with a sodium source.

In the 1950s and 1960s, the two principal routes to STPP were from thermal acid (exploited in the United States and Europe) and from WPA by chemical purification, which has been termed the *wet salts* process (exploited in Europe and by Blockson/Olin in the United States). In the *wet salts* process, WPA first has any excess sulfate reduced with phosphate rock; then sodium hydroxide/bicarbonate is added, raising the pH to the point where many impurities precipitate out. The acid is then filtered and concentrated, and further sodium is added to make the sodium phosphate liquor that is fed to the STPP plant. The *wet salts* process is a crude purification process but perfectly acceptable for technical and industrial rather than food grade products. The relative cost of STPP made from thermal acid comes down to the relative cost of phosphorus and wet process acid; in turn, this comes down to the relative cost of electricity and sulfur. From the earliest days, most phosphorus plants have been established where good-quality rock and cheap power (usually hydroelectricity but occasionally, as for Albright & Wilson in the nineteenth century and northern European producers in the early twentieth century, cheap coal) coincided. Although in the 1950s sulfur suffered at one stage poor availability and therefore was highly priced, generally, wet process acid will always be significantly cheaper. Furthermore, because of the relatively large volumes of WPA going to fertilizer manufacture, it would usually be straightforward to take a relatively small quantity for the detergent market.

The rise of demand for STPP and therefore phosphoric acid was coincident with the rise of the Israeli phosphate industry. Negev Phosphates was formed in 1952 to exploit the phosphate reserves of that region; previously, rock was imported from Jordan into Haifa to make phosphate fertilizers. Twenty-one deposits were found, and the most important that are mined today are Oron, Arad, and Nahal Zin, all of which are relatively close to the Dead Sea. Also in the 1950s, Israel developed and expanded its chemical industry based on the brine from the Dead Sea. The Israel Mining Industries (IMI)–Institute for Research and Development invented a hydrochloric acid route to phosphoric acid in the late 1950s, which of necessity incorporated solvent extraction. IMI also developed a solvent extraction process for the purification of phosphoric acid made via sulfuric acid. These and related invention were protected with a series of patents by Avraham M. Baniel, Ruth Blumberg, and others from 1961 [27]. Hydrochloric acid was chosen because of the nearby source of chloride from the Dead Sea operations (a pipeline transferred magnesium and calcium chloride brine to the production site near Arad). A simplified reaction is shown in the following equation:

$$Ca_3(PO_4)_2 + 6HCl \rightarrow 2H_3PO_4 + 3CaCl_2 \tag{1.5}$$

The calcium chloride is soluble; unreacted solids are separated from the phosphoric and hydrochloric acids and calcium chloride mixture; the phosphoric acid and hydrochloric acid are then separated by solvent extraction from the calcium chloride. The phosphoric and hydrochloric acids are then separated by distillation. The hydrochloric acid route was implemented at Arad but was very corrosive and led to a difficult commissioning and operation.

The phosphoric acid so produced was not particularly pure by the standards of even the following decade (e.g., it was brown, not colorless, and had a high 0.5% iron content); however, it was clear and could be used instead of thermal acid for liquid fertilizer, animal feed phosphates, and detergent phosphate manufacture. In another installation in Haifa, Israel, acids were purified, as well as being separated from calcium chloride. Even as projects for the largest phosphorus furnaces ever, globally, were launched in the 1960s, the writing was on the wall.

In 1966, Albright & Wilson investigated some developments made by IMI including the use of salt (sodium chloride) as a source of sodium instead of sodium bicarbonate (soda ash) or sodium hydroxide, generating hydrochloric acid as a coproduct, and the cleaning of WPA by a solvent identified by IMI for use in a simple procedure (diisopropyl ether, DIPE). This latter process was further developed by Albright & Wilson in conjunction with IMI and an engineering proposal put forward to the Albright & Wilson board as an alternative to further phosphorus production (the project under consideration in Newfoundland).

FIGURE 1.9 Albright & Wilson solvent extraction demonstration plant. (Courtesy of Hugh Podger, Brewin Books Limited, and Solvay.)

The findings revealed in a Dutch patent [28] using methyl isobutyl ketone (MIBK) to separate nitric acid from phosphoric acid inspired the Albright & Wilson team led by Alan Williams to develop the outline of a solvent extraction process. A second research team examined the use of acetone in the purification of WPA. All work was stopped at the end of 1970 because of the financial situation within Albright & Wilson due in large part to the Newfoundland project. As difficulties continued to mount at Newfoundland, work was restarted on the purification of WPA with the construction and commissioning of a demonstration plant for the process using MIBK (see Figure 1.9). A larger pilot plant was installed in December 1972 for which the mixer–settlers were provided by Davy Powergas.

On April 1, 1974, commissioning commenced on a plant to produce 40,000 tons per year P_2O_5 as purified acid on Albright & Wilson's Whitehaven site in Cumbria, England. The coproduct acid was transferred to their Humberside fertilizer factory in railcars. The process was code-named *MO*. *MO* referred to Marchon Oldbury. The Whitehaven site was established by two Austrians, Fred Marzillier and Frank Schon in the 1940s, hence Marchon. Unlike the rather conservative Albright & Wilson, Marchon was very entrepreneurial; under the site was an anhydrite mine, a source of valuable sulfur in the 1950s, and so Marchon made sulfuric acid, WPA, surfactants, and STPP. The use of a code name for a project is not unusual in the chemical industry generally; in this case, it is indicative of the secrecy that surrounded this technology not just in Albright & Wilson but in all the other companies working on this process.

Unfortunately, IMI claimed Albright & Wilson had infringed its patents and served a writ in 1976. The case was settled out of court with Albright & Wilson purchasing 3 years' supplies of phosphate rock from Israel. In 1979, Albright & Wilson had extended their PWA plant with a second train of 95,000 tons per year P_2O_5 and modified the process so that it produced two product grades, *salts* grade for STPP production and technical grade for other phosphate salts.

Also in 1979, Albright & Wilson had added a WPA plant (known as F5) to produce strong acid directly from the filter. The technology (HDH—hemidihydrate) was provided by Fisons and is now available from Yara. The product acid was separated from calcium sulfate hemihydrate, which was recrystallized to gypsum (dihydrate) in order to recover the residual phosphoric values. This product was further concentrated to the strength required for the purification process. Economically, the Whitehaven operation relied on good rock prices from Morocco and good fertilizer grade acid (underflow from the PWA plants) prices; environmentally, it depended on being permitted to discharge gypsum from the site into the Irish Sea. In 1984, Albright & Wilson commissioned a further modification, termed underflow extraction (known as UFEX); this process substituted 77% sulfuric acid for phosphoric in the underflow. This meant that up to 97% of the feed WPA could be extracted as purified acid, thus breaking the link with fertilizer acid prices. Another innovation was the installation of a Kühni column (by Kühni AG, Switzerland, now Sulzer Chemtech) instead of more mixer–settlers. Although less flexible than a series of mixer–settlers, the column had a much smaller footprint and required less power input (from pumps), less leakage paths for solvent, and less equipment.

In the early 1980s, Albright & Wilson became concerned about a surplus capacity of phosphorus; the Newfoundland plant was producing more reliably, but the Whitehaven PWA plants were displacing internal thermal acid demand. Albright & Wilson considered acquiring a number of US companies and in 1985 completed the purchase of the Industrial Chemicals Group of the Mobil Corporation with plants in Fernald, Ohio, and Charleston, South Carolina (the latter originally owned by Virginia Carolina). This acquisition both provided an outlet for phosphorus and opened up other prospects. Unfortunately, the Albright & Wilson phosphorus price was higher than that from Monsanto, the company supplying Mobil. In the late 1970s, Mobil had acquired the know-how of the Budenheim PWA process and had been in preliminary talks with Olin, the leading US producer of STPP based on WPA at that time. A study group concluded that a PWA plant, in partnership with either TexasGulf at Aurora, North Carolina, or IMC in Florida as well as the closure of the Newfoundland plant, should be pursued. Serious discussions between Albright & Wilson and TexasGulf (at that stage a subsidiary of Elf Aquitaine of France) were held in 1986, the relative merits of the Albright & Wilson and Budenheim processes evaluated and a plant size of 75,000 tons per year P_2O_5 as purified acid agreed. Olin joined the party, contributing $16 m, and the plant scope was increased to produce 120,000 tons per year P_2O_5 with 50,000 tons per year P_2O_5 destined for Olin's phosphate plants at Joliet, Illinois. The plant was officially opened in 1990. In 1991, Olin withdrew from the partnership and closed its facilities at Joliet, and Albright & Wilson took over its US phosphate business.

Meanwhile, Rhône–Poulenc was developing a new plant in Geismar, Louisiana. Coincidently, both Rhône–Poulenc and Purified Acid Partners (PAP comprising

TABLE 1.1
US Phosphorus Plants in the Early 1980s

Company	Location	Capacity 1000 Tons P_4/Year
Electro-Phos	Pierce, FL	22
FMC	Pocatello, ID	137
Mobil (Albright & Wilson)	Charleston, SC	Mothballed
	Mt. Pleasant, TN	Mothballed
	Nicholas, FL	7
Monsanto	Columbia, TN	135
	Soda Springs, ID	95
Occidental	Columbia, TN	45
Stauffer (R-P)	Mt. Pleasant, TN	45
	Silver Bow, MT	42
	Tarpon Springs, FL	23
TVA	Muscle Shoals, AL	Mothballed
Total		566

Albright & Wilson, TexasGulf, and Olin) used the same engineering, procurement, and construction company (EPC), Jacobs Engineering, and although the design teams were kept separate and every effort made to maintain confidentiality, it is possible to see many similarities between the two plants (and a later plant for Astaris at Soda Springs, Idaho). The rationale for the Rhône–Poulenc project was straightforward. Rhône–Poulenc had acquired Stauffer Chemicals US phosphate business, which made phosphorus through to phosphate derivatives; however, the phosphorus plants were getting old, electricity prices were rising, and the environmental challenges of phosphorus production were beginning to become financially important. Furthermore, the market in the United States for STPP was in decline and had fallen some 40% since the late 1960s. The status of phosphorus plants in the early 1980s is summarized in Table 1.1, within 20 years every plant except the Monsanto plant at Soda Springs would be closed.

Monsanto and FMC held the view at that stage in the early 1990s that a phosphorus-based business model still worked, although that would change by the turn of the century. Rhône–Poulenc closed its phosphorus plants in the early 1990s.

Although many phosphate companies carried out work on the solvent extraction route to purified phosphoric acid in the 1960s and 1970s (see Table 1.2) and many patents were granted, only four processes could be considered commercially significant in the last quarter of the twentieth century. We have looked at two, the IMI process (purification of WPA via sulfuric acid) and the Albright & Wilson process.

To categorize broadly, the IMI process was probably the first real industrial scale commercial process. (Toyo Soda in Japan could claim to be the first commercial solvent extraction plant that operated a process from 1962 to 1968 at Tonda with a nominal capacity of 5000 tpa P_2O_5 producing both technical and food grade acids. HCl was used to attack the rock with butanol as solvent under license from IMI. In 1970, a new plant was started up with a capacity of 18,000 tpa P_2O_5 based on WPA via sulfuric acid.) The IMI process had examples in Israel, Spain (although later abandoned),

TABLE 1.2
Selection of PWA Solvent Extraction Patents between 1960 and 1978

Company	Patent	Issue
Albright & Wilson	US3947499	1976
	US3914382	1975
	US3912803	1975
Allied Chemical Corp	US3723606	1973
Arad Chemical Industries	IS32320	1968
Armour Agricultural Chemical	US3408161	1968
Azote et Produits Chimiques	GB1357614	1971
Chemische Fabrik Budenheim	GB1344651	1974
Canadian Industries	US3298782	1967
Central Glass Co.	GB1296668	1972
Cominco	US3388967	1968
Dow Chemical	US3449074	1969
FMC	US3684438	1972
	US3410656	1968
GIULINI	GB1350293	1974
Goulding	GB1342344	1974
IMI	US3304157	1967
	US3311450	1967
	US3573005	1971
Produits Chimiques Ugine Kuhlmann	US4108963	1978
Monsanto	US3684439	1972
	US3479139	1969
Montedison	GB1172293	1969
NKK	JP7215458	1971
Occidental Petroleum	US3694153	1972
Office National Industriel de l'Azote	US3497329	1970
Produits Chimiques Pechiney–Saint-Gobain	US3397955	1968
	US3366448	1968
	US3607029	1971
Prayon	US3970741	1974
Produits Chimiques et Metallurgiques du Rupel	GB1323743	1973
St. Paul Ammonia Products	US3375068	1968
Stamicarbon NV	US3363978	1968
Susquehanna Western Inc	US3359067	1967
Tennessee Corporation	US3318661	1967
	US3367738	1968
Toyo Soda Manufacturing	US3920797	1975
	US4154805	1977
	US3529932	1970
Typpi Oy	GB1129793	1968

Bilt and Gujarat, India, and Coatzacoalcos, Mexico. The Coatzacoalcos plant has been operating since 1971 during which time it has been extended and the product purity improved. Initially, the site was owned by Fertilizantes Fosfatados Mexicanos (FFM). FFM merged with Guanos y Fertilizantes de Mexico (Guanomex) to form the Mexican government-owned Fertimex in 1977. In 1992, Troy Industries acquired the facility and in 1994 formed a joint venture with Albright & Wilson (A&W Troy). Albright & Wilson completed purchase of the Troy interest in 1998; in 2001, the Coatzacoalcos site became part of Rhodia and is now in the Innophos family.

The Albright & Wilson MIBK-based process can claim, today, to be used on the largest PWA plant in the world at Aurora, North Carolina, and has had, perhaps, the greatest commercial impact on the industry as a whole. The most prolific is the Prayon process.

Société Anonyme Métallurgique de Prayon at Engis, near Liège, Belgium, was formed in 1882. Prayon was initially a zinc producer; a sulfuric acid plant built in 1888 was a natural adjunct. Prayon's WPA processes, the most important being the dihydrate route, began their development from the Dorr Company work in the 1930s. In the 1940s, Prayon started to develop continuous reaction and filtration stages. Dissatisfied with a German *Gropel* filter, Prayon developed its own tilting pan design. In 1953, it licensed its first plant to Smith and Douglas in Streator, Illinois. This licensing policy has continued alongside development of its technology. Today, over half the world's WPA plants are Prayon technology. In the 1970s, Prayon developed and patented [29] its own solvent extraction process, using a mixture of DIPE and tributyl phosphate (TBP) as solvent and its own design of equipment. Two plants were built in Belgium, the first at Puurs in 1976 (33,000 tpa P_2O_5 mothballed in 1992) and the second at Engis in 1982 (initially 40,000 tpa P_2O_5). Subsequently, two 25,000 tpa P_2O_5 plants were built in Korea and Indonesia, although not without operational difficulties.

In 1987, production commenced at a new plant at Cajati, 240 km from São Paulo, Brazil; the plant is known as Fosbrasil. Originally a four-partner joint venture between Monsanto, Quimbrasil, Prayon, and Société Belge d'Investissement, the plant used Prayon technology to produce 25,000 tpa P_2O_5 of technical grade acid. Subsequently, ownership has changed due to divestments and both capacity increased (to 90,000 tons per year P_2O_5) and purity improved.

In 1992, Prayon and Rhône–Poulenc formed a joint venture Europhos. Through the 1960s and early 1970s, the French government had brought a diverse range of small French chemical companies (including Coignet) into two combines; Rhône–Poulenc was one. A number of patents were granted for solvent extraction processes that had been assigned to a variety of companies; however, in the end, three plants of importance can be credited to the French, with Louis Winand as their champion. The first was a pilot plant at Les Roches de Condrieu near Lyon, France; the second was half a plant at Rouen, France; the third was initially rated at 70,000 and subsequently was expanded to 90,000 tons per year P_2O_5 plant at Geismar, Louisiana. One object of the Europhos joint venture was to rationalize production so Puurs was mothballed and Les Roches closed. (The Rouen plant was integrated with an STPP plant so did not have much of the postpurification equipment needed to produce purified acid products. The output was sodium phosphate liquor and a small quantity of raffinate that was piped to a fertilizer plant.)

In 1997, Prayon, Budenheim, and Office Cherifien des Phosphates (OCP, Morocco) entered a joint venture, Euro Maroc Phosphore (Emaphos), and agreed to build a 130,000 tons per year P_2O_5 plant at Jorf Lasfar, Morocco, with Prayon technology. In 1998, Prayon bought Rhône–Poulenc out of the Europhos joint venture. Albright & Wilson and OCP had considered a joint venture in 1994, but Albright & Wilson backed away, as it did in South Africa and in 1999 in Brazil.

Around 2000, there were a series of acquisitions and divestments. The nonpharmaceutical divisions of Rhône–Poulenc were bundled into a new corporation, Rhodia, which included its phosphate operations in Europe and the United States. Rhodia then purchased Albright & Wilson. Monsanto spun out its phosphate businesses into a new company, Solutia. FMC did likewise, creating a stand-alone phosphate division. Then Solutia and FMC formed a joint venture, Astaris. Both the Rhodia/Albright & Wilson and FMC/Solutia deals required Federal Trade Commission (FTC) approval to meet antitrust legislation. The FTC ruled that Rhodia should sell its share of the Aurora facility, which was now owned by PotashCorp (PCS), and that Astaris should sell its Augusta site to Prayon. Timing was everything; at one stage in this period, Albright & Wilson was in negotiations to buy Solutia; had that deal gone through, the US industrial phosphate landscape would have been very different.

The FMC/Solutia deal was complicated by a \$100 m project for a new PWA plant at Soda Springs, Idaho. FMC's nearby phosphorus plant at Pocatello was old and the Environment Protection Agency had ordered improvements to the plant's emissions. FMC had a good STPP business in Green River, Wyoming: the phosphorus source was reasonably local as was the sodium source, and the customer, Procter & Gamble, had a local plant to process the STPP to detergent blends for sale. FMC evaluated whether to continue with the phosphorus plant, buy acid from Aurora, or build a new PWA plant themselves. FMC had worked on PWA in the United States and had a number of patents; FMC also had a 25,000 tons per year P_2O_5 PWA plant in Huelva, Spain. The Spanish plant originally used IMI technology and then migrated to its own, US-developed technology. Commencing late 1997, FMC started to work up a project scope for an 80,000 tons per year P_2O_5 PWA plant. The plant was to be located at the Agrium site and operated on FMC's behalf by them; initially, Agrium was to supply WPA at the required feed strength of approximately 52% P_2O_5 and receive the raffinate return from FMC for fertilizer manufacture (monoammonium phosphate [MAP] and diammonium phosphate [DAP]). In order to meet this new demand from FMC, Agrium in turn embarked on a number of projects itself. Just prior to the meetings where FMC and Solutia signed the joint venture in April 1999, Agrium announced to FMC that its capital program was not large enough to afford all the modifications necessary to supply the FMC plant. FMC agreed to take responsibility for some of these areas in return for a reduction in WPA price. The changes were incorporated into the project scope and the capital estimates rose from an initial ballpark of \$50 m to \$85 m. As well as taking on an additional engineering scope, FMC had taken on additional commissioning and operational complexity. The project proceeded with Jacobs Engineering and was completed on program in May 2001, a month after the FTC approved the FMC/Solutia joint venture. Commissioning the plant was always going to be challenging for several reasons. Firstly, there were two big plants to commission (the PWA plant and the significantly expanded WPA plant), essentially at the same time. Secondly, the remote location of the plant and its weather were a hindrance—it takes time to gather additional personnel and equipment in remote places, so the rate of making changes that is a feature of commissioning is slowed; PWA plants like to be warm and stable, whereas it is often cold in Idaho. Thirdly, PWA plants are difficult to commission; the difference between success and failure is small; there are several examples where considerably greater resource and time than originally planned have been required to get a PWA plant up to a respectable rate and some where the plant has been quietly abandoned. In this case just about everything that could go wrong in commissioning did—the WPA feed acid was unstable, and utilities, plant, and equipment broke; nevertheless, some acid was produced. Agrium and Astaris struggled on and 2 years later had managed to improve output and reliability somewhat. Unfortunately, a key plank of the business case, the Procter & Gamble contract, was lost to Rhodia. Astaris chose to close the plant. Shortly afterwards, in October 2003, Solutia sued FMC for fraud over the alleged technology failure seeking more than $322 m in damages. The case was eventually settled "on the court steps" for a relatively modest sum. Astaris, like Rhodia, carried out a major restructure closing plants and 2 years later was acquired in November 2005 by ICL. Rhodia sold its phosphate operations: in the United States, including Coatzacoalcos, Mexico, to Bain Capital, creating Innophos; in the United Kingdom, it closed Whitehaven phosphate and PWA plants and sold the rest of the business to Thermphos (which was spun out of Hoechst [Knapsack historically] and was itself the subject of A&W's takeover intentions in 1998); in France, the Rouen PWA/STPP unit was closed; in Belgium, the Rieme WPA plant was sold to Nilephos. Meanwhile with the closure of Astaris' Idaho plant, PCS added a third train to its Aurora plant at a cost of $70 m in 2003 adding a further 80,000 tpa P_2O_5 capacity. The plant was expanded again in 2006 at a cost of $73 m, taking total capacity to 327,000 tons per year P_2O_5 making it the largest PWA plant in the world. Figure 1.10 is a photograph of the PWA plant at Aurora; the first two trains are incorporated in the plant on the right, and the third and fourth are in the center and left of the photograph.

Thermphos was (before closing in 2013) the only producer of phosphorus in Europe but had no access to PWA, so formed a joint venture with Kemira GrowHow in 2007. The joint venture, Crystalis, established a 30,000 tons per year P_2O_5 plant at Kemira's Siilinjarvi site in Finland. The plant combined Kemira's WPA pretreatment technology and Albright & Wilson's solvent extraction and crystallization technologies. Albright & Wilson's earliest solvent extraction patents described crystallization as a final purification step to make, for example, high-purity grades for the cola (carbonated soft drinks) and pharmaceutical markets [30]. In the 1990s, Albright & Wilson installed a batch crystallization unit at Whitehaven as part of the PWA plant; under Rhodia, this unit was relocated to the Widnes, United Kingdom, site and converted to a continuous plant. Kemira had also worked on purification via crystallization and had patents in this field [31].

In the early twenty-first century, the phosphorus and PWA/phosphate businesses in the Americas and Europe are relatively consolidated and stable. Areas of potential growth and development are North Africa, the Middle East, China, and Kazakhstan. In North Africa OCP, Morocco has developed from a modest, French-managed,

FIGURE 1.10 Purified phosphoric acid plant Aurora, North Carolina. (Courtesy of PotashCorp, Saskatoon, Saskatchewan, Canada.)

phosphate mining operation in the 1920s to the leading supplier of fertilizer in the world; it is said that the annual fertilizer price negotiations between OCP and Indian customers effectively set the global market price of fertilizer and phosphate rock. It is integrated forward through the Emaphos joint venture to PWA. Other North African countries, Tunisia and Algeria, for example, utilize their phosphate reserves, but the next big operation will be the Maaden project in Saudi Arabia. In 2013, projects are underway for a PWA and STPP plant.

Unlike the single phosphorus plant in the United States (Monsanto, Idaho) and Europe (Thermphos, Vlaardingen), China has many phosphorus furnaces ranging from small plants with very simple designs and little environmental protection to at least one large unit based on Uhde technology. These plants are under three major pressures: competition for electricity—China's generation capacity cannot keep up with overall demand; China's environmental legislation is on a par with international standards and many furnaces simply fail to meet these standards; consequently, the Chinese government ordered the closure of small, old plants; overcapacity and rising labor costs mean that it cannot be economic to keep plants idle or running at low rates. It would seem that China is ripe for PWA plants; however, the economics have not yet been compelling enough for a major switchover. Nevertheless, in 2007, Wengfu Group Co. Ltd. brought the first PWA plant in China on stream. The plant and process designs were carried out by Bateman Litwin (Litwin had a license for IMI PWA in the 1960s); Bateman utilized their pulse column technology in the solvent extraction section of the plant. The plant is rated at 100,000 tons per year P_2O_5.

Having not yet fully embraced PWA technology, China has the opportunity to evaluate an emerging process, kiln phosphoric acid (KPA). Sections of the KPA process have similarities with the feed preparation for a phosphorus furnace and the hydrator section of a thermal phosphoric acid plant. In essence, phosphate rock, silica, and green petroleum coke are dried, milled, formed into agglomerates in a balling drum, dried again, and then fired in a rotating kiln. The kiln solids are maintained above 1180°C, the temperature at which the phosphate reduction reaction commences. Phosphorus pentoxide is recovered as a strong phosphoric acid from the kiln off-gases and is more pure than WPA but carrying heavy metals. It is likely that the recovered acid could be purified to a food grade standard without the need for a full PWA plant. The KPA process does not require sulfuric acid; the electricity demand is light (400 kWh/te P_2O_5) compared to phosphorus production and the by-products are claimed to have utility rather than being a waste to handle. On the other hand the process is challenging—to the kiln and in terms of process control. Nevertheless, the process is deemed sufficiently mature, and potentially profitable, that Minemakers Limited of Australia have agreed a license with the process developers, JDCPhosphate, Inc., to implement the technology on the Wonarah project, in Australia's Northern Territory.

When the progress of the business and technology of phosphorus, phosphoric acid, and phosphates is considered, it is clear that several different components are required.

Inventors and explorers make the first critical steps, for example, Brandt with the phosphorus discovery and Gahn with the realization that bones were a far more concentrated source of phosphorus than urine; the inventors are often commercially incompetent or uninterested and secretive.

Developers are needed to apply discoveries, getting them out of the laboratory, whether to develop manufacturing processes or useful products; examples would include the match in the nineteenth century that gave the phosphorus industry a push or STPP in the twentieth. Of course, without customers, there can be no sales, and there can be no sales without salesmen. Again, developers can gain market share with new processes like the continuous reaction and filtration of WPA or the unit processes of PWA. But the inventors and technology developers require business managers to bring the whole enterprise together, and no amount of innovative technology can combat poor business management.

Finally, one must not discount good fortune: Albright & Wilson took account of the fish toxicity of phosphorus, yet it proved to be more toxic than anticipated and was one of many contributing factors to the closure of the plant at Long Harbour. Knapsack (Hoechst) operated a good PWA process, but when the plant burnt down in 1987, it was not rebuilt and had long-term business consequences. Finally, the level of difficulty of the commissioning of the Astaris plant at Idaho could so easily have happened at either Aurora or Geismar—the margin between success and failure is fine on PWA plants.

1.2 CHEMISTRY AND PROCESS OVERVIEW

1.2.1 Introduction

In this section, we shall consider the chemistry of the whole industry starting with phosphate rock. Simplified reaction equations are presented for the principal phosphate products; these equations form the basis of different product costs in the economics section of this chapter. The later chapters will look at the chemistry of different products and processes in sufficient detail to inform plant designs. For a comprehensive treatment of the chemistry and biochemistry of phosphorus and its products, the reader is referred to Corbridge [64].

An overview of the industry is depicted in Figure 1.11. The diagram commences with phosphate rock as the source of phosphorus. The majority of the world's rock, about 30 m tons per year, is converted to phosphoric acid through the reaction with sulfuric acid. The majority of this acid, around 90%, is further converted to fertilizers, principally ammonium phosphates, through reaction with ammonia, but also superphosphates through the reaction of the phosphoric acid with more rock. The minority of this acid is purified further; as acid, commercial grades vary in purity from a technical grade suitable for the manufacture of technical phosphate salts, through food grade, to pharmaceutical and semiconductor grades. Food grade acid is also used to make food grade phosphates. A minority of the world's rock is reacted to form elemental phosphorus; this requires coke, silica, and a lot of electrical energy (12–14 MWh per ton of phosphorus). A small proportion of rock is sold as a cheap fertilizer (also known as phosphate flour or direct application rock—DAR).

Phosphorus has several allotropes: white (this is the most common form and is also referred to as yellow phosphorus as this is its usual appearance—the color is derived from dissolved organic compounds), red, violet, and black. Diphosphorus, P_2, only exists in gaseous form. All commercial production is directed to make white phosphorus (total worldwide production is about 1,000,000 tons per year P_4). Red amorphous phosphorus (RAP), a product itself in matches, ammunition, phosphor bronze, metal phosphides, and pigments, is made by heating white phosphorus in a small agitated reactor to over 250°C.

Phosphine (PH_3), a highly toxic, flammable gas, may be produced by the reaction of metal phosphides with water or the treatment of RAP with water at about 250°C or the alkaline hydrolysis of white phosphorus. Phosphine is a product itself and is used to produce catalysts and mining and extraction agents; reaction with formaldehyde and either hydrochloric or sulfuric acids gives tetrakis (hydroxymethyl) phosphonium chloride (THPC) or sulfate (THPS); both are used as flame retardants for cotton and cotton-rich fabrics. Two other important phosphorus derivatives are phosphorus trichloride (PCl_3) and phosphorus pentasulfide (P_4S_{10}); a minor derivative is phosphorus sesquisulfide (P_4S_3). All are processed further to a wide range of phosphorus derivative products; well-known examples include glyphosate, an herbicide originally invented by Monsanto, and bisphosphonates, which are used in treatments for osteoporosis. In general, the processes of phosphorus derivatives are more chemically complex and far more hazardous but of a smaller scale than phosphate processes.

The overlap between phosphates and phosphorus occurs with thermal phosphoric acid and its related products. As was discussed in the history section, prior to commercial solvent extraction–based purification plants, all pure phosphoric acid came from the hydration of phosphorus pentoxide (P_2O_5) itself formed by burning phosphorus in phosphorus burners. Many years ago, the economics in some parts of the world permitted the use of thermal acid in fertilizers. The plants that produce thermal acid are capable of varying its concentration depending upon

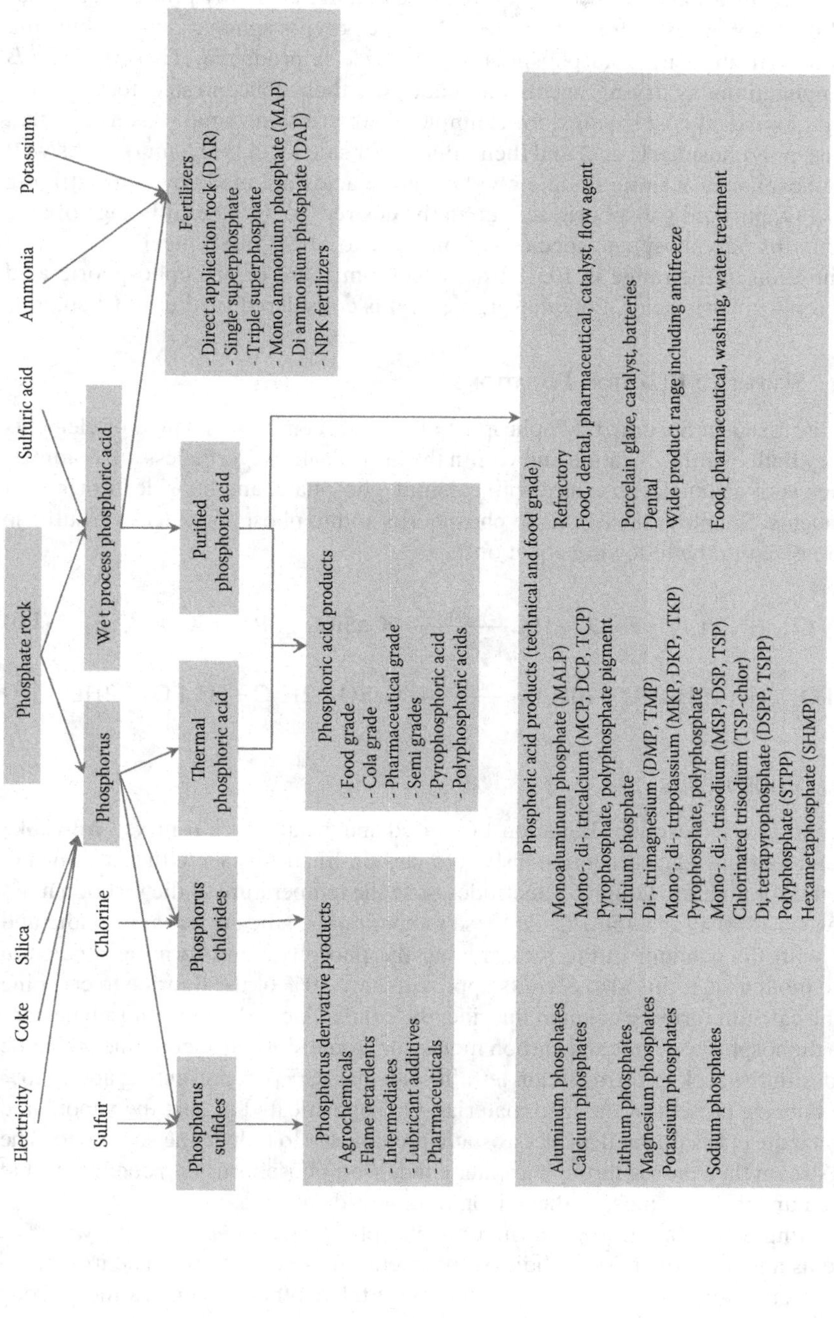

FIGURE 1.11 Industry overview.

the amount of water allowed into the system; the usual commercial concentration is 85% H_3PO_4 (61.6% P_2O_5) with a range of 75%–92% H_3PO_4. As the water addition is decreased, higher concentrations up to 116% H_3PO_4 (84% P_2O_5) may be produced. Higher concentrations to 118% and even above that are normally produced by adding solid phosphorous pentoxide to strengthen the polyphosphoric acid. By burning phosphorus in dry air, solid phosphorus pentoxide is produced. These *polyacids* have applications as drying agents and catalysts; their concentrated forms allow potential logistical cost savings, for example, there are some small businesses that purchase polyphosphoric acid and then dilute it for sale local to the market. As well as the phosphorus burning route, polyphosphoric acids are made by concentrating 85% H_3PO_4 purified phosphoric acid up to the desired grade. The full range of acid that contains polyphosphate species is from 95% to 118%, but typically the lowest concentration in the range is 105% (the eutectic mixture of orthophosphoric acid and pyrophosphoric acid). Polyphosphoric acid is considered further in Chapter 3.

1.2.2 Simplified Reaction Equations

As is discussed in Section 1.2.7, phosphate rock is a complex mélange of elements that vary both by mine location and within the mine itself. Nevertheless, all commercial rock is of an apatite structure with calcium, phosphate, and fluoride as principal components. Simplified reactions to phosphorus and to phosphoric acid via sulfuric acid are given in the following equations:

$$Ca_{10}(PO_4)_6F_2 + 9SiO_2 + 15C \xrightarrow{1400°C} 9CaSiO_3 + 3P_2 + 15CO + CaF_2 \quad (1.6)$$

$$Ca_{10}(PO_4)_6F_2 + 10H_2SO_4 + 20H_2O \xrightarrow{80°C} 10CaSO_4 \cdot 2H_2O + 6H_3PO_4 + 2HF \quad (1.7)$$

1.2.3 Phosphorus

In the phosphorus reaction (Equation 1.6), dried and ground rock is mixed with coke and sand and introduced continuously to a carbon-lined furnace either as a blend, briquettes, or nodules. Graphite electrodes raise the temperature of the reaction mass to between 1400°C and 1500°C. At these temperatures, silica is strongly acidic and reacts with the calcium in the rock freeing the phosphate and forming a calcium silicate molten slag; this also permits approximately 80% of the fluoride to combine with the calcium forming calcium fluoride; the coke reduces the rock phosphate content to diphosphorus vapor and carbon monoxide; some silica also combines with the fluorides in the rock to form silicon tetrafluoride gas (SiF_4), representing about 20% of the fluoride present in the feed material. On sophisticated plants, the vapors and entrained dust pass through an electrostatic precipitator to reduce the solid load. The vapor stream then passes through a water condenser; phosphorus (P_4) condenses and is stored under water; most of the silicon tetrafluoride goes into solution.

Iron impurities in the rock form iron phosphide (*ferrophosphorus*), which is tapped as a melt, cooled, and solidified for sale to the steel industry. The iron phosphide reactions are set out in Equations 1.8 and 1.9; other impurities in the rock feed, including sulfur, magnesium, and aluminum (which is also introduced with the coke), follow similar reactions and together account for a loss of around 8% of the phosphorus content of the feed material when that remaining in the calcium silicate slag is included:

$$Fe_2O_3 + 3C \rightarrow 2Fe + 3CO \tag{1.8}$$

$$8Fe + P_4 \rightarrow 4Fe_2P \tag{1.9}$$

In practice, the relative amounts of rock, silica, and coke, together comprising the *furnace burden*, are calculated according to two ratios: The first is the desired silica/lime ratio (SiO_2/CaO), usually between 0.8 and 1.0, and the second is the phosphate/carbon (P_2O_5/C) ratio, usually between 2.3 and 2.6; in both cases, these ratios depend upon the analysis of both rock and coke, the need to provide more carbon than calculated from stoichiometry, to account for losses, and plant experience. Electricity consumption is 12.5–14 MWh per metric ton of phosphorus, which represents around 50% of the full cost. The vapor stream from the condenser is predominantly carbon monoxide together with a small quantity of hydrogen (resulting from moisture entrained in the air in the feed materials), phosphine and phosphorus particles, and other minority components; on more complex plants, this stream is used to provide heat to dry the rock or process it in a number of ways, or fed to a local power plant; if fully recycled, this stream can reduce the energy cost by close to 20%. The molten slag is waste; it is run off and solidifies. It has been used as road base in the past; however, the presence of radionuclide makes this problematic now. The smaller more simple plants with no precipitators rely on the condenser to knock out the dust, and without carbon monoxide recovery, everything goes up the stack. Observing these plants one can see a large flame emerging from the main vent stack due to burning carbon monoxide and hydrogen, as well as billowing white fumes from the conversion of small particles of phosphorus to phosphoric acid; dust with measurable radioactivity is dispersed in the vicinity. Figure 1.12 shows a flow sheet for phosphorus production.

1.2.4 Phosphoric Acid

In the phosphoric acid reaction, ground rock and sulfuric acid are co-fed to a reactor or series of reactors containing phosphoric acid and calcium sulfate. The most common reaction holds the reaction temperature at 70°C–80°C and produces the dihydrate form of calcium sulfate, which is usually referred to as gypsum. The reaction mass is fed continuously to a filter; the filter cake is washed with water in a countercurrent fashion, and the resultant weak acid is returned to the reactor to control both the solid content and the P_2O_5 content of the acid; the filter cake is then transferred either to the gypsum stack or to the sea if the plant is located nearby suitable open water. It is possible to treat the gypsum, principally by washing and drying, to the extent that it is suitable for processing to plasterboard. This is practiced in Belgium and Japan but is usually uneconomic compared with mining virgin gypsum. From the filter, the acid is pumped to storage. At this stage the acid concentration is approximately 30% P_2O_5; for use in fertilizer or further processing, the acid is concentrated

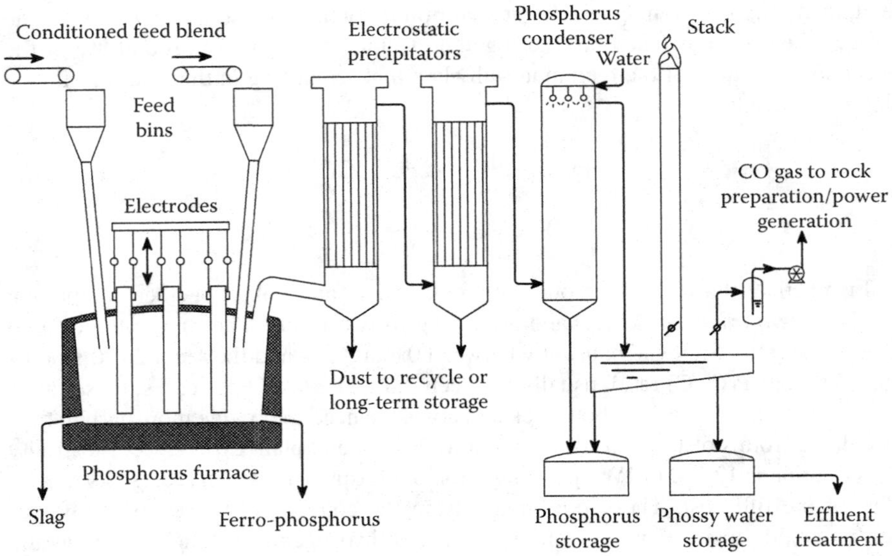

FIGURE 1.12 Phosphorus plant flow sheet.

to 40% and 54% P_2O_5. Phosphoric acid concentrators utilize low-pressure steam; steam is provided by dedicated boilers, power plant on site, or usually, from the heat recovery steam generator of the sulfuric acid plant that supplies the sulfuric acid. In addition to gypsum, a phosphoric acid plant generates a complex phosphatic sludge particularly in the concentration and storage stages and fluoride-based waste. The sludge must be worked off into solid fertilizer products to recover the value of the phosphate it contains. On some plants, the fluoride-based waste is processed to fluosilicic acid (H_2SiF_6) for sale in the production of aluminum and fluoridation of water.

1.2.5 Fertilizers

Fertilizers provide a range of nutrients to support plant growth; the principal macronutrients are nitrogen (N), phosphorus (P), and potassium (K); other important components are aluminum (Al), calcium (Ca), magnesium (Mg), and sulfur (S). Fertilizers also contain micronutrients, for example, zinc (Zn) and molybdenum (Mo); whereas the concentration of principal nutrients is measured in weight percent, micronutrient concentration is measured in parts per million (ppm). Fertilizers are sold with specific NPK ratios reflecting the different requirements of different crops at different stages in their growth cycle and the depletion of key nutrients in the soil. NPK is short for weight percentage nitrogen (%N), phosphorus pentoxide (%P_2O_5), and potassium oxide (%K_2O). For example, a 1000 kg big bag of 28-14-14 fertilizer would contain 280 kg nitrogen, 140 kg phosphorus pentoxide, and 140 kg potassium oxide. Fertilizer manufacturers react or blend several different materials to make different fertilizers depending on the final product form, liquid or solid; on product sophistication, for example, a simple bulk blend or a double-coated horticultural product; and on the price and availability of the raw materials. Single superphosphate (SSP), triple superphosphate (TSP), MAP, and DAP are the main phosphatic fertilizer chemicals and together with finely ground phosphate rock (DAR), defluorinated rock, and partially acidified rock are used either as products themselves or as components of NPK fertilizers. Phosphoric acid is utilized in NPK production and nitric acid in the production of nitric phosphates. The reactions to SSP, TSP, MAP, and DAP are shown in the following equations:

$$Ca_{10}(PO_4)_6F_2 + 7H_2SO_4 \rightarrow [3Ca(H_2PO_4)_2 + 7CaSO_4] + 2HF \quad (1.10)$$

$$Ca_{10}(H_2PO_4)_6F_2 + 14H_3PO_4 \rightarrow 10Ca(H_2PO_4)_2 + 2HF \quad (1.11)$$

$$NH_3 + H_3PO_4 \rightarrow NH_4H_2PO_4 \quad (1.12)$$

$$2NH_3 + H_3PO_4 \rightarrow (NH_4)_2HPO_4 \quad (1.13)$$

SSP is also known as normal, regular, or ordinary superphosphate. SSP is a mixture of monocalcium phosphate and calcium sulfate and all the other nonvolatile components of feed material. TSP, MAP, and DAP contain lower levels of impurities than SSP and higher P_2O_5 levels. A key component of the value of a fertilizer is its *available* phosphate content, which is less than its total phosphate content. Available P_2O_5 is measured by standard procedures; one measures citrate-soluble P_2O_5, and the other water-soluble P_2O_5; a total P_2O_5 measurement completes the analysis.

SSP production is relatively straightforward. Sulfuric acid is diluted and cooled and pumped continuously to a mixing vessel; usually this is a TVA cone mixer that has no moving parts; mixing is achieved by the swirl energy of the pumped acid. Phosphate rock is crushed and milled and conveyed to the same mixing vessel. The reacting slurry is fed via a mixing conveyor onto a moving bed in the *den*; sometimes the mixing vessel is part of the den assembly; the slurry becomes more solid forming a cake during a retention time of the order of 30 min. At the end of the moving bed is a cutter that breaks up the cake. The superphosphate is then sent either directly to storage for curing, typically for 2 weeks, or is mixed with other materials, for example, clay, limestone, potash, or ammonium sulfate. This mix, together with water, is then fed to a granulator and then to classification and storage. Off-gases, principally hydrogen fluoride and silica fluoride fumes, are scrubbed. An even simpler process is known as *run of pile* (ROP); diluted sulfuric acid is pumped to a TVA conical mixer and contacts the milled rock. The reaction slurry then falls to a conveyor and is conveyed for curing. Figure 1.13 shows a simplified flow sheet of superphosphate manufacture.

The manufacture of TSP, or more usually granular triple superphosphate (GTSP), commences with the reaction of phosphoric acid and milled phosphate rock in one or more reactors. Feed acid concentration is typically 54% P_2O_5; different concentrations are possible, both higher and lower and simply affecting the water balance. The reaction slurry overflows to a blunger (a mixing vessel with rotating arms) where it is mixed with recycled final product fines and milled oversize material. The mixed material is then dried in a rotary drier and screened. The product that meets specification is transferred to storage and packing.

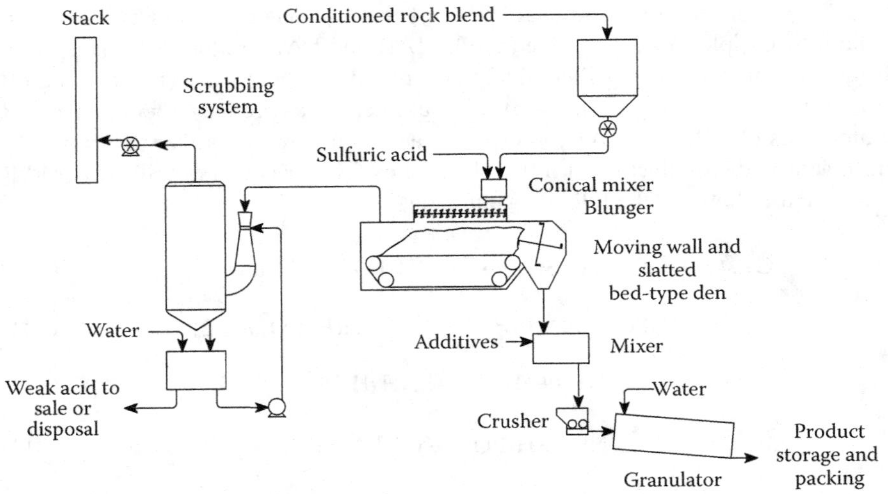

FIGURE 1.13 SSP plant flow sheet.

There are several different process plant designs to manufacture ammonium phosphates; many modern plants either follow or are based on the TVA patents, US2729554, issued in 1956 [32], and US3954942, issued in 1976 [33]. These claimed firstly a continuous process that could use anhydrous ammonia without undue creation of unavailable P_2O_5 values and secondly the concept of a pipe reactor further reducing production plant complexity. In recent years, patents have been granted for derivative designs; in the main, the newer designs have the objective either of incorporating additional components into the ammonium phosphate, such as micronutrients (e.g., US7497891, granted to the Mosaic Company and issued in 2009 [34]) or improving the energy efficiency of the process. In the Mosaic process, phosphoric acid and micronutrients are fed to two agitated prereactors; one reactor feeds the preneutralizer where the acid reacts with ammonia, and the second feeds the pipe cross reactor (PCR). In the PCR, phosphoric acid and ammonia react at high temperature producing a molten ammonium phosphate, which is sprayed into the granulator forming rough ammonium phosphate particles. A rolling bed of particles progresses along the granulator; ammonia is sparged into the bottom of the bed; preneutralizer liquor is sprayed onto the bed; milled recycle material is returned to the granulator and incorporated in the bed. The rolling bed makes the particles substantially spherical; they are then transferred to the drum drier and dried in hot air at 250°C–320°C. A dry, hot product at 80°C–90°C is then screened; oversize is milled and recycled with fine material; in specification material is cooled and sent to packing. The molar ratio of ammonia to phosphoric acid controls the ratio of MAP and DAP, 2:1 being the theoretical ratio for 100% DAP. The ratio also affects solubility; the solubility curve is shown in Figure 6.3 [35,36]. Consequently, the molar ratio chosen in the preneutralizer is 1.4, corresponding to the maximum solubility for the corresponding temperature and therefore most easily pumped. The molar ratio is increased to 1.85–1.95 in the granulator by sparging

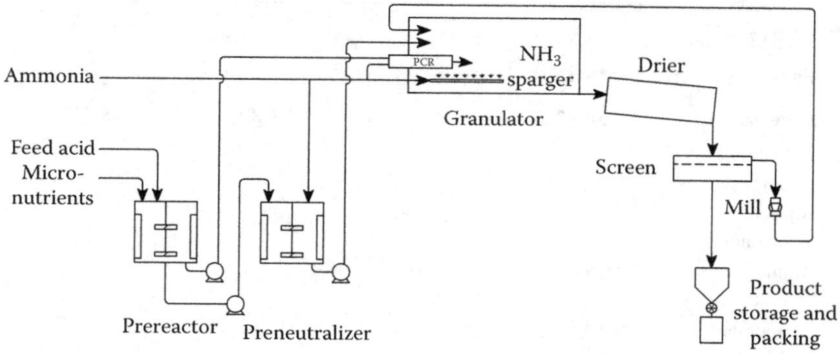

FIGURE 1.14 Mosaic ammonium phosphate flow sheet based on US7497891.

with ammonia, where minimum solubility promotes solidification. Impurities such as iron, aluminum, and magnesium phosphates account for the difference with the theoretical ratio. Sulfur may be incorporated in the product being added either via a spray of elemental sulfur or by the addition of solid ammonium or calcium sulfate. Figure 1.14 shows a simplified flow sheet of the Mosaic process.

The flow sheet for mixed NPK fertilizers is very similar to that in Figure 1.13. Two additional liquor feeds are added, sulfuric acid and nitrogen solution, nitrogen solution being a variable liquid mixture of ammonium nitrate, ammonia, and water. Solid feeds may be added: SSP, enhanced or TSP, ammonium sulfate, and potash or potassium sulfate. Using a software model, a formulation is chosen depending on the desired NPK product ratio, P_2O_5 availability, and raw material prices. Table 1.3 shows fertilizer raw materials, with their weight percent N, P_2O_5, and K_2O giving both a typical analysis figure and a range.

Other fertilizers include liquid fertilizers, some of which are formulated in the field, and nitric fertilizers where nitric acid is used to acidulate phosphate rock or superphosphates.

1.2.6 Purified Phosphoric Acid

WPA is insufficiently pure for uses other than fertilizers. The principal impurities are measured in weight percent. The most elementary purification is the reduction of hydrogen fluoride values; this is done to permit the production of animal feed grade dicalcium phosphate. Chemical purification, also known as the *wet salts* process, whereby the wet acid pH is raised to pH 9 with either sodium carbonate or sodium hydroxide to precipitate metal phosphates, is described in Section 2.2 of Chapter 2 and is still practiced. Chemical purification does not provide a sufficiently pure phosphate for many applications, in fact anything other than technical grade sodium phosphates. All purified acid is now manufactured by solvent extraction, which is discussed in detail also in Chapter 2. Most technical, food, dental, pharmaceutical, and the least demanding semiconductor specifications are satisfied by solvent-extracted acid; the most pure grades require further purification with crystallization or filtration.

TABLE 1.3
Fertilizer Raw Materials

Materials	Formula	%N	$\%P_2O_5$	$\%K_2O$
Anhydrous ammonia	NH_3	82.2 82.2–82.2		
Aqueous ammonia	NH_3	20.0–25.0		
Ammonium nitrate	NH_4NO_3	33.5 33.0–35.0		
Ammonium sulfate	$(NH_4)_2SO_4$	20.5 20.0–21.0		
Urea	$(NH_2)_2CO$	45.0 45.0–46.0		
Nitric acid	HNO_3	22.2 22.2–22.2		
Nitrogen solutions[a]		44.0 30.0–55.0		
MAP[b]	$NH_4H_2PO_4$	11.0 11.0–13.0	53.5 48.0–61.0	
DAP[c]	$(NH_4)_2HPO_4$	18.0 18.0–21.0	46.0 46.0–53.0	
Phosphate rock (75 BPL)[d]	$Ca_{10}(PO_4)_6F_2$		34.3 28.0–38.0	
Phosphoric acid	H_3PO_4		54.0 28.0–54.3	
SSP			19.0 19.0–20.0	
TSP			45.0 44.0–46.0	
Potash	KCl			60.0 60.0–63.0
Potassium sulfate	K_2SO_4			50.0 48.0–53.0

[a] Nitrogen solutions have varying proportions of ammonium nitrate, ammonia, urea, and water.

[b] There are a range of commercial MAP grades including 11-(51-55)-0 and small quantities at 12-61-0.

[c] Nearly all commercially DAP is 18-46-0.

[d] BPL is the traditional way of expressing the P_2O_5 content of rock as tricalcium phosphate $Ca_3(PO_4)_2$, for example, 75 BPL is $75 \div 2.185 = 34.3\%P_2O_5$.

1.2.7 Derivative Phosphates

Derivative phosphates of commercial importance include phosphates of aluminum, ammonia, calcium, lithium, potassium, and sodium. The final product specification governs the quality of both the phosphoric acid and the metal salt or oxide reactants. Typically food grade products require food grade raw materials, similarly technical grade products; however, careful attention to the analysis of the raw materials and the product specification does permit the use of raw materials that might not in all respects meet either a food or technical grade specification. For example, a phosphoric acid meeting a food grade specification in all impurities other than sodium could be used in the production of food grade sodium phosphates. (Strictly speaking, there is no sodium specification for food grade acid;

it falls within the alkali phosphate specification. The test for this specification requires that an acid sample dissolved in alcohol remains clear. If the levels of sodium or other alkali phosphates are too high the sample is cloudy and fails the test. In this example the decision to use the acid would be based on analysis of the principal minor impurities.)

Most commercial aluminum phosphate is monoaluminum phosphate; its equation is shown as Equation 1.14. With heating, the monophosphate is transformed to aluminum meta- and polyphosphates as well as berlinite, $AlPO_4$:

$$Al(OH)_3 + 3H_3PO_4 \rightarrow Al(H_2PO_4)_3 + 3H_2O \tag{1.14}$$

The principal use for aluminum phosphate is as a refractory binder. Other applications, such as paint pigment or piezoelectric crystals, and other aluminum phosphate forms are discussed in Chapter 6.

Technical and food grade ammonium phosphates are made as well as fertilizer grades, although at much higher purity. The process is based on continuous vacuum crystallization. The chemical equations, however, are identical to those for fertilizer. The main uses are flame retardants and specialty fertilizers.

There are six commercial calcium phosphate chemicals that are produced utilizing thermal or purified phosphoric acid. The monocalcium and dicalcium phosphates are produced in both the hydrated and anhydrous forms (MCP, MCPa, DCP2, DCPa): tricalcium phosphate (which is available in a number of forms or mixtures—the most common commercial grade is hydroxyl apatite, $Ca_5(OH)(PO_4)_3$) and calcium pyrophosphate (CPP). The calcium phosphates are discussed in detail in Chapter 5. Equations 1.15 through 1.20 summarize the reactions; all the equations assume calcium hydroxide (*slaked lime*) as the calcium source (calcium oxide, *lime*, is usually purchased and slaked on site); for mono- and dicalcium phosphates, operating companies also use calcium carbonate as an alternate, the choice governed primarily by supply price and in some cases impurity considerations. The wet slaking process allows the rejection of other minerals present in the lime:

$$2H_3PO_4 + Ca(OH)_2 \rightarrow Ca(H_2PO_4)_2 \cdot H_2O + H_2O \tag{1.15}$$

$$Ca(H_2PO_4)_2 \cdot H_2O \xrightarrow{\text{Heat}} Ca(H_2PO_4)_2 + H_2O \tag{1.16}$$

$$H_3PO_4 + Ca(OH)_2 \xrightarrow{\leq 38^{\circ}C} CaHPO_4 \cdot 2H_2O \tag{1.17}$$

$$H_3PO_4 + Ca(OH)_2 \xrightarrow{\geq 38^{\circ}C} CaHPO_4 + 2H_2O \tag{1.18}$$

$$3H_3PO_4 + 5Ca(OH)_2 \rightarrow Ca_5(OH)(PO_4)_3 + 9H_2O \tag{1.19}$$

$$2CaHPO_4 \cdot 2H_2O \xrightarrow{\text{Heat}} Ca_2P_2O_7 + 5H_2O \tag{1.20}$$

Calcium phosphates have a range of applications, and the foremost are food and dental uses.

Lithium phosphate is made for use in porcelain glazes, in polymer intermediates, and as a catalyst in the isomerization of propylene oxide. Lithium iron and lithium iron magnesium phosphates are used in emerging battery technology and have some advantages over lithium ion batteries; at present, there are several routes to these products, which are briefly discussed in Chapter 6. The simple lithium phosphate reaction is shown in the following equation:

$$H_3PO_4 + 3LiOH \rightarrow Li_3PO_4 + 3H_2O \tag{1.21}$$

Di- and trimagnesium hydrated phosphates (DMP3, TMP8) are produced in small quantities often in conjunction with dicalcium phosphate for dental applications: either magnesium phosphate acts as a stabilizer for the dicalcium phosphate when it is mixed in with fluorides in toothpaste formulations. Other uses include antacids, laxatives, and food additives. The magnesium phosphate reactions are shown in Equations 1.22 and 1.23; either magnesium oxide or magnesium hydroxide can be used as the magnesium source:

$$H_3PO_4 + Mg(OH)_2 \rightarrow MgHPO_4 + 2H_2O \tag{1.22}$$

$$2H_3PO_4 + 3Mg(OH)_2 \rightarrow Mg_3(PO_4)_2 + 6H_2O \tag{1.23}$$

There are five potassium phosphate products of commercial interest: mono-, di-, and tripotassium phosphates (MKP, DKP, TKP); tetrapotassium pyrophosphate (TKPP); and potassium tripolyphosphate (KTPP). Potassium phosphates have a wide range of applications, many in liquid formulations; cleaners and antifreeze are two examples. The potassium phosphate reactions are shown in the following equations:

$$H_3PO_4 + KOH \rightarrow KH_2PO_4 + H_2O \tag{1.24}$$

$$H_3PO_4 + 2KOH \rightarrow K_2HPO_4 + 2H_2O \tag{1.25}$$

$$H_3PO_4 + 3KOH \rightarrow K_3PO_4 + 3H_2O \tag{1.26}$$

$$2K_2HPO_4 \xrightarrow{\text{Heat}} K_4P_2O_7 + H_2O \tag{1.27}$$

$$2K_2HPO_4 + KH_2PO_4 \xrightarrow{\text{Heat}} K_5P_3O_{10} + 2H_2O \tag{1.28}$$

Commercially, sodium phosphates remain the largest group of derivative phosphates with volumes greater than the combined volume of calcium, potassium, and ammonium phosphates. Although it is currently in decline, STPP has since the 1950s been the largest volume sodium phosphate. There are at least 10 commercial sodium phosphates (excluding hydrates): mono-, di- and trisodium phosphates (MSP, DSP, TSP); chlorinated trisodium phosphate (TSP-chlor); sodium aluminum phosphate (SALP); SAPP, also known as disodium pyrophosphate (DSPP); tetrasodium pyrophosphate (TSPP); sodium trimetaphosphate (STMP); STPP; and sodium hexametaphosphate (SHMP). Hexametaphosphate is a generic term used erroneously to describe polymeric phosphates of varying chain lengths (more correctly termed glassy phosphates). True hexametaphosphate is a cyclic compound with six phosphorus atoms as is trimetaphosphate (three phosphorus atoms), a polymerized phosphate of varying chain length. Sodium carbonate (*soda ash*—Na_2CO_3) and sodium hydroxide (*caustic soda*—NaOH) are both sodium sources of commerce; indeed, the volumes consumed in the production of STPP have in the past been sufficient to affect the pricing of these raw materials at a national level. The sodium phosphate reactions are shown in the following equations using sodium hydroxide:

$$H_3PO_4 + NaOH \rightarrow NaH_2PO_4 + H_2O \tag{1.29}$$

$$H_3PO_4 + 2NaOH \rightarrow Na_2HPO_4 + 2H_2O \tag{1.30}$$

$$H_3PO_4 + 3NaOH \rightarrow Na_3PO_4 + 3H_2O \tag{1.31}$$

$$Na_3PO_4 + \frac{1}{4}NaOCl + 11H_2O \rightarrow Na_3PO_4 \cdot \frac{1}{4}NaOCl \cdot 11H_2O \tag{1.32}$$

$$16H_3PO_4 + Na_2CO_3 + 6Al(OH)_3 \rightarrow 2NaAl_3H_{14}(PO_4)_8 \cdot 4H_2O + CO_2 + 11H_2O \tag{1.33}$$

$$2NaH_2PO_4 \xrightarrow{225°C–250°C} Na_2H_2P_2O_7 + H_2O \tag{1.34}$$

$$2Na_2HPO_4 \xrightarrow{>450°C} Na_4P_2O_7 + H_2O \tag{1.35}$$

$$3NaH_2PO_4 \xrightarrow{>500°C} (NaPO_3)_3 + 3H_2O \tag{1.36}$$

$$NaH_2PO_4 + 2Na_2HPO_4 \xrightarrow{>450°C} Na_5P_3O_{10} + 2H_2O \tag{1.37}$$

$$(n-2)NaH_2PO_4 + 2Na_2HPO_4 \xrightarrow{>800°C} Na-(NaPO_3)_n - ONa + nH_2O \tag{1.38}$$

1.2.8 Phosphate Rock

Phosphorus as phosphate has been applied to crops through the use of manure for thousands of years. In the nineteenth century, *guano*, from the Kichwa language of the Andes, meaning *the droppings of birds*, from Chile was used extensively as a fertilizer. In the second half of the nineteenth century, mined and subsequently acidulated phosphate rock was used. Today, although trials are still carried out on alternative sources, all phosphoric acid is made from phosphate rock. Phosphate rock resources are not inexhaustible so efforts to close the phosphorus cycle are important and are discussed in Chapter 7.

The geochemists tell us that the Earth's crust comprises 0.1%–0.35% P_2O_5 and has a mass of the order of 10^{20} tons suggesting a P_2O_5 content of $1.0–3.5 \times 10^{17}$ tons. Much of this occurrence is inaccessible whether on land or in ocean beds. In 2010, the

TABLE 1.4
Phosphate Rock Mined and Remaining Reserves

	Production		
Mine	**2009**	**2010**	**Reserves**
China	60,200	65,000	3,700,000
United States	26,400	26,100	1,400,000
Morocco and Western Sahara	23,000	26,000	50,000,000
Saudi Arabia[a]		11,600	534,000
Russia	10,000	10,000	1,300,000
Tunisia	7,400	7,600	100,000
Jordan	5,280	6,000	1,500,000
Brazil	6,350	5,500	340,000
Egypt	5,000	5,000	100,000
Israel	2,700	3,000	180,000
Syria	2,470	2,800	1,800,000
Australia	2,800	2,800	82,000
South Africa	2,240	2,300	1,500,000
Algeria	1,800	2,000	2,200,000
Togo	850	800	60,000
Canada	700	700	5,000
Senegal	650	650	180,000
Other countries	8,620	9,500	620,000
World total	166,000	176,000	66,000,000

After the USGS Mineral Commodities Survey 2011.

[a] The Ma'aden company's Al Jamid mine commenced operation in 2011 with an expected output of 11.6Mt per year and has reserves of 534Mt; source http://www.maaden.com.sa/eng/phosphate_project.htm.

worldwide production was 176 million tons and reserves 65 billion tons [37]. Table 1.4 shows the estimates of world production and reserves for the major national producers. Figure 1.15 shows a graph of the output of the world and the United States since 1900.

The totals in Table 1.4 suggest a resource life of 375 years; however, this assumes neither further growth in fertilizer demand nor the realization of new phosphate resources. World population growth is directly linked to food supply. Food supply is directly linked to the availability of fertilizer and consequently phosphate rock. Therefore world population growth drives phosphate rock demand.

Phosphate rock is found in two forms, igneous and sedimentary, the latter providing most of the world's supply. The largest igneous rock deposits currently mined are in Russia, South Africa, and Brazil. Geologists tell us that millions of years ago igneous deposits were eroded and washed into the seas settling to the seabeds. Plants and animals incorporated and concentrated phosphates into their body structures and in time became part of the ground and fossilized. Over millions of years of Earth movement, geological phosphate ore bodies were formed and these are the

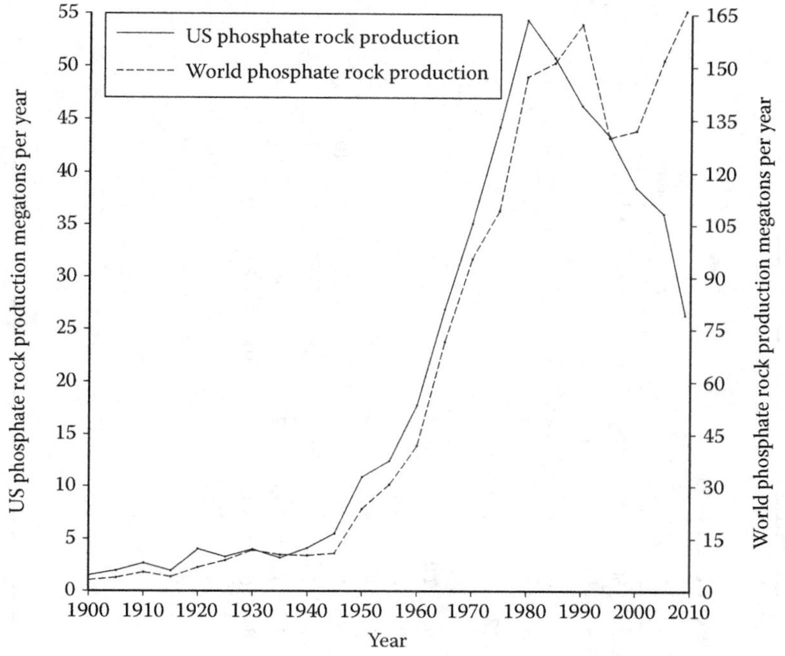

FIGURE 1.15 The US and world phosphate rock production 1900–2009 data from USGS.

sedimentary rocks now mined. Some resources are inaccessible or uneconomic, others are economic but require underground mining, and others require only the removal of a relatively shallow overburden.

Of the nearly 200 different forms of phosphate rock, all but a handful are in the apatite group. The three forms of interest are fluorapatite, chlorapatite, and hydroxylapatite, now conventionally referred to as apatite-(CaF), apatite-(CaCl), and apatite-(CaOH). Fluorapatite and its partially carbonated or hydroxylated variants are the major ore. The unit cell formula $Ca_{10}(PO_4)_6F_2$ is used in simplified equations; however, a more generalized formula is $M_{10}X_2(RO_4)_6$ where M = Ca, Pb, Na, K, Sr, Mn, Zn, Cd, Mg, Fe^{2+}, Al, C (as CO_3), H_2O, and the rare earths especially Ce; X = F, OH, Cl, and Br; and $RO_4 = PO_4$, AsO_4, VO_4, SO_4, and SiO_4 [38]. The phosphate ore body is always complex; the phosphate ion is often accompanied by a good proportion of the periodic table, if only in trace amounts. Trace impurities are relatively unimportant for fertilizer production but must be considered carefully when contemplating purifying phosphoric acid. Phosphate deposits vary in composition around the world and even within a local mine, hence the need to analyze the rock on a regular basis as part of a wet process acid plant control strategy. Nevertheless, the impurity profiles of rock from different regions are broadly similar and understood in the industry. Table 1.5 shows a selection of phosphate rock from different locations with typical analyses. It should be noted that these are beneficiated rock, that is, mined phosphate ore that may have been washed and separated from gangue by flotation, sometimes calcined, sized, and concentrated up to a marketable product. The concentration

TABLE 1.5
Typical Impurity Analyses of Phosphate Rock from Different Locations

	%												ppm	
Rock Source	**P_2O_5**	**CaO**	**SiO_2**	**Fe_2O_3**	**Al_2O_3**	**MgO**	**CO_2**	**SO_3**	**F**	**Na_2O**	**K_2O**	**C[a]**	**Cl**	**As**
Algeria	34.5	54.5	2.4	0.3	0.25	0.45	1.3	1.6	4.0	0.45				
Australia[b]	32.0	45.0	14.5	0.65	0.5	0.1	1.2	0.3	3.4	0.15				
Brazil	35.5	47.5	0.4	1.7	0.2	0.05	1.7		2.5	0.1				
China[c]	33.8	44.6	9.88	1.7	0.98	0.21	1.12	0.31	2.79					
Christmas Is	37.5	50.5	0.6	1.4	1.5	0.1	2.3		1.3	0.4				
Egypt	28.5	45.5	8.3	0.9	0.2	0.15	5.0	2.0	2.9	1.0				
Israel	32.0	51.0	1.6	0.2	0.1	0.2	6.0	2.0	4.0	0.6				
Jordan	34.0	52.0	3.2	0.12	0.1	0.15	4.2	1.3	4.0	0.4	0.02			
Morocco[d]	34.0	52.0	2.4	0.15	0.2	0.2	4.2	1.2	3.9	1.2	0.8	0.2	200	10
Morocco[e]	33.0	53.0	2.0	0.2	0.3	0.35	3.1	2.1	4.4	0.7	0.06	<0.1	600	2
Nauru[f]	38.0	52.4	0.16	0.18	0.34	0.29	3.3	0.28	2.75	0.2		0.4	32	
Togo	35.5	50.5	5.0	1.6	1.1	0.12	1.8	0.3	4.1	0.2	0.04	0.04	375	

[a] Organic carbon.
[b] Duchess deposit, Phosphate Hill, Queensland.
[c] *World Survey of Phosphate Deposits* 4th edn. British Sulphur Corporation Limited, 1980.
[d] Khourigba.
[e] Youssoufia–calcined.
[f] Now mined out.

process for igneous rock phosphates is more extensive because the primary ore may contain as little as 10% apatite.

There are many considerations when selecting a phosphate rock supply for the production of wet process acid. Obviously, price is a consideration, as is security and volume of reasonably consistent supply. For example, the purchase of small quantities of phosphate rock from different suppliers might be cheapest in terms of rock purchase; however, if the supplies vary greatly in impurity analysis, this may lead to a poorly controlled acid plant, consuming more sulfuric acid than necessary and generating increased P_2O_5 losses. Therefore, for the business as a whole, this might not be the best strategy.

In terms of analysis, the first consideration in rock selection is P_2O_5 content as this is linked directly to the amount of phosphoric acid that will be produced; rock price is linked primarily to P_2O_5 content—in practice, this is often referred to as bone phosphate of lime (BPL) content for historical reasons (BPL is equivalent to 2.183 times P_2O_5). Wet process acid plants have P_2O_5 efficiencies in the range of 85%–98% depending primarily on the rock quality and assuming good plant management. The second consideration is the calcium content, always expressed as CaO, as this governs the amount of sulfuric acid that is required to acidulate the rock. The CaO/P_2O_5 weight ratio is a common measure of rock quality; for pure apatite, the ratio is 1.32; commercially available rock may have a CaO/P_2O_5 ratio up to 1.6. The third consideration is the grindability of the rock. Different rocks require different finenesses to maintain similar reaction kinetics; an existing plant of fixed size needs to maintain similar reaction rates to maintain output. The grinding or milling of rock is potentially the third or fourth highest variable cost, as an electricity charge, and influences plant profitability.

Other impurity considerations are as follows:

a. Silica, SiO_2: Broadly speaking, silica is present either as quartz, which reacts very slowly; reactive silica, which reacts with fluorine to form sodium or potassium silicofluoride and may cause blockages on cool surfaces such as pipework and heat exchangers, and hamper filtration; or residual silica remaining in solution or present as a colloid in the acid, which can be a problem in solvent extraction.
b. Iron, Fe_2O_3; aluminum, Al_2O_3; and magnesium, MgO: Iron and aluminum form phosphate sludges as the acid is concentrated to sale concentrations contributing to P_2O_5 losses and maintenance requirements. As P_2O_5 concentrations increase, high iron contents contribute to increased acid viscosity. Aluminum affects crystal habit helpfully; high aluminum content is associated with good filtration rates. Aluminum also reacts with fluoride ions to form aluminofluoride complexes. Above certain levels, magnesium concentration is linked to increased product acid viscosity and to the formation of struvite, $MgNH_4PO_4 \cdot 6H_2O$, when making ammonium phosphates. (Human kidney stones also comprise struvite.) In the industry, two ratios are used to describe these impurities, the iron and aluminum (I&A) ratio and the minor element ratio (MER). The lower the ratios, the more pure the rock. In practice, it is very difficult to manufacture the standard 18-46-0 NPK fertilizer when using an acid with an MER greater than 0.1; a working limit of 0.085

is used when contemplating fertilizers with nitrogen content greater than or equal to 18%. I&A and MER are defined in the following equations:

$$I\&A = \frac{(\%Al_2O_3 + \%Fe_2O_3)}{P_2O_5} = 0.08 - 0.10 \quad (1.39)$$

$$MER = \frac{(\%Al_2O_3 + \%Fe_2O_3 + \%MgO)}{P_2O_5} = 0.08 - 0.20 \quad (1.40)$$

c. Carbonate, CO_2: Carbonate is a nuisance as it creates foam as carbon dioxide is released from the reaction mass. The volume of gas in the slurry must be controlled in order to control the reaction and is achieved with antifoaming agents. In turn, these agents can cause problems downstream in solvent extraction units.
d. Sulfate, SO_3: The sulfate content of rock simply reduces the amount of sulfuric acid required for acidulation.
e. Fluoride, F: Fluoride may be as high as 4% of the rock analysis. During the phosphoric acid reaction, the fluoride ions released from the rock form hydrofluoric acid (HF), which in turn reacts with reactive silica, forming silica tetrafluoride, SiF_4. Silica tetrafluoride is then hydrolyzed to hydrofluosilicic acid, H_2SiF_6, a strong acid that participates in the acidulation of the rock. The silica hexafluoride ion reacts with sodium and potassium ions to form sodium and potassium silica hexafluoride, Na_2SiF_6 and K_2SiF_6, precipitates that can cause blockages. These reactions are shown in Equations 1.41 through 1.44. Insufficient reactive silica in the reaction mass leads to HF remaining in solution and the evolution of some hydrogen fluoride gas, both of which are very corrosive to metals. The fluoride levels in acid intended for further treatment on a purification plant must not be too high; the design and sizing of the concentration, solvent stripping, and defluorination process steps are all affected by the fluoride levels in the feed acid, which must be managed from corrosion, blockage, and emission standpoints. Therefore, acid soluble silica may be added to achieve a reasonable fluoride balance. In general, acid intended for purification via solvent extraction should have a slight fluoride, rather than silica, excess to ameliorate the solvent extraction step:

$$4HF + SiO_2 \rightarrow SiF_4 + 2H_2O \quad (1.41)$$

$$3SiF_4 + 2H_2O \rightarrow 2H_2SiF_6 + SiO_2 \quad (1.42)$$

$$2Na^+ + SiF_6^{2-} \rightarrow Na_2SiF_6 \quad (1.43)$$

$$2K^+ + SiF_6^{2-} \rightarrow K_2SiF_6 \quad (1.44)$$

f. Sodium, Na_2O, and potassium, K_2O: Sodium and potassium react with hydrofluosilicic acid to form sodium and potassium silica hexafluoride.

g. Organic carbon, C: *Organics* (meaning complex organic compounds) cause foaming and may form solids and discoloration of the acid. Organics can be reduced significantly by calcining the rock as part of the beneficiation process. Alternatively, chemical oxidation is possible but expensive and is usually restricted to food grade acids after solvent extraction. Organic levels much above 500 ppm will adversely affect solvent extraction processes.
h. Chloride, Cl: Chloride content is linked to corrosion. In general, the wet acid process is very corrosive requiring the use of graphite, plastics, and high-alloy steels; even with these materials, particularly the steels, high chloride content (above about 0.1%) is problematic.
i. Arsenic, As, and cadmium, Cd: Heavy metals are toxic and include arsenic, cadmium, and lead, Pb. Both fertilizers and phosphoric acids have heavy metal limits in their specifications. Acid intended for fertilizer use does not normally undergo any particular treatment to reduce heavy metal content; therefore, the heavy metal levels in the rock must be such that permit fertilizer production within specification. Heavy metals are specifically removed from phosphoric acid going to food grade applications. Clearly, the cost of this removal is linked to the quantity to be removed.

With a representative analysis of a prospective rock, it is possible to make a preliminary assessment of the variable cost of production.

Equation 1.45 [39] is empirical and gives an estimate of sulfuric acid consumption per ton of P_2O_5 product acid produced out of the filter:

$$S=\left(\frac{1.732\text{CaO}}{P_2O_5-0.02\text{CaO}}-1.225\frac{SO_3}{P_2O_5}+0.062\right)\frac{100}{100-\%\text{SL}} \tag{1.45}$$

where

S is the tons 100% H_2SO_4 per ton of P_2O_5 produced out of the filter
CaO is the weight fraction of CaO in the rock
P_2O_5 is the weight fraction of P_2O_5 in rock
SO_3 is the weight fraction of SO_3 in rock
%SL is the percent of soluble P_2O_5 losses in filtration and spillage

For example, for the Morrocan Khourigba rock in Table 1.5 and assuming 1.5% soluble P_2O_5 losses,

$$S=\left(\frac{1.732\times0.52}{0.34-0.02\times0.52}-\frac{1.225\times0.012}{0.34}+0.062\right)\frac{100}{100-1.5}=2.79 \tag{1.46}$$

So, if producing phosphoric acid from Khourigba rock, the plant would consume 2.79 tons 100% sulfuric acid per ton of phosphoric acid expressed in tons P_2O_5. The rock requirement must take account of both insoluble and soluble P_2O_5 losses. Insoluble losses include *unattacked losses*, representing phosphate rock that has not been effectively acidulated possibly due to encrustation by calcium sulfate during the reaction, and *lattice losses*, which include $CaHPO_4$, dicalcium phosphate, losses (also known as citrate-soluble losses), and metal phosphates coprecipitating with the calcium sulfate. As an estimate, unattacked losses might amount to 0.5% and lattice losses to 3%. Rock consumption, R, in tons per ton of phosphoric acid expressed in tons P_2O_5 is calculated from

$$R = \frac{100}{P_2O_5\left(\%REC\right)} \tag{1.47}$$

where

P_2O_5 is the weight fraction of P_2O_5 in rock
%REC is the overall P_2O_5 recovery

Thus, for the same Khourigba rock,

$$R = \frac{100}{0.34 \times 95} = 3.10 \tag{1.48}$$

So, if producing phosphoric acid from Khourigba rock, the plant would consume 3.10 tons of rock per ton of phosphoric acid expressed in tons P_2O_5.

In practice, a company considering a new rock supply, whether for a new plant or an existing one, will always carry out laboratory-based trials. From these trials, more accurate estimates can be made of rock and sulfuric acid consumption, P_2O_5 losses, calcium sulfate (*gypsum*) production, and filterability. These data allow the sizing of new plant or rate comparisons on existing plant.

1.2.9 Wet Process Acid

There are a number of WPA processes. These processes may be categorized by licensor and the calcium sulfate hydrate, for example, the Prayon Mark IV dihydrate process or the Yara hemihydrate and HDH processes. In essence, all the processes comprise four steps: rock preparation (grinding), reaction, filtration, and concentration; some processes, operating with a suitable rock, claim to be able to avoid the first and last steps. Although the desired product is phosphoric acid, in terms of design and control, it is more helpful to think of these processes as crystallization processes to produce calcium sulfate with a minimal P_2O_5 content and a valuable phosphoric acid by-product. Consequently, the $CaSO_4$–P_2O_5–H_2O system has been studied in considerable detail [25]. Figure 1.16 shows the system in the area of interest. In this region, the three forms of calcium sulfate are dihydrate, $CaSO_4 \cdot 2H_2O$; α-hemihydrate,* α-$CaSO_4 \cdot ½H_2O$; and anhydrite II, $CaSO_4$ II. Both the acidulation of phosphate rock and the attendant dilution of sulfuric acid are exothermic. As reaction temperatures approach 75°C, the rate of formation of α-hemihydrate increases and dihydrate decreases [40] with consequent decrease in filtration efficacy. Furthermore, the higher the temperature, the greater the corrosive effect of

* Indicates an overstoichiometric and variable content of water of crystallization.

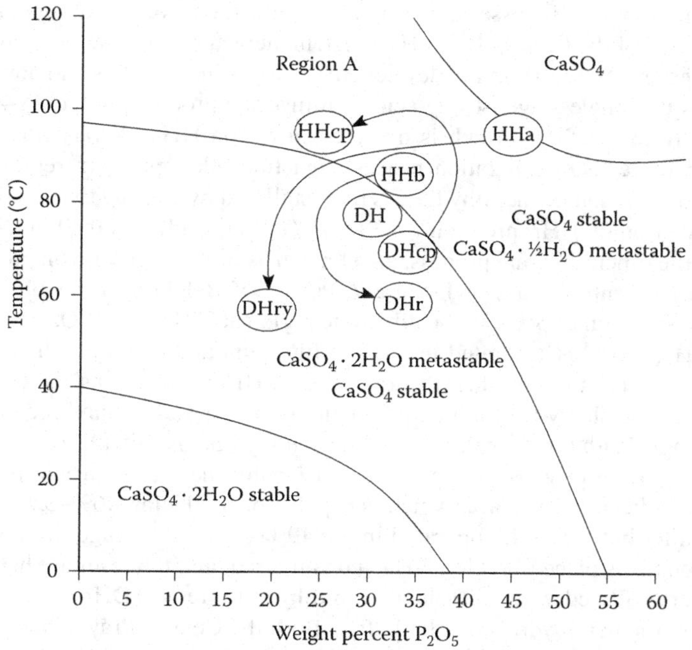

FIGURE 1.16 Equilibrium diagram for the $CaSO_4$–P_2O_5–H_2O system. (Adapted with permission from Dahlgren, S.-E., Fertilizer materials, calcium sulfate transitions in superphosphate, *J. Agric. Food Chem.*, 8(5), 411–412. Copyright 1960 American Chemical Society.)

liquid and vapor phase in this reaction. Consequently, an operating region DH, as shown in Figure 1.16, is chosen for dihydrate processes. The reaction is controlled just below the dihydrate–hemihydrate equilibrium line in the range of 70°C–80°C at 28%–32% P_2O_5 concentration. The principal advantages of the dihydrate process are its simplicity, reliability, and flexibility, particularly to handle a wide range of rock quality. Its principal disadvantages are that its product acid is only 28%–30% P_2O_5 concentration and its overall P_2O_5 efficiency is lower than other processes (around 93%–97% compared to 95%–99%).

The acid feed to DAP plants is typically 40%–42% P_2O_5; merchant grade acid, that is, WPA sold to either fertilizer or purified acid producers, is typically 52%–54% P_2O_5. When the hemihydrate processes were originally developed, one key selling point was their facility to produce acid at or close to this concentration by utilizing the heat of reaction rather than requiring a concentration step. These early plants proved problematic, and a typical product concentration for nondihydrate processes now is 40%–42% P_2O_5 [41]. Concentrating an acid from a dihydrate plant from 30% to 42% is possible with either hot water or low-grade steam in one step. The capital cost of the concentration section balances the capital cost of the more complex nondihydrate processes. The lower P_2O_5 efficiency is balanced by better availability compared with nondihydrate processes, so overall the dihydrate process is perhaps the best compromise; certainly, it is the most prolific process worldwide.

The nondihydrate processes are variations on the theme of operating in the hemihydrate region of the $CaSO_4$–P_2O_5–H_2O system, depicted as region A in Figure 1.16. The simplest example, originally developed by Fisons in the 1960s and now licensed by Yara, is the single-stage, strong acid, hemihydrate process producing 40%–50% P_2O_5 acid from the filter, which is operated in region HHa. A variation of this is the hemihydrate recrystallization process, reaction takes place in region HHb in Figure 1.16, and then the hemihydrate is recrystallized by cooling the reaction slurry to dihydrate (region DHr) producing a 30%–32% P_2O_5 acid from the filter. The third variant is the dihemihydrate process, also known as the Central Prayon process [42], developed by Central Glass of Japan and Prayon of Belgium in the 1960s; in this process, reaction takes place in the dihydrate region at 32%–35% P_2O_5 concentration (region DHcp); acid is separated in the first filtration, and then the dihydrate slurry is converted to high-quality hemihydrate (region HHcp) by raising the temperature with steam. The slurry acid is filtered again and water washed and recycled to the reaction stage. With three weeks' aging, the hemihydrate is suitable as an alternative to natural gypsum of which there is none in Japan. The fourth variant is the HDH process, also licensed by Yara, which is capable of producing 46%–52% P_2O_5 acid from the filter but is usually operated in the 40%–43% P_2O_5 range. In this process, the reaction takes place in region HHa and is filtered, and the resultant hemihydrate is then recrystallized with sulfuric acid to dihydrate (region DHry), thus releasing P_2O_5 lost in the hemihydrate crystal lattice. Both the Central Prayon and HDH processes claim very high P_2O_5 efficiencies (98.5%–99%).

Wet process acid plants face two main environmental challenges, the control of fluoride-based vapors and the management of by-product calcium sulfate, which is always referred to as gypsum. Every ton of P_2O_5 product compares to approximately 5 tons of gypsum. If the phosphate rock is suitable, containing low levels of radionuclides and heavy metals, and the gypsum is sufficiently washed and treated to reduce P_2O_5 and fluoride levels, then, as with the Central Prayon process, the gypsum may enter the commercial gypsum market. Unfortunately, these conditions are rarely met, and so the gypsum is stored local to the plant in *stacks*, which must be managed to avoid water runoff, preventing pollutants entering local water courses. Minimizing P_2O_5 levels in gypsum and maximizing P_2O_5 output amount to the same thing and depend on good control of the crystallization process (in that good crystals filter and wash well so making for a good separation between crystal and acid).

The crystallization process is the third of three reactions involving calcium and sulfate ions:

1. The first, the ionization of sulfuric acid, is very fast, measured in seconds and depending on the dispersion of the acid in the reaction mix.
2. The second, separating calcium ions from the phosphate rock, is not so fast, measured in minutes, and can be blocked if calcium sulfate crystallizes on rock particles, coating them and protecting them from acidulation.
3. The third reaction, between calcium and sulfate ions to form calcium sulfate, depends on temperature and the concentrations of calcium, sulfate, and P_2O_5. It is the slowest reaction and is measured in hours. Typically, for industrial systems, it is 4–6 h.

Equation 1.49 [39] relates these variables:

$$\left[Ca^{2+}\right]\left[SO_4{}^{2-}\right]^{(1.25-0.01t)} = 0.46^{\left(\frac{2.4-\log t}{0.912}\right)\frac{P_2O_5}{100}} \quad (1.49)$$

where

Ca^{2+} is the concentration Ca^{2+}, weight percent
$SO_4{}^{2-}$ is the concentration of $SO_4{}^{2-}$, weight percent
t is the temperature, °C
P_2O_5 is the concentration of P_2O_5, weight percent

As with all crystallization processes, as the solubility product (in this case calcium and sulfate ion concentration product) reaches the solubility limit, the liquid phase starts to become supersaturated and crystals of calcium sulfate begin to form on suitable nuclei; as the solubility product continues to increase, the liquid phase reaches the supersaturation limit and spontaneous crystallization occurs. Spontaneous crystallization produces large quantities of tiny crystals, which are very difficult to filter and contribute to increased viscosity of the reaction mass. Consequently, the design and operation of the plant will seek to remain in the saturated zone. The saturation and supersaturation curves of the CaO/SO_4 system may be derived from Equations 1.50 and 1.51 [39] and are expressed in weight percent CaO and SO_4, for a system at 75°C and 30% P_2O_5 concentration:

$$(\%CaO)\times(\%SO_4) \simeq 0.83 = K_S \quad (1.50)$$

$$(\%CaO)\times(\%SO_4) \simeq 1.30 = K_{SS} \quad (1.51)$$

The rate of crystallization is a linear function ranging from zero on the saturation curve up to 105 kg/h on the supersaturation curve (for the 75°C and 30% P_2O_5 system) as shown in the following equation:

$$Q = \varphi(K_{SS} - K_S) \quad (1.52)$$

where

Q is the quantity of gypsum, kg slurry/h/m³ of reactor volume, crystallized without spontaneous formation of nuclei
φ is the crystallization mass transfer constant equivalent to 214 kg/m³ for a 25% solids by volume slurry in the 75°C and 30% P_2O_5 system
K_{SS} is the solubility product of the supersaturated solution
K_S is the solubility product of the saturated solution

The simplified reaction (Equation 1.2) indicates that an excess of sulfuric acid drives the reaction. Too great an excess will lead to spontaneous crystallization. The optimum excess, as measured in the reaction slurry liquor, is in the range of 2%–3% SO_4 [40].

The size and choice of reactor type depend very much on the anticipated rock type and analysis, the desired output, and the design philosophy of the licensor.

As a general guideline, using Equations 1.50 through 1.52, we may generate 214 × (1.30 − 0.83) = 100 kg/h gypsum per m^3 of reactor volume and still remain in a region of stable crystal growth. Gypsum specific gravity is 2.32, so 100 kg/h occupies 100/2320 = 0.043 m^3/h. Most systems operate with 25% volume basis slurry; therefore, the total reaction volume is 4 × 0.043 = 0.17 m^3/h. For a 1 m^3 reactor volume, the residence time would be 5.8 h. Specific reaction volumes for dihydrate systems vary from 1 to 2 m^3/ton of P_2O_5/day and for hemihydrate systems 5–6 m^3/ton of P_2O_5/day. For example, a Jacobs Engineering–designed plant in Florida producing 1000 tons P_2O_5/day has a total slurry volume of 1600 m^3 [43]; in Jordan, Prayon converted a plant originally designed for 1500 tons P_2O_5/day with a specific reactor volume of 1 m^3/ton of P_2O_5/day to an operation working at 1.9 m^3/ton of P_2O_5/day with a 2530 m^3 reaction volume (the original design had not achieved full capacity for a variety of reasons [44]). Large plants produce over 2000 tons P_2O_5/day.

Reactors are designed to provide sufficient time for rock dissolution, gypsum crystal growth, and disengagement of foams. The addition of both rock and sulfuric acid is managed in different ways but always with the goal of achieving rapid and efficient dispersion in the reaction mass and so avoids local regions of supersaturation and the consequent formation of many very small crystals. Given the dilution of sulfuric acid is responsible for most of the heat load, a localized excess could also lead to high temperatures and the formation, in a dihydrate system, of small hemihydrate crystals. In both cases, the consequence will be poor filtration, reduced capacity, and increased P_2O_5 losses.

The approach adopted by Prayon, licensor for more than 50% of dihydrate plants worldwide, is to specify a multicompartment reactor; the standard Prayon Mark IV plant has six compartments each rectangular of 245 m^3 capacity with a single agitator. Figure 1.17 shows a simplified flow sheet of the reaction section. The reactor design is expandable by adding more compartments; this might be specified for a plant required to process igneous rock, which is normally less reactive than sedimentary rock. Typically, the compartments are constructed from concrete, then lined with rubber, and finally covered with graphite bricks. Metal plates are usually placed on the reactor floor under the agitators to minimize local brick erosion. The slurry moves from one compartment to the next either by flowing over the dividing wall or

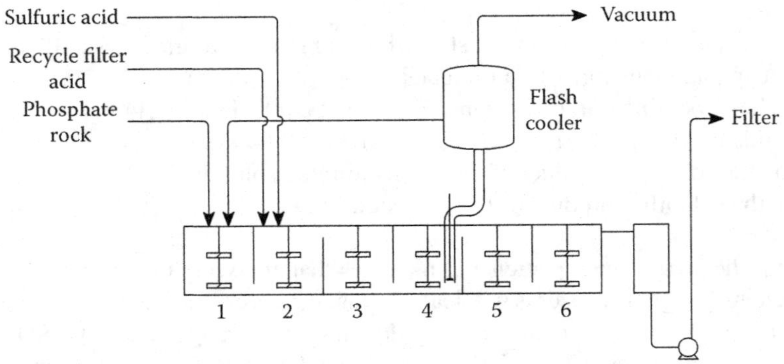

FIGURE 1.17 Prayon Mark IV reaction flow sheet.

through an opening at the base of the wall. Rock is introduced to compartment 1 and mixed with the reaction slurry by the single agitator in compartment 1. The slurry then flows continuously into compartment 2 to which is added sulfuric acid and weak phosphoric acid from the filter. The slurry continues to flow to compartments 3 and 4; sometimes sulfuric acid will also be added in these compartments in an attempt to limit SO_4 excess and so control supersaturation. From compartment 4, one portion of the slurry passes to the first *digestion* compartment 5 to allow crystal development; the other portion is drawn into the low-level vacuum flash cooler (LLFC) [45]. Water is evaporated under vacuum, cooling the slurry, which is then returned to compartment 1. From compartment 5, the slurry passes to compartment 6 and onto filtration.

The Rhône–Poulenc process was originally conceived as a single reactor with evaporative cooling. As well as the main agitator, several small helices are installed around the reactor both to disperse sulfuric acid and aid heat transfer at the liquid surface; air is sucked through the reactor vapor space aiding evaporation and carrying heat away. After a number of years, Rhône–Poulenc modified their process, introducing a second reactor and naming the combination the Diplo process. Figure 1.18 shows a simplified flowsheet of the Rhône–Poulenc Diplo reactor. Just like the Prayon process and most others, the reactor construction is concrete, rubber lined, and graphite brick lined. Other *single* tank technologies include Dorr–Jacobs, Siape, and Badger–Raytheon.

The benefits claimed for the Rhône–Poulenc process include a more simple design and higher recirculation rate, leading to better gypsum crystal quality and therefore better filtration. Certainly one or two agitators and one or two vessels are much simpler than eight or more compartments and agitators. The recirculation rate achieved by a large agitator within the reactor is generally higher than that possible with a pumped circulation. A high recirculation rate aids the maintenance of the sulfate level at the desired control point and so avoids local supersaturation and poor gypsum crystal quality. The weakness of the single- or double-reactor approach, with blown air cooling, is that the cooling surface area available is limited to the diameter of the reactor; this is not a problem for small- and medium-sized reactors. As plant reactor

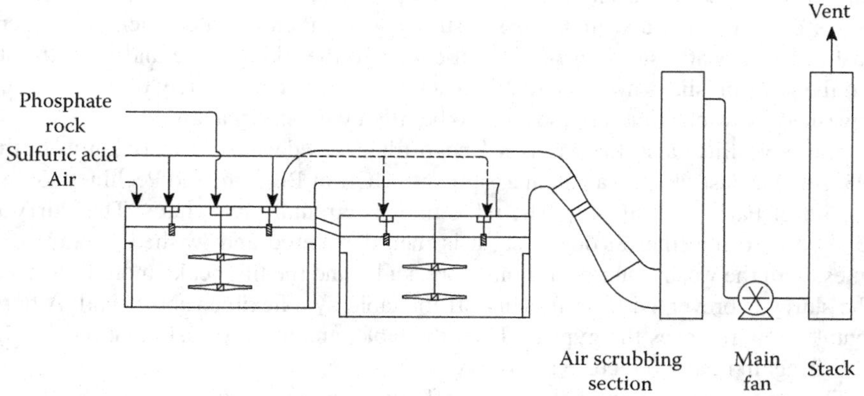

FIGURE 1.18 Rhône–Poulenc Diplo reactor flow sheet.

size increases, the volume/diameter ratio is such that there comes a point when there is insufficient heat transfer area for the increased cooling demand. Cooling air is normally at ambient temperature, so for plants built in hot climates, either the air must be cooled or the reduced temperature difference, compared to cooler climates, accepted. A flash cooler does not depend on local air temperature.

There are many advantages and disadvantages of one wet process reactor technology versus another, and there is insufficient space in this book to address them all. Each case must be evaluated on its own merits, which also include capital cost and program.

There are three filter types in use on all wet process plants to carry out the separation of phosphoric acid from the calcium sulfate crystal: the tilting pan filter, the table filter, and the continuous belt filter. All filters are open to atmosphere and therefore require ventilation to remove fumes, usually just at the slurry feed stage; they all depend on a suitable cloth, supported on a grid system, to retain the gypsum (or hemihydrate) crystals; they all use vacuum to aid liquor flow through the filter cake, and all of them use water washing to minimize P_2O_5 losses; they are all available in a variety of standard sizes ranging 10–300 m^2. Typically, clean water is sprayed onto the filter cake at the last stage before leaving the filter; this slightly acidified water is then reused at another filter stage where the cake is carrying more acid, and so the filter water picks up more acid and is eventually pumped to the reactor. Overall, the wet acid process requires water for evaporation and for water of crystallization. The water balance is maintained by the water addition at the filtration stage.

The tilting pan filter is also known as the Prayon or Bird–Prayon filter. Prayon commenced the development of the filter in the 1950s and has been granted several patents [46] over the years for continued developments, most recently in 2004 [47]. The drawings in the 2004 patent are clear; however, a photograph of a Prayon filter undergoing maintenance may aid the reader's comprehension (Figure 1.19). Prayon licensed Bird to make the filter in the United States. The filter comprises a number of trapezoid filter pans, arranged in a rotating ring. The slurry flows into a pan, the pan moves along its circular path, and the free acid liquor drains through the filter cake and passes through a radial tube to a central collection point and then to the filtered acid tank; at the same time, the next pan is being filled with slurry. The pan moves further along and undergoes a strong wash, then dewater, then subsequent washes and dewater steps; eventually, it comes to the tilting stage and the pan tilts and the gypsum slides away. In the final stage, the pan is completely overturned and is washed. Now clean, it moves round to be filled with slurry again.

The table filter also known as a Ucego filter was developed in Belgium in the 1950s and is described in a patent assigned to UCB of Belgium and Pechiney–Saint-Gobain of Paris in 1961 [48]. The table filter is circular and rotates. The slurry is pumped onto the filter at one zone; it is then dewatered and washed a number of times, with the wash water becoming more acidic and the filter cake more P_2O_5-free. The slurry is prevented from flowing off the table by a flexible rubber wall. A horizontal screw removes the gypsum from the table, and after a wash, that part of the table is again ready to receive the slurry.

The continuous moving belt filter is offered by a number of suppliers, for example, Eimco [49] (the reference is another patent with clear drawings; Figure 1.20

FIGURE 1.19 Tilting pan filter. (Courtesy of Solvay.)

FIGURE 1.20 Belt filters on WPA. (Courtesy of Solvay.)

shows a photograph of two belt filters utilized on WPA). The belt is made of reinforced rubber and is shaped by rollers to form side walls and the channels to allow the liquid to pass below the cloth that sits on the belt. The belt moves horizontally in a straight line; slurry is added, and at various stages along the length of the filter water wash, weak acid wash and vacuum are applied; at the end of the filter, the gypsum falls off and the cloth passes over a support roller. On the return toward the drive roller, the cloth and belt are separated from each other so that they can both be washed.

The tilting pan and table filter tend to be used for larger duties; even for large dihydrate plants, there may be only one of this type of filter. For smaller duties, the continuous belt filter is often the cheaper option.

Filter size is chosen on the basis of filtration trials, which can be done on small test equipment. Typically, as part of a rock trial in the laboratory or pilot plant samples of the reaction, the slurry will be filtered on the filtration test rig and an estimate made of the surface area required to achieve the production of 1 ton P_2O_5 per day for different filtration times. The filtration times are measured in minutes and standardized to 90 s; P_2O_5 losses are measured and recorded. Typical specific filtration rates range 3–9 tons P_2O_5/m^2/day and depend on the gypsum crystal quality, which in turn depends on rock type and quality and the sulfate content of the filtrate; other factors include whether flocculants are added to the acid to aid filtration and temperature.

The acid leaving the filter contains 1%–2% solids comprising fine gypsum crystals and sodium and potassium silicofluorides. The acid is normally held for some time in intermediate storage before transfer to the concentration section. Usually, the acid temperature drops in the storage and more solids are precipitated, which are returned to the reactor. This operational problem varies in severity with the acid impurity composition and occasionally is addressed with a separation device such as a centrifuge or clarifier step before concentration.

For a dihydrate plant, the filter acid is usually concentrated to 40%–42% P_2O_5 to be suitable for fertilizer manufacture. For wider sale as merchant grade acid, it is concentrated to 52%–56% P_2O_5. It is possible to concentrate from 28% to 56% P_2O_5 in one step; however, doing so leads to a high level of solid precipitation on heat exchange surfaces. In turn, this leads to frequent production interruptions for cleaning. As a result, a filter acid is often concentrated to 40%–42% P_2O_5 and then clarified; the solid-rich underflow is either returned to reaction or filtered, usually on a small belt filter. Figure 1.21 depicts a flow sheet of a two-stage concentration section with clarifiers. The clarified acid is then transferred forward to the 52%–56% P_2O_5 concentrator. At this stage, the acid may be further clarified or simply allowed to stand in a final product storage tank. The concentration step comprises feeding the acid into a circulating flow into a concentrator vessel. The concentrator always operates under vacuum, usually achieved with steam ejectors, and is usually constructed from carbon steel with a rubber lining. Below the concentrator body, a large axial flow pump maintains circulation to the concentrator heat exchanger, which is either graphite or exotic alloy tubular construction. The head developed by the pump is relatively small—just enough to overcome resistance due to flow

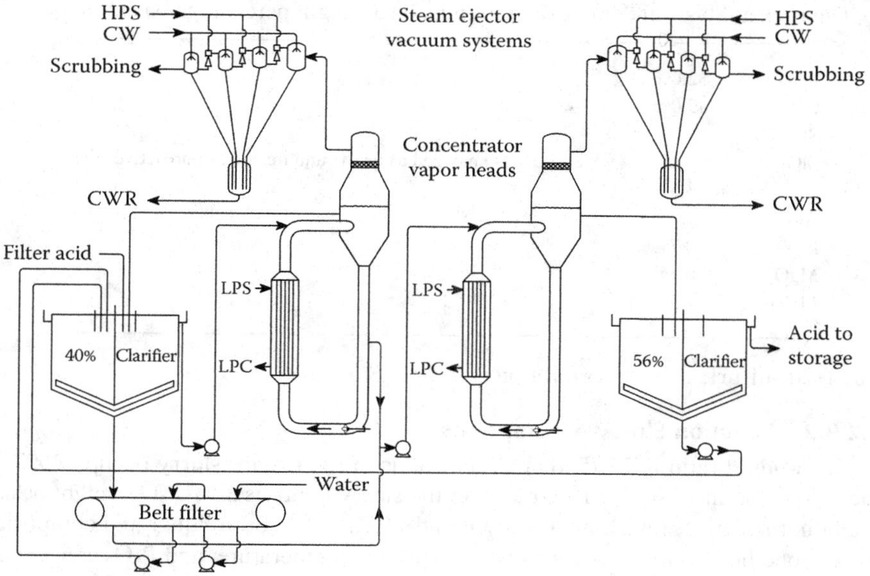

FIGURE 1.21 Phosphoric acid concentrator flow sheet. *Note*: HPS, high pressure steam; LPS, low pressure steam; LPC, low pressure condensate; CW, cooling water; CWR, cooling water return.

through the heat exchanger. The shell to tube side temperature difference is kept as low as possible (less than 5°C) to slow down the rate of scaling as solids are always precipitated. It is possible to consider this operation as a crystallizer and to recycle solids to increase the available surface area for crystal growth, in turn producing larger crystals that are easier to filter. Precipitation remains a maintenance challenge throughout the industry and requires regular cleaning of storage tanks, ship's holds, and tanker cars as well as frequent attention to heat exchange surfaces.

This section on wet process acid will now conclude with an example heat and mass balance and calculations to predict acid quality based on rock analysis.

1.2.9.1 Mass Balance

For any chemical process, a mass balance forms the basis for plant design, process control, and product costs. The following mass balance draws heavily on the example in Becker [39]. There are many approximations in a generalized mass balance of this process because it is chemically complex and the interactions and solubilities of different minor components vary between different rocks. That is why it is essential to carry out laboratory trials for any prospective rock. Nevertheless, the following balance is a reasonable approximation useful as the basis for a plant model.

The rock analysis under consideration is, on a weight percentage basis, as follows:

P_2O_5	32.0%
CaO	49.0%
SO_3	2.0%
SiO_2	3.0% (of which 1.5% is analyzed as quartz and treated as unreactive silica)
F	3.6%
Al_2O_3	0.7%
Fe_2O_3	0.7%
MgO	0.5%
Moisture	1.5%

The feed sulfuric acid concentration is 98% H_2SO_4.

1.2.9.2 Reaction Slurry Assumptions

Solid-to-liquid ratio is 25:75, that is, the solids in the reactor slurry occupy 25% of the slurry volume; P_2O_5 concentration in the slurry liquor is 30% P_2O_5 weight basis; the liquor specific gravity, at 30% P_2O_5 and 75°C, is 1.3. Both liquor and composite slurry specific gravity vary with impurity profile, temperature, and P_2O_5 concentration. Equation 1.53 relates P_2O_5 concentration with specific gravity for wet process acid. Real-life tests will vary a little in the following equation:

$$SG = a + bP + cP^2 + dP^3 + eP^4 \tag{1.53}$$

where

SG is specific gravity
P is P_2O_5 content

A specific gravity of 1.3 and 30% P_2O_5 concentration imply a P_2O_5 content of 1300 × 30% = 390 kg/m³ in the slurry liquor. The solid (gypsum) specific gravity is 2.32.

P_2O_5 loss assumptions (weight percentage of P_2O_5 in rock feed), as follows:

Unattacked losses (due to coating, etc.)	0.5%		
Lattice losses (due to formation of dicalcium phosphate, etc.)	3.0%		
Total insoluble losses	3.5%		
Filtration losses (P_2O_5 lost with gypsum)		1.0%	
Spillage, maintenance, etc.		0.5%	
Total soluble losses		1.5%	
Total P_2O_5 losses			5.0%

1.2.9.3 Mass Balance Calculations

Phosphate rock consumption, *R*, in tons per ton P_2O_5 produced is calculated using Equation 1.48:

$$R = \frac{100}{0.32 \times 95} = 3.29 \tag{1.54}$$

Sulfuric acid consumption, S, in tons 100% H_2SO_4 per ton P_2O_5 produced is calculated using Equation 1.45 and accounts for the SO_3 content of the feed rock and soluble losses:

$$S = \left(\frac{1.732 \times 0.49}{0.32 - 0.02 \times 0.49} - \frac{1.225 \times 0.02}{0.32} + 0.062 \right) \frac{100}{100 - 1.5} = 2.76 \tag{1.55}$$

Gypsum cake production, in tons of solids per ton of feed rock, is calculated using Equation 1.56. *IS* represents the weight percentage of insolubles in the feed rock, principally, unreactive silica (quartz):

$$G = \frac{172}{56}(\mathrm{CaO} \times 0.98) + IS \tag{1.56}$$

$$G = \frac{172}{56}(0.49 \times 0.98) + 0.015 = 1.49 \tag{1.57}$$

Reactor slurry production, V_S, in m^3 per ton of rock, is calculated using Equation 1.58. Reactor slurry production depends on cake production, G; the solid content, usually taken as 25%; and the specific gravity of gypsum:

$$V_S = \frac{G}{0.25 \times 2.32} \tag{1.58}$$

$$V_S = \frac{1.49}{0.58} = 2.57 \tag{1.59}$$

A proportion of the product acid is recycled from the filter with the water wash. The amount of P_2O_5 in the recycle acid equates to the difference between the P_2O_5 content of the slurry and the P_2O_5 produced from the phosphate rock attack. In other words,

P_2O_5 recycled (tons per ton of rock) = P_2O_5 in slurry − P_2O_5 from rock

P_2O_5 in slurry (t/t rock) = (slurry volume − volume of solids) × concentration of P_2O_5 in acid

P_2O_5 from rock (t/t rock) = P_2O_5 in rock − insoluble losses

$$P_2O_5 \text{ slurry} = 2.57 \times (1 - 0.25) \times 0.390 = 0.752 \tag{1.60}$$

$$P_2O_5 \text{ rock} = 0.32 \times (0.32 \times 0.035) = 0.309 \tag{1.61}$$

$$P_2O_5 \text{ recycled} = 0.752 - 0.309 = 0.443 \tag{1.62}$$

$$\frac{P_2O_5 \text{ recycled}}{P_2O_5 \text{ produced}} = \frac{0.443}{0.309} = 1.43 \tag{1.63}$$

Process water requirements comprise water of crystallization for the calcium phosphate dihydrate (gypsum), water for cooling the reaction mass by vaporization, and dilution water to adjust the P_2O_5 concentration of liquor phase of the reaction slurry. Water of crystallization, W_C, is calculated using the following equations:

$$W_C = 0.643(\text{CaO} - 0.70\text{SO}_3) - 0.003 \quad (1.64)$$

$$W_C = 0.643(0.49 - 0.70 \times 0.02) - 0.003 = 0.303 \quad (1.65)$$

Water for vaporization cooling, W_{VAP}, allows the reaction heat, equivalent to 795.3 MJ/ton of 100% H_2SO_4, to be balanced by heat of evaporation of water, which is taken as 2344 MJ/ton, as shown in the following equation:

$$W_{VAP} = \frac{795.3 \times S}{2344 \times R} \quad (1.66)$$

$$W_{VAP} = \frac{795.3 \times 2.76}{2344 \times 3.29} = 0.285 \quad (1.67)$$

Water of dilution, W_D, is the quantity of water required to adjust the P_2O_5 concentration of the slurry liquor, in dihydrate plants to 28%–32% P_2O_5. Soluble impurities in the rock substitute water. The effect of these impurities in assessing W_D is best assessed experimentally; however, a good approximation is possible, for a given rock, using dissolution factors. These may be calculated by comparing various phosphate rock analyses with the produced acid; values from Becker are used in the succeeding text. The following relationship is used:

$$\text{Impurity in rock} \times \frac{P_2O_5 \text{ in acid}}{P_2O_5 \text{ in rock}} \times \text{Dissolution factor} \quad (1.68)$$

The total weight of 1 m^3 of product acid is 1.3 tons. The weight of P_2O_5 in the acid is 30% × 1.3 = 0.390 tons. The weight of P_2O_5 in rock, less lattice losses, is 0.32 × 0.97 = 0.310 tons. So in tons/m^3,

Phosphoric acid content as H_3PO_4 is

$$\frac{390 \times 2 \times 98}{142} \times 10^{-3} = 0.5383 \quad (1.69)$$

2.5% H_2SO_4 in solution is

$$0.025 \times 1.300 = 0.0325 \quad (1.70)$$

0.7% Al_2O_3 in the rock is 80% solubilized as Al^{3+}:

$$\left(0.007 \times \frac{2 \times 27}{102}\right) \times \frac{0.390}{0.310} \times 0.8 = 0.0037 \quad (1.71)$$

0.7% Fe_2O_3 in the rock is 90% solubilized as Fe^{2+}:

$$\left(0.007\times\frac{2\times55.8}{159.6}\right)\times\frac{0.390}{0.310}\times0.9=0.0056 \tag{1.72}$$

0.5% MgO in the rock is 100% solubilized as Mg^{2+}:

$$\left(0.005\times\frac{24.3}{40.3}\right)\times\frac{0.390}{0.310}\times1.0=0.0038 \tag{1.73}$$

3.6% F^- in the rock is 50% solubilized as H_2SiF_6:

$$\left(0.036\times\frac{144.1}{6\times19}\right)\times\frac{0.390}{0.310}\times0.5=0.0286 \tag{1.74}$$

Typically, 4–6 kg calcium stays in solution, taking 5 kg in this case:

$$\frac{40}{56}\times5\times10^{-3}=0.0036 \tag{1.75}$$

Therefore, the total weight of components is 0.616 tons.

Water content in 30% P_2O_5 product acid is

$$1.300-0.616=0.684 \tag{1.76}$$

Water of dilution per ton of rock, W_D

$$W_D=0.684\times\frac{0.310}{0.390}=0.54 \tag{1.77}$$

Total process water, W_T, is the sum of water of crystallization, water for vaporizing cooling, and water of dilution less the water content of the phosphate rock and the sulfuric acid, as expressed in the following equations:

$$W_T=W_C+W_{VAP}+W_D-W_R-W_S \tag{1.78}$$

$$W_R=0.015 \tag{1.79}$$

$$W_S=\frac{2.76}{3.29}\times0.02=0.017 \tag{1.80}$$

$$W_T=0.30+0.28+0.54-0.015-0.017=1.09 \tag{1.81}$$

The volume of the recycle acid is best deduced from the measurement of the density of a real sample due to slight shrinkage. Shrinkage is of the order of 0.5%. Recycled acid comprises a P_2O_5 component and the total process water, W_T;

recycled acid volume, V_{RA}, in m^3 per ton of rock, is calculated using the following equations:

$$V_{RA} = \left(\frac{P_{REC}}{SG} + W_T \right) \times 0.995 \tag{1.82}$$

$$V_{RA} = \left(\frac{0.443}{0.390} + 1.09 \right) \times 0.995 = 2.21 \tag{1.83}$$

The P_2O_5 content of the recycle acid, $P_{REC\%}\%P_2O_5$, is given by the following equations:

$$P_{REC\%} = \frac{P_{REC}}{(P_{REC}/P) + W_T} \% \tag{1.84}$$

$$P_{REC\%} = \frac{0.443}{\frac{0.443}{0.30} + 1.09} \times 100 = 17.3\% \tag{1.85}$$

The total plant water demand is greater than the total process water demand. If wet rock milling is used, compared to dry milling or no milling, there is a water requirement, although this forms part of the process water that is recycled. A more significant water requirement is the water retained in the gypsum filter cake. Residual moisture for easily filtered gypsum is 20%–25% but can be as high as 50% for difficult filtrations. Furthermore, there is a water requirement depending on whether the gypsum is dry or wet stacked. In dry stacking, the gypsum is conveyed by belt or truck to the stack. In wet stacking, the gypsum is slurried to a 15%–30% by weight concentration and pumped either to the stack or to sea. If the slurry is pumped to sea, seawater is used. If the slurry is pumped to stack, process water is used; in turn, this water is recycled from the stack to the plant, and with it, some of the P_2O_5 originally lost with the gypsum during filtration. Overall, P_2O_5 efficiency is improved by 1%–2%. For 25% residual moisture, water in the filter cake, W_F, in tons per ton of rock is given by

$$W_F = \frac{0.25}{0.7} \times G = \frac{0.25}{0.7} \times 1.49 = 0.53 \tag{1.86}$$

Water is added to the process via the filter wash; this flow, W_{FF}, is the sum of the total process water, W_T, and the water in the filter cake, W_F, in tons of water per ton of rock as shown in the following equation:

$$W_{FF} = W_T + W_F = 1.09 + 0.53 = 1.62 \tag{1.87}$$

From the equations earlier, it is possible to build a preliminary model of the flows around the reaction and filtration sections of a WPA plant and, given a rock analysis, to estimate a product acid quality.

1.2.10 Thermal Acid

Phosphoric acid manufactured by the thermal route is known as *thermal acid*. Phosphorus is burned in the presence of air in a combustion chamber, so forming P_2O_5 vapor; the flame temperature is in the range of 1800°C–2000°C. The P_2O_5 vapor, at 500°C–1000°C, passes into a second vessel, the hydrator, where it contacts water and recycling acid, thus forming thermal phosphoric acid. Two chemical reactions occur in this process; both are exothermic, the phosphorus oxidation strongly so. These reactions are shown in the following equations:

$$P_4 + 5O_2 \rightarrow 2P_2O_5 \quad \Delta H = -3012\,\text{kJ/mol} \tag{1.88}$$

$$P_2O_5 + 3H_2O \rightarrow 2H_3PO_4 \quad \Delta H = -188\,\text{kJ/mol} \tag{1.89}$$

The simplest plants have no energy recovery; however, more sophisticated plants are able to recover at least 60% of the heat of reaction as high-pressure steam [50,51]. Steam is raised either directly in jackets around the combustor or indirectly, using hot oil to transfer the heat from the combustor to a steam raising heat exchanger. Another process option is to coproduce solid P_2O_5. Dry air is used in the combustion section; the P_2O_5 vapor passes to a condensing tower and a proportion of the vapor flow condenses forming solid P_2O_5. The solid P_2O_5 is withdrawn from the tower with a screw conveyor and bagged for sale or mixed with water to form phosphoric acid. The vapor passes to wet scrubbers, which form phosphoric acid. Careful control of the amount of water introduced to the process either as water vapor in the combustion air or as water for hydration and scrubbing dictates the concentration of product acid. A simplified flow sheet is shown in Figure 1.22. Figure 1.23 shows a photograph of the thermal phosphoric acid plant at Břeclav, Czech Republic, operated by Fosfa. The combustion chamber is on the left of the photograph, the hydrator at the center, and the electrostatic precipitator (an alternative to wet scrubbers) to the right, exhausting to the vent stack.

Thermal acid is pure to the extent that impurities are measured in ppm (weight). Consequently, care must be taken with the impurity levels of the hydrating water, and it is not unusual for the feed water to a thermal acid plant to be treated to some degree. Another consideration is corrosion. Most of the plant is constructed with 316L stainless steel, which has good corrosion resistance to phosphoric acid up to 60°C. Above this temperature, the corrosion rates start to increase and the corrosion products dissolve into the acid potentially affecting its impurity profile. For the high-temperature sections of the plant, different approaches are used to manage corrosion, but at the low-temperature areas, typically when the product acid is formed, transferred, and stored, poor temperature control will lead to corrosion and affect product quality. However, the most important impurity that requires reduction to meet the specifications for thermal acid is arsenic, As. Arsenic is next to phosphorus in the periodic table and behaves in a similar fashion to phosphorus in the phosphorus furnace and is the principal

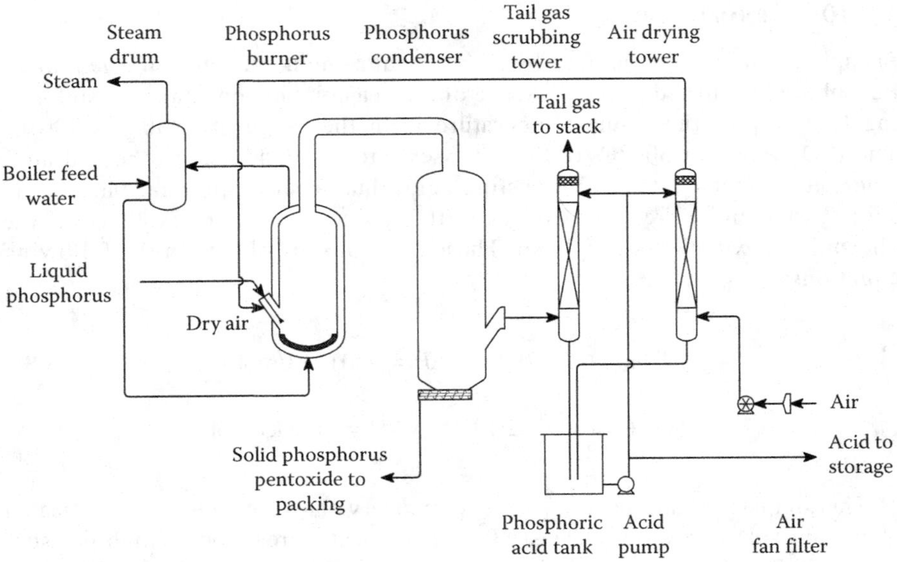

FIGURE 1.22 Thermal acid flow sheet.

FIGURE 1.23 Thermal acid plant. (Courtesy of Fosfa.)

impurity in phosphorus. Arsenic is poisonous; typical specifications permit concentrations less than 1 ppm in phosphoric acid. Arsenic is removed by precipitation and filtration using hydrogen sulfide gas, H_2S, or sodium hydrosulfide, NaHS, or sodium sulfide, Na_2S. Other processes using copper and electrolysis are no longer practiced industrially. Hydrogen sulfide gas is very toxic, and care is required in its handling and managing if it evolves as a consequence of using either sodium sulfides. Prior to treatment, arsenic levels are typically 10–20 ppm and posttreatment 0.1 ppm.

1.2.11 Kiln Process Acid (KPA)

The origins of the modern KPA process go back to patents granted in the nineteenth century to convert phosphate rock in a blast furnace either to phosphorus or phosphoric acid [53]. Waggaman [54] and coworkers carried out studies into the production of phosphoric acid in a fuel-fired furnace in the 1920s. Guernsey and Yee were granted a patent in 1922 to produce phosphoric acid from phosphate rock, sand, and coal at temperatures below that at which a fluid melt is obtained; any equipment capable of handling the solid mix at about 1300°C, such as an internally fired rotary kiln, would suffice [55]. Further study was carried out by Pike in the 1930s [56–58] and two commercial scale blast furnaces were built, one by Victor Chemical at Nashville, Tennessee, in 1929, the other by Coronet Phosphate Co., at Pembroke, Florida [59]. The potential operating cost advantage of the blast furnace compared to the electric furnace was well understood, although the greater gas volumes lead inevitably to larger equipment and greater capital cost per ton of acid. Favorable hydroelectricity rates awarded to the electric furnaces undermined the commercial advantage of the blast furnaces and they were closed and dismantled.

In later decades the kiln process was investigated by a number of organizations: in the 1960s by FMC [60], in the 1970s by Olin [61], and in the 1980s by Occidental Research [62]. The FMC process achieved a good degree of heat integration (a headache with the blast furnaces of the 1930s) but did not conquer the problem of the kiln charge material melting. The Occidental Research process, also known as the Hard process after one of the inventors, achieved stability in a 0.84 m diameter by 9.1 m long pilot scale rotary kiln [63] in tests run in December 1981 and May 1982, the breakthrough being that additional silica addressed the melting problem.

Subsequently, the process has further developed and is known as the improved Hard process. There is a better understanding of the kiln operating temperature range, the specification for the petroleum coke, optimum size distribution of the feed materials and feed ratios, reaction kinetics, and the nature and utility of the by-product aggregate.

1.3 ECONOMICS

In terms of raw materials, the economics of the whole phosphate industry rest on the prices of phosphate rock, sulfur, electricity, and, for ammonia-based fertilizers, ammonia, the price of which is strongly linked to natural gas prices.

Phosphate rock is supplied by the industry for use within the industry. The price of rock is therefore set by the industry; there is some price influence from energy costs for mining and government taxation. Taxation is usually a stable cost component; however, in 2008, China set a 20% export tax on fertilizers with the aim of conserving Chinese P_2O_5 resources. One of the consequences of this tax was a spike in phosphate rock and sulfur prices worldwide. Prices subsequently settled down and broadly follow fertilizer demand, which itself is linked to food demand and therefore ultimately population growth. The price of phosphate rock, like any other item of commerce, is a matter for negotiation between buyer and seller and will depend on many factors of which P_2O_5 content is the most important. Various bodies monitor and publish prices.

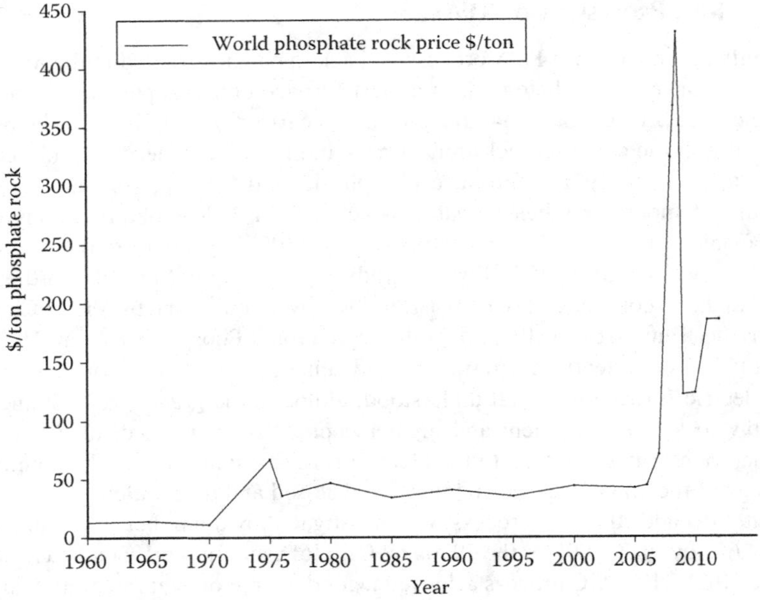

FIGURE 1.24 World phosphate rock prices 1960–2012 data from World Bank GEM Commodities database.

Figure 1.24 shows phosphate rock prices from 1960 to 2012 plotted from The World Bank: Global Economic Monitor (GEM) Commodities.

Approximately 90% of all sulfur is converted to sulfuric acid. Approximately 65% of sulfur is recovered from oil and natural gas operations and coal for power stations and is known as recovered sulfur; 1% is mined; 22% is recovered as sulfuric acid from metal smelting operations. The annual worldwide production of sulfuric acid is 200 million tons H_2SO_4 of which 65% goes to produce phosphoric acid. The recovered sulfur supply is inelastic in that it is a fixed by-product of refineries and coal desulfurization; therefore, prices are demand led. As the phosphate industry takes such a large proportion of the output, it may be argued that it has some degree of price control. Sulfur and sulfuric acid prices are monitored and reported; nine benchmark prices are Vancouver, Middle East, Mediterranean, North Africa, China, India, Black Sea, Tampa, and Benelux. Each location is preceded by either FOB, meaning free on board, or CFR, meaning cost and freight; these are two of several International Commercial (INCO) terms for supplying goods; other terms may be agreed between buyer and seller. Each price is given in US dollars per metric ton and specifies either a spot purchase or contract supply. Most phosphoric acid producers take sulfur and produce sulfuric acid as they are able to use the by-product steam making phosphoric acid and other products as well as generating electricity from the steam. Of course, some producers are conveniently located near to smelting plant and are able to negotiate prices for sulfuric acid that take account of not generating by-product steam and electricity.

Historically, the location of phosphorus plants has depended on the access to a local phosphate source and coal, which in turn permits an advantageous electricity price; obvious examples include the phosphorus plants that were built in Tennessee and the Western States of the United States. With increasing societal demands for electricity, prices are driven to a relatively uniform level worldwide, in many cases controlled by government, although again electricity is a commodity like any other and buyers and sellers come to individual agreements. For the purpose of cost comparison, below average prices from the United States are used.

Ammonia prices are linked back to natural gas prices, and the phosphate industry, in general, must compete with other users for supplies. Ammonia and fertilizer prices are collated by several organizations including Fertilizer Week and the National Agricultural Statistics Service of the US Department of Agriculture.

1.3.1 Production Costs of Phosphorus and Phosphoric Acid

An illustration of the relative costs of phosphoric acid via the wet route or the thermal route is given in the succeeding text. A typical product cost sheet is structured as in the following; it lists all the cost components, usages, and cost per unit and comprises raw materials, in order of importance; chemical services, which do not form part of the product but are essential in its processing (e.g., defoamer or the solvent in a solvent extraction plant); utilities such as steam, water, and nitrogen; direct labor, meaning personnel required to operate, maintain, develop, and supervise the plant; maintenance, which is usually budgeted at 1%–5% of the capital or replacement cost of the plant depending on the technology; and indirects, which include administrative overheads, depreciation, taxes, and insurance.

Table 1.6 displays a cost sheet for sulfuric acid as the majority of phosphoric acid enterprises include a sulfuric acid plant. The wet process acid must be purified to compete

TABLE 1.6
Sulfuric Acid Plant Cost Sheet

Sulfuric Acid Plant		Capacity 2400 t/Day H_2SO_4		330 Days/Year	
	Usage/t H_2SO_4	Price	Unit	\$/t H_2SO_4	%
Raw materials					
Sulfur	0.333	100	\$/t	33.3	96%
Catalyst				1.0	3%
Utilities					
Electricity cost	40	0.07	\$/kWh	2.7	8%
Electricity credit	–65	0.07	\$/kWh	–4.4	–13%
Steam credit	–1.2	10	\$/t	–12.0	–35%
Water				2.0	6%
Direct labor	40 people	50,000	\$	2.5	7%
Maintenance	3.5% capital cost	1,750,000	\$	2.2	6%
Administration			\$	1.0	3%
Depreciation			\$	6.3	18%
				35	100%

with thermal acid so a generalized cost sheet is included (Table 1.9) for a purified acid plant; the impact of the different solvent extraction technologies on these costs will be considered in more detail in the next chapter. The key learning point is that the cost of purified acid is dominated by the cost of the feed acid (and in this situation the return acid credit), followed by maintenance, administration and depreciation costs. The cost impact of a functioning solvent extraction technology is relatively low. Tables 1.7 through 1.11 show the cost sheets for a wet phosphoric acid plant, a phosphoric acid concentrator plant, a purified acid plant, a phosphorus plant, and a thermal acid plant. The phosphate rock input costs are on the same cost per ton of P_2O_5 basis, as well as identical utility costs.

Comparing the total production cost of purified phosphoric and thermal acids, it is obvious that the latter is far more expensive, nearly double, for the same P_2O_5, utility, labor, and overhead rates. Consequently, as well as phosphoric acid, any phosphate derivatives made from thermal acid will be much more expensive than those from purified acid. In the past, the business case for phosphorus manufacture was based on access to local rock, coal, and cheap electricity, including hydroelectricity. Unfortunately, if, as in China at the beginning of the twenty-first century, the domestic demand for electricity grows, and as a result of local weather conditions, the hydroelectricity literally dries up in dry seasons, then phosphorus production must face the full commercial rate for electricity as well as having output limited by its restricted electricity supply. A similar argument applies for coke. Consequently, in fully developed economies, the only justification for a phosphorus plant is to meet a market

TABLE 1.7
WPA Plant Cost Sheet

Phosphoric Acid Plant		Capacity 1000 t/Day P_2O_5		330 Days/Year	
	Usage/t P_2O_5	Price	Unit	\$/t P_2O_5	%
Raw materials					
Phosphate rock[a]	3.29	158	\$/t	519.8	76%
Sulfuric acid[b]	2.76	35	\$/t	95.7	14%
Additives				5.0	1%
Utilities					
Electricity	150	0.07	\$/kWh	10.1	1%
Steam			\$/t		
Water	5.33	0.25	\$/t	1.3	0.2%
Gypsum disposal				2.5	0.4%
Direct labor	50 people	50,000	\$	7.6	1%
Maintenance	3.0% capital cost	3,000,000	\$	9.1	1%
Administration			\$	2.4	0.4%
Depreciation			\$	30.3	4%
				684	100%

[a] Phosphate rock of 32% P_2O_5 concentration, usage includes total losses (5%).
[b] 98% H_2SO_4 used, calculation based on 100% H_2SO_4.

TABLE 1.8
Phosphoric Acid Concentrator Plant Cost Sheet

Phosphoric Acid Concentrator Plant		Capacity 1000 t/Day P_2O_5		330 Days/Year	
	Usage/t P_2O_5	Price	Unit	\$/t P_2O_5	%
Raw materials					
Phosphoric acid[a]	1.005	684	\$/t	687.2	96%
Additives				1.0	0.1%
Utilities					
Electricity	25	0.07	\$/kWh	1.7	0.2%
Steam	1.8	10	\$/t	18	3%
Cooling water	50	0.05	\$/t	2.5	0.3%
Direct labor	10 people	50,000	\$	1.5	0.2%
Maintenance	2.5% capital cost	250,000	\$	0.8	0.1%
Administration			\$	2.4	0.3%
Depreciation			\$	3.0	0.4%
				718	100%

[a] Based on 30% P_2O_5 filter acid concentrating up to 59% P_2O_5 and accounting for precipitation and processing losses.

TABLE 1.9
Purified Phosphoric Acid Plant Cost Sheet

Purified Phosphoric Acid Plant		Capacity 250 t/Day P_2O_5		330 Days/Year	
	Usage/t P_2O_5	Price	Unit	\$/t P_2O_5	%
Raw materials					
Phosphoric acid[a]	1.454	718	\$/t	1044.1	98%
Return acid credit[b]	−0.428	574	\$/t	−245.9	−23%
Solvent	0.5	1.8	\$/kg	0.9	0.1%
Chemicals				50	5%
Utilities					
Electricity	150	0.07	\$/kWh	10.1	1%
Steam	3.5	10	\$/t	35.0	3%
Cooling water	20	0.05	\$/t	1.0	0.1%
Process water	1	0.9	\$/t	0.9	0.1%
Direct labor	30 people	50,000	\$	18.2	2%
Maintenance	2.5% capital cost	2,250,000	\$	27.3	3%
Administration			\$	9.6	0.9%
Depreciation			\$	109.1	10%
				1060	100%

[a] Based on 59% P_2O_5 concentrated, low sulfate, acid feed, 2.5% losses.

[b] Return acid credit is negotiated and ranges 0%–100% of the value of the feed acid, here it is 80%.

TABLE 1.10
Phosphorus Plant Cost Sheet

Phosphorus Plant		Capacity 110 t/Day P_4		330 Days/Year	
	Usage/t P_4	Price	Unit	\$/t P_4	%
Raw materials					
Phosphate rock[a,b]	9.9	123	\$/t	1222.0	45%
Silica[c]	0.24	1.5	\$/t	0.4	0%
Coke	2.4	90	\$/t	216.0	8%
Electrodes				130	5%
Utilities					
Electricity	13,250	0.07	\$/kWh	887.8	33%
Water	1	0.9	\$/t	0.9	0%
Direct labor	30 people	50,000	\$	41.3	2%
Maintenance	2.5% capital cost	1,500,000	\$	41.3	2%
Administration			\$	15.0	1%
Depreciation			\$	165.3	6%
				2720	100%

[a] Based on 25% P_2O_5 rock and 8% P_4 losses.

[b] Rock price same as wet acid case based on P_2O_5 content.

[c] Silica requirement accounting for SiO_2 content of phosphate rock (25% SiO_2).

TABLE 1.11
Thermal Phosphoric Acid Plant Cost Sheet

Thermal Phosphoric Acid Plant		Capacity 180 t/Day P_2O_5		330 Days/Year	
	Usage/t P_2O_5	Price	Unit	\$/t P_2O_5	%
Raw materials					
Phosphorus	0.437	2720	\$/t	1188.6	92%
Water	0.4	0.9	\$/t	0.4	0%
Chemicals				20	2%
Utilities					
Electricity	10	0.07	\$/kWh	0.7	0%
Water	1	0.9	\$/t	0.9	0%
Direct labor	20 people	50,000	\$	16.8	1%
Maintenance	2.5% capital cost	625,000	\$	10.5	1%
Administration			\$	15.0	1%
Depreciation			\$	42.1	3%
				1295	100%

demand for phosphorus and its organic derivatives. There is now only one phosphorus furnace in the United States, in Idaho, and one in Europe, in the Netherlands. On the other hand, the business case for the wet acid route to purified acid is also obvious; that is why there are several purified acid plants around the world.

REFERENCES

1. I. Asimov, *Asimov on Chemistry*, Anchor Books, New York, 1975.
2. S. White and D. Cordell, *Peak Phosphorus: The Sequel to Peak Oil* [Online]. Available at: http://phosphorusfutures.net/peak-phosphorus (Accessed April 08, 2013).
3. J. Bostock, Tran., Pliny the Elder, The Natural History, Book XVII. *The Natural History of the Cultivated Trees*, Chapter. 8—The proper mode of using manure [Online]. Available at: http://www.perseus.tufts.edu/hopper/text?doc=urn:cts:latinLit:phi0978.phi001.perseus-eng1:17.8 (Accessed April 08, 2013).
4. J. Emsley, *The Shocking History of Phosphorus: A Biography of the Devil's Element*, New edition, Pan Books, London, U.K., 2001.
5. J. Wisniak, Phosphorus-from discovery to commodity, *Ind. J. Chem. Technol.*, 12, 108–122, 2005.
6. M. E. Weeks, *Discovery of the Elements 1933*, Kessinger Publishing, Whitefish, MT, 2003.
7. R. E. W. Maddison, Studies in the life of Robert Boyle, F.R.S, Part V. Boyle's Operator: Ambrose Godfrey Hanckwitz, F.R.S, *Notes Rec. R. Soc. Lond.*, 11(2), 159–188, 1955.
8. R. Boyl, *A Paper of the Honourable Robert Boyl's, Deposited with the Secretaries of the Royal Society, October 14, 1680 and Opened Since His Death; Being an Account of His Making the Phosphorus, etc*, Royal Society of London, London, U.K., 1753.
9. R. Boyle, T. Snowden, and N. Ranew, *The Aerial Noctiluca, or, Some New Phenomena and a Process of a Factitious Self-Shining Substance: Imparted in a Letter to a Friend Living in the Country*, Printed by Tho. Snowden, London, U.K., and are to be sold by Nath. Ranew, 1680.
10. R. Watt, *Bibliotheca Britannica; or, A General Index to British and Foreign Literature*, Printed for A. Constable and Company, Edinburgh, Scotland, 1824.
11. E. B. R. Prideaux, A textbook of inorganic chemistry, in *Phosphorus,* vol. 4, Part II, J. Newton Friend, Eds. pp. xxviii+238. C. Griffin & Co., Ltd., London, U.K., 1934. 18s., *J. Soc. Chem. Ind.*, 53(35), 746–748, 1934.
12. J. Walker, *The True History of the Invention of the Lucifer Match,* ed. M. Heavisides, Heavisides & Son, Stockton-on-Tees, U.K., 1909.
13. M. G. L. Banks, *The Manchester Man*. Hurst and Blackett, London, U.K., 1876.
14. R. E. Threlfall, *The Story of 100 Years of Phosphorus Making, 1851–1951*, Albright & Wilson, Oldbury, England, 1951.
15. W. H. Waggaman, *Phosphoric Acid, Phosphates, and Phosphatic Fertilizers*, Reinhold Pub. Corp., New York, 1952.
16. T. Parker, Electrical furnace, US Patent 482586, September 1892.
17. H. Podger, *Albright & Wilson, The Last 50 Years: The Story of the Growth, Struggles and Disintegration of a Long-Established Member of the International Chemical Industry*, Brewin, Studley, England, 2002.
18. A. M. Smith, *Manures and Fertilisers*, Thomas Nelson, London, U.K., 1952.
19. E. Farber, *History of Phosphorus*, Project Gutenberg, 2010.
20. P. Lesher, A load of guano: Baltimore and the growth of the fertilizer trade, *NM*, 18, 121–128, 2008.
21. C. D. Wright, *The Phosphate Industry of the United States,* Government Printing Office, Washington, DC, 1893.

22. T. W. Coslett, Treatment of iron or steel for preventing oxidation or rusting, US Patent 870937, November 1907.
23. W. H. Allen, Process for rust proofing metal, US Patent 1206075, November 1916.
24. F. R. G. Richards, Treatment of iron or steel for preventing oxidation or rust, US Patent 1069903, August 1913.
25. A. V. Slack, *Phosphoric Acid, Part I*, Marcel Dekker, New York, 1968.
26. S. G. Nordengren, Manufacture of phosphoric acid, US Patent 1776595, September 1930.
27. A. M. Baniel and R. Blumberg, Process for the recovery of phosphoric acid from aqueous reaction mixtures produced by the decomposition of tricalcium phosphate with hydrochloric acid, US Patent 3304157, February 1967.
28. A. H. De Rooij and J. Elmendorp, Process for recovering phosphoric acid from aqueous solutions containing nitric acid and phosphoric acid, Swiss Patent 1063248, March 1967.
29. E. W. Pavonet, Method for purifying phosphoric acid, US Patent 3970741, July 1976.
30. T. A. Williams and F. M. Cussons, Purification of phosphoric acid, US Patent 3912803, October 1975.
31. J. Aaltonen, S. Riihimäki, P. Ylinen, and A. Weckman, Process for production of phosphoric acid by crystallization of phosphoric acid hemihydrate, US Patent 6814949, November 2004.
32. F. T. Nielsson, Ammoniation of superphosphate, US Patent 2729554, January 1956.
33. F. P. Achorn and J. J. S. Lewis, Granular ammonium phosphate sulfate and monoammonium phosphate using common pipe-cross-type reactor, US Patent 3954942, May 1976.
34. L. A. Peacock, Method for producing a fertilizer with micronutrients, US Patent 7497891, March 2009.
35. H. L. Thompson, P. Miller, P. M. Johnson, I. W. McCamy, and G. Hoffmeister, PILOT PLANTS. Diammonium phosphate, *Ind. Eng. Chem.*, 42(10), 2176–2182, October 1950.
36. B. Wendrow and K. A. Kobe, The alkali orthophosphates. Phase equilibria in aqueous solution, *Chem. Rev.*, 54(6), 891–924, December 1954.
37. US Geological Survey, 2011, Mineral commodity summaries 2011: US Geological Survey, 198 pp [Online]. Available at: http://minerals.usgs.gov/minerals/pubs/mcs/2011/mcs2011.pdf (Accessed April 09, 2013).
38. J. R. Van Wazer, *Phosphorus and Its Compounds Chemistry*, Vol. I, Interscience Publishers, New York, 1958.
39. P. Becker, *Phosphates and Phosphoric Acid: Raw Materials, Technology, and Economics of the Wet Process*, 2nd edn., Marcel Dekker, New York, 1989.
40. M. Jamialahmadi and H. Müller-Steinhagen, Crystallization of calcium sulfate dihydrate from phosphoric acid, *Dev. Chem. Eng. Miner. Process.*, 8(5–6), 587–604, 2000.
41. J. H. Wing, *The Hemi Era in Phosphoric Acid*, Presented at the American Institute of Chemical Engineers Clearwater Convention, Clearwater Beach, FL, 2006, p. 24.
42. Prayon displays its phosphate technology and operations, *Phosph. Potass.*, 174, 38–43, August 1991.
43. Reactors, agitators and filters for phosphoric acid plants, *Phosph. Potass.*, 174, 23–37, August 1991.
44. P. A. Smith, Successful debottlenecking at JPMC, *Phosph. Potass.*, 192, 30–34, August 1994.
45. S. V. Houghtaling, Manufacture of phosphoric acid, US Patent 4188366, February 1980.
46. A. L. Davister, Multiple cell filter with baffles, US Patent 4330404, May 1982.
47. S. Kurowski, Continuous filtration device with pivoting cells, US Patent 0089599, May 2004.
48. A. H. Parmentier, Filter with a horizontal rotating table, US Patent 3262574, July 1966.
49. S. S. Davis, Horizontal vacuum filter, US Patent 3426908, February 1969.

50. W. Klemm, B. Kuxdorf, P. Luhr, U. Thummler, and H. Werner, Process for making phosphorus pentoxide and optionally phosphoric acid with utilization of the reaction heat, US Patent 4603039, July 1986.
51. H. Rosenhouse and J. F. Shute, Heat recovery in the manufacture of phosphorus acids, US Patent 4713228, December 1987.
52. S.-E. Dahlgren, Fertilizer materials, calcium sulfate transitions in superphosphate, *J. Agric. Food Chem.*, 8(5), 411–412, May 1960.
53. J. Van Rumbeke, Process of making phosphoric acid, US Patent 540124, May 1895.
54. W. H. Waggaman, Manufacture of phosphoric acid by the volatilization process, *Ind. Eng. Chem.*, 16(2), 176–179, February 1924.
55. E. W. Guernsey and J. Y. Yee, Process of producing phosphoric acid, US Patent 1422699, July 1922.
56. R. D. Pike, Volatilization of phosphorus from phosphate rock I—Experiments in crucibles and rotary kiln, *Ind. Eng. Chem.*, 22(3), 242–245, March 1930.
57. R. D. Pike, Volatilization of phosphorous from phosphate rock II—Experiments in volatilization of phosphorus and potash in a blast furnace, *Ind. Eng. Chem.*, 22(4), 344–349, April 1930.
58. R. D. Pike, III—Calculation of performance of a blast furnace for volatilization of phosphorus and potash, *Ind. Eng. Chem.*, 22(4), 349–354, April 1930.
59. S. D. Gooch and W. H. Waggaman, Production of phosphorus and phosphoric acid, US Patent 1888896, November 1932.
60. W. C. Lapple, Recovery of phosphorus values and cement clinker from a phosphatic ore, US Patent 3235330, February 1966.
61. W. C. Saeman, Phosphorus recovery feed control method, US Patent 3558114, January 1971.
62. R. A. Hard and J. A. Megy, Process for reducing phosphate ore, US Patent, 4351809 September 1982.
63. J.A. Megy, A credible alternative to the WPA process, *Fertil. Int.*, 424, 81–89, May–June 2008.
64. D. E. C. Corbridge, *Phosphorus: Chemistry, Biochemistry and Technology*, 6th ed., CRC Press, Boca Raton, FL, 2013.
65. Furne, L. F., *Les merveilles de l'industrie, ou Description des principales industries modernes*, vol. 4, Jouvet et Cie., Paris, France, 1860.

2 Purification of Phosphoric Acid

2.1 INTRODUCTION

The purification of phosphoric acid means many things to different people. To those in the fertilizer business, it may simply mean filtration to remove solids, especially those of complexes that have been deliberately formed, or perhaps partial defluorination. On the other hand, to those supplying phosphoric acid to the semiconductor industry, it may mean ultrafiltration of already highly purified acid.

In this chapter, all the common purification processes are discussed; they are gathered into three groups: the pretreatment processes such as desulfation and crude defluorination; the core purification processes of which solvent extraction is the most common by far; and the posttreatment processes such as defluorination, decolorization, and ultrafiltration.

Individual process operations are used at different stages of purification on different plants, for example, dearsenication processes are sometimes located at the early stages of a purified acid plant, sometimes late. *Crude* is used to differentiate those very similar processes located before core purification from those after; the equipment is often near identical but the operation slightly different. Depending on the feed acid quality and the desired product, some processes are used in isolation. Table 2.1 summarizes phosphoric acid purification processes.

Taken together, the processes in Table 2.1 have the goal of producing a purified acid from a crude wet process acid made by the reaction of phosphate rock and sulfuric acid. The manufacture of purified acid via the hydrochloric acid route has the same ultimate goal, and there are many similarities in the intermediate processes, but it is different and is addressed as one of the Israeli Mining Institute (IMI) processes.

Prior to the emergence of the purification of wet process phosphoric acid (WPA) via solvent extraction, the only alternative to thermal acid in the manufacture of sodium tripolyphosphate (STPP) was the product of a chemical purification process also known as the *wet salts* process. In this process, impurities are caused to precipitate by the addition of sodium hydroxide or carbonate, producing sodium orthophosphate liquor sufficiently pure for the production of STPP and other technical grade sodium phosphates.

In addition to the purification processes, there are several ancillary processes that do not of themselves sufficiently purify the acid but are nevertheless essential components of the overall purification plant. In both the pre- and posttreatment stages, the acid undergoes concentration steps. Typically, WPA is available at 28%–30% P_2O_5 or 42% P_2O_5, depending on whether the dihydrate or hemihydrate route is employed, or 54% P_2O_5; the ideal acid concentration feed to most solvent extraction

TABLE 2.1
Phosphoric Acid Purification Processes

Process	Process Goal	Effect
Pretreatment		
Desulfation	Reduce SO_4 through addition of calcium (or barium) source, e.g., phosphate rock or limestone.	SO_4 in concentrated phosphoric acid reduced to 0.5% from 2% to 4%.
Crude defluorination	Reduce F through addition of silica, followed by concentration or blowing the acid with an air/steam mixture.	F reduced from 2%–5% to 0.15%–0.5%, Mg and K reduced through concentration, possible to recover H_2SiF_6 for sale.
Crude dearsenication (sulfiding)	Heavy metal removal, especially arsenic, through precipitation with sulfide.	Arsenic reduced from 50–100 to <5 ppm, cadmium from 50 to <1 ppm.
Organic reduction	Removal of organic species through carbon treatment.	TOCs reduced to less than 250 ppm or even below 100 ppm depending on downstream process.
Solvent extraction	Reduce both anionic and cationic impurities through solvent extraction and release by addition of water or sodium hydroxide if the purified product is required as a sodium phosphate solution.	Most impurities reduced to <1 ppm for food grade acid. For most solvent extraction processes, SO_4 is the exception achieving typically 5000–100 ppm purifications, respectively. Sodium hydroxide may be deliberately added into the process with residual levels being greater than 100 ppm.
Posttreatment		
Dearsenication (sulfiding)	Heavy metal removal, especially As, through precipitation with sulfide.	As reduced from 50–100 ppm to <3 ppm.
Decolorization	Removal of organic species to produce a clear acid.	TOC reduced to <10 ppm, color to water white (<5 APHA/Hazen/Pt–Co color scale) by adsorption or oxidation.
Defluorination	Reduce F through distillation or hot air stripping.	F reduced to <10 ppm at product acid concentration; sales concentration usually 85% H_3PO_4.
Crystallization	Reduce impurities through freeze crystallization.	Taking food grade acid as feed; crystallization is capable of reducing most impurities from ppm levels to <50 ppb.
Ultrafiltration	Reduce impurities through ultrafiltration.	Production of close to food grade acid from WPA.

processes is 58%–59% P_2O_5; clearly therefore, the supplied acid must undergo a concentration step. After solvent extraction, the acid concentration in some processes is 45% P_2O_5, from some processes significantly less, whereas the sales concentration is 61.6% P_2O_5; again a concentration step is required. Concentration is often combined with other operations such as defluorination; in the pretreatment operation, solids precipitation presents a significant operational challenge; therefore, these steps require careful attention. Many of the crude purification processes remove impurities by precipitation; consequently, solid/liquid separation equipment is required, and a wide range of filters and centrifuges may be used.

Solvent extraction processes must recover solvents from acids leaving the plant, whether it is product or raffinate acid. Most of the solvents in use have a molecular weight of about 100 and are volatile, forming azeotropes with water. Consequently, acid leaving the solvent extraction section is stripped with steam to remove the soluble or entrained solvent. Organophosphorus extractants, generally tributyl phosphate (TBP), are also used as solvents but have very high boiling points and must therefore be recovered by other means. Over time and use, the solvent may degrade, forming variant chemical species; it may also become contaminated by dissolving organic compounds from the feed acid; these compounds are removed in a solvent treatment section.

Purification processes distribute the impurities in the feed acid between the product(s) and a by-product. In most purification processes, the by-product is a usable acid; for most PWA plants, it goes to a fertilizer plant; however, in some cases where there is no immediate fertilizer outlet, the by-product becomes a waste to be minimized and the plant utilizes a *total exhaustion* process. The *total exhaustion* process has 95%–97% P_2O_5 efficiency. The remaining 3%–5% does not *disappear* but is either returned to the WPA plant reactor, used to make a small quantity of fertilizer, or neutralized and landfilled. Some PWA plants are designed with one final product, whether salts (for STPP), technical, or food grade; others are designed for two or even all three. The latter are more flexible, better able to deal with fluctuating markets and feed acid quality but are more complex and may be more expensive to operate. Meanwhile, for those PWA plants with a fertilizer grade acid outlet, the raffinate stream must not be treated as a dump for the impurities, as the fertilizer feed acid must meet certain specifications to make acceptable fertilizer. For example, a WPA plant might produce an acceptable acid for fertilizer manufacture that is close to the limit for cadmium and/or uranium; in the PWA process, most of these species distribute to the raffinate stream to go to the fertilizer plant. Clearly, that acid stream would be out of specification; therefore, either careful blending with the rest of the WPA is required or further cleaning of the raffinate stream is needed. Figure 2.1 illustrates this problem: a WPA plant produces an acid with a particular metal impurity of 90 ppm against a fertilizer feed specification of 100 ppm; 50,000 tons are sent to the PWA plant that reduces the concentration to 1 ppm in the product acid but returns 50% of the feed acid that now has an impurity content of 179 ppm. This stream must then be blended with four times the feed to the PWA plant to achieve the fertilizer specification. This problem arises in all purification plants and units within these plants. A crystallization plant converting food grade acid into semiconductor grade acid inevitably produces by-products with higher impurities than the feed *food grade* acid; the art is ensuring that these by-products are saleable and do not become an embarrassment.

The term *split* is in common use in the industry to describe the ratio between the feed, the raffinate, and the product, and it is similar to the common term yield. The definition of split becomes more complex when applied to different PWA plants

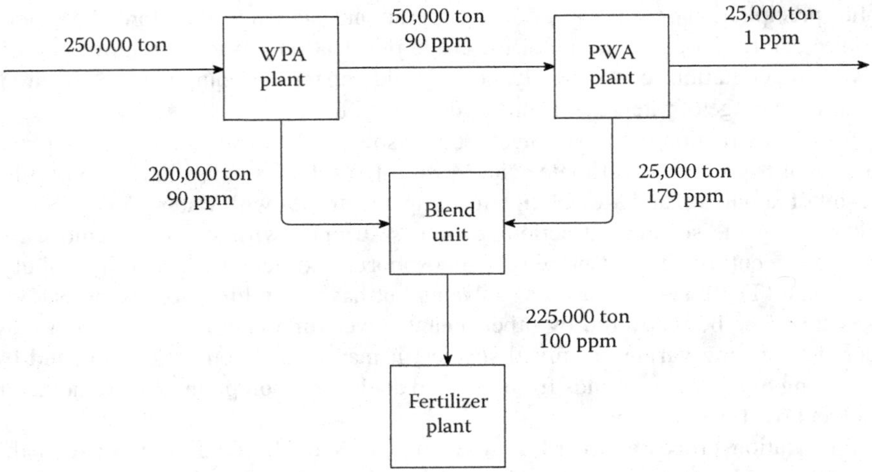

FIGURE 2.1 Balancing impurities between WPA, PWA, and fertilizer plants.

and is explained later. PWA plants have splits in the range 50%–80%, depending on the PWA technology, the feed acid quality, and the outlets. Generally speaking, in chemical engineering the higher the yield, the better, and on PWA plants, a high yield, or split, implies a lower process cost within the plant boundary. But the economic basis of the plant depends on several other factors that include the cost of the feed acid to the plant, including any pretreatment costs, and the credit for the raffinate return, and these and other costs can easily outweigh the benefit of a higher split.

The specification, which includes maximum impurity levels, concentration, and appearance, of a specific acid is a matter for agreement between the supplier and the receiver. All suppliers publish sales specifications for different acid grades and will, by agreement, supply a detailed chemical analysis of specific batches or sales deliveries. The sales specifications frequently include various abbreviations to signify that the product meets the standard of a particular organization, for example, USP-NF stands for United States Pharmacopeia National Formulary, and products used in healthcare products sold in the United States must meet these standards. Table 2.2 lists the most common standards organizations that issue standards applying to phosphoric acid and its phosphate derivatives.

The standards of the organizations in Table 2.2 have much in common with each other and are much less specific than the purchase specifications of individual corporations. This is because individual corporations often have a specific process or market application that may be sensitive to a specific impurity. This impurity might affect product taste, color, or performance. The common factors in the standards include the following:

Concentration—phosphoric acid for food, pharmaceutical, and semiconductor grades is a minimum 75% H_3PO_4. (Most sales from producers are at 85% and range 75%–91.5%.)

TABLE 2.2
National and International Standards Organizations for Phosphoric Acid

Abbreviation	Organization
ACS	American Chemical Society
BP	British Pharmacopeia
CAC/JECFA	Codex Alimentarius Commission/Joint Expert Committee Food Additives
EP	European Pharmacopeia
FCC	Food Chemicals Codex
JP	Japanese Pharmacopeia
NF	National Formulary
SEMI®	Global industry association for the micro- and nanoelectronics industries
USP	United States Pharmacopeia

Appearance—clear, water white. *Water white* is an old-fashioned term dating back to the nineteenth century. Appearance is now measured by both light absorbance and platinum–cobalt (Pt–Co) color scale tests. The Pt–Co color scale test is defined by ASTM D 1209-05 [1] and is often referred to in the industry as the American Public Health Authority (APHA) scale or HAZEN number (after the chemist who proposed the test)
Taste—tasteless
Odor—odorless
Nitrates—not more than 5 mg/kg (5 ppm)
Chlorides—not more than 200 mg/kg (200 ppm) as chlorine
Sulfates—not more than 1500 mg/kg (1500 ppm)
Fluoride—not more than 10 mg/kg (10 ppm)
Arsenic—not more than 3 mg/kg (3 ppm)
Lead—not more than 4 mg/kg (4 ppm)
Heavy metals (as Pb)—total heavy metal analysis adds up to not more than 10 mg/kg (10 ppm). *Heavy metals* is a loose and common term in the industry and implies impurities that are often metals, often toxic to humans, and are often removed by precipitation with sulfide. The *heavy metals* include As, Ba, Cd, Cu, Fe, Hg, Mo, Pb, and Zn. Of these, the elements of greatest concern because of both the usual impurity profile in the feed acid and the effectiveness of the various purification processes are As, Pb, and Cd

Within a phosphate business, or multiproduct site, various acid grades are produced either for sale as acid or for further processing to derivative phosphates. A broad categorization is given in Table 2.3 based on concentration and range of impurity for each group of acid. To amplify on the categorizations: *filter* acid is the acid from the filter immediately after the reaction system on a WPA plant; *fertilizer* or *merchant* acid is acid ready for processing on a fertilizer plant, such as a MAP/DAP plant, and has undergone some purification and concentration, and it is a marketable

TABLE 2.3
Phosphoric Acid Group Categorization

Acid Group	Concentration	Impurity Level (ppm)
Filter acid	28% P_2O_5	5,000–25,000
Fertilizer/merchant acid	42%–54% P_2O_5	10,000–50,000
PWA feed acid	54%–59% P_2O_5	5,000–25,000
Raffinate acid	25%–45% P_2O_5	20,000–50,000
Technical acid	50%–61.6% P_2O_5	500–5,000
Food acid	61.6% P_2O_5	0.5–250
Cola/pharma acid	61.6% P_2O_5	0.5–100
Semi (LCD) acid	61.6% P_2O_5	0.1–1
Semi (semi) acid	61.6% P_2O_5	0.01–0.1

product; *PWA feed* acid is filter or fertilizer acid that has undergone sufficient treatment that it is ready to go into a PWA plant; *filter, fertilizer, merchant,* and *PWA* feed acids are usually referred to collectively as green acid (because it is often that color), especially in the context of purification; *raffinate* acid is that stream rejected from most PWA plants and is suitable for use in fertilizer processes; *technical* acid is a partially purified acid suitable for processing to derivative phosphates for the technical markets; *food grade* acid is suitable for use directly in food applications or for the manufacture of food grade products; and *cola/pharma grade* acid is of a slightly higher purity than a general food grade acid and can be used as such—cola grade reflects addition requirements made by suppliers of cola drinks, similarly pharma and pharmaceutical industry customers. Acid for the semiconductor business ranges in purity depending upon the application. For liquid crystal displays, the purity requirements are similar to those for food grade acid. For the manufacture of semiconductors, a very high purity is required with most impurities below 50 ppb.

Table 2.4 shows indicative analyses of various acids. The reality is that regardless of the acid, its analysis constantly changes; this is a function of the changing composition of the feed rock for WPA and changes in the purification chemistry in response to small changes in the WPA; in addition, two acid samples from the same bottle frequently have different *analyses* when tested in different laboratories. The first four acids are wet process acids from four different locations; the first three are made from sedimentary rock and the fourth, Kola, from igneous rock. As the world's largest rock producer, Morocco markets several different rocks as well as acids; this example at 59.3% P_2O_5, 5240 ppm sulfate, and 1300 ppm fluoride has undergone defluorination, desulfation, and concentration and is ready to go directly to a solvent extraction system. It is notable that the iron content is lower than the other two acids from sedimentary rocks. The acid from North Carolina has negligible levels of arsenic but relatively high levels of aluminum, iron, and titanium and has not undergone crude defluorination; the designer of a PWA plant might consider allowing this acid to go straight to solvent extraction or might consider a further concentration step in conjunction with crude defluorination and clarification to remove precipitates. The Idaho acid is typical of those from the *Western States* with relatively high

TABLE 2.4
Indicative Analyses of Various Phosphoric Acid Grades

	Morocco	North Carolina	Idaho	Kola	Raffinate	Technical	Food	Thermal	LCD	Semi
% H_3PO_4	82	78	73	73	62	76	85	85	85	85
% P_2O_5	59.3	56.6	53	53	45	54.8	61.6	61.6	61.6	61.6
ppm										
SO_4	9600	7100	20,300	19,260	13,300	2100	78.0	1.0	1.9	0.9
F	1300	4060	8,100	1,010	25	310	5.1	4.0		
Cl	60	44		507	0					
Al	221	3050	7,700	2,400	560	1.0	1.0	1.3	0.49	0.05
As	5	0	17	12	13	2.5	0.4	0.2	0.05	0.03
B	19		40	0	48	18.1	17.1	0.0	0.46	0.04
Ba									0.46	0.04
Ca	500	1029	68		215		3.5	3.0	1.44	0.15
Cd	47	29	125	29	119	0.5	0.2	0.3	0.04	0.04
Co									0.04	0.03
Cr	374	230	634	19	946	1.1	0.5	0.4	0.20	0.04
Cu	19	2	58	96	48	0.5	0.1	0.0	0.05	0.03
Fe	1421	8729	5,760	0	3,594	6.7	1.0	6.7	1.80	0.09
K	480	1344	422	1,152	1,214	73.8	49.2	4.9	1.20	0.12
Li										
Mg	5160	6950	2,880	7,010	13,100	2.9	0.5	0.4	0.20	0.05
Mn	10	58	86	365	25	0.5	0.5	0.4	0.10	0.05
Mo	21	14		8	53	30.8	4.8	1.0	0.10	0.05
Na	1740	650	140	890	1,670	25,800	100	11.3	2.00	0.24

(continued)

TABLE 2.4 (continued)
Indicative Analyses of Various Phosphoric Acid Grades

	Morocco	North Carolina	Idaho	Kola	Raffinate	Technical	Food	Thermal	LCD	Semi
Ni	75	33	163	36	190	0.5	0.5	0.4	0.20	0.04
Pb	1	0	1	1	3	0.3	0.1	0.1	0.10	0.04
Sb						1	0.8	5.2	0.40	0.26
Si	58	59	557		147					
Ti	120	864	125	720	304	4.9	0.5	0.1	0.20	0.05
U	72	91	192	0	182	0.01	0.01	0.09	0.05	0.02
V	264	48	1,248	67	668	0.5	0.5	0.1	0.04	0.02
Zn	797	518	2,208	19	2,016	64.3	1.0	1.3	2.00	0.05
TOC	48	144	384	144	121					

levels of chromium, vanadium, and zinc; some rock from this region also has high organic levels. Chromium and zinc may be tackled with both sulfide precipitation and solvent extraction. Vanadium, as well as molybdenum, is a challenging impurity that forms complexes with phosphoric acid and is less amenable to purification by solvent extraction than other elements. Acid from Kola rock is high in some impurities but low in many compared with acids from sedimentary rocks. The raffinate acid is typical of the acid returning to the fertilizer plant from the PWA plant after solvent extraction of the Moroccan acid. Different PWA technologies produce different raffinates (or underflow as it is also known in the industry). The technical grade acid is a typical analysis of acid suitable for producing industrial sodium phosphates such as STPP; the sodium content is very high and reflects the use of sodium hydroxide in the solvent extraction process; alternatives include water and ammonia. If a PWA plant is adjacent to a customer plant for, say, STPP or ammonium phosphate fertilizer, then either sodium or ammonia will be chosen as a consequence; both are more effective than water and are accounted for in the interplant transfer value. The impurities in the thermal acid are generally low, less than 5 ppm. Calcium, iron, and sodium are notable exceptions; it is possible that calcium is introduced with dilution water in making the acid from phosphorus pentoxide vapor, the iron is a result of pick up from the plant, and the sodium is introduced if sodium bisulfide is used in the dearsenication process.

2.2 CHEMICAL PURIFICATION

The purification of a chemical by precipitation and separation of impurities is a standard technique applied from the laboratory to the plant scale. STPP is made by reacting either sodium hydroxide or sodium bicarbonate with phosphoric acid, producing sodium ortholiquor—a solution containing both MSP and DSP, which is then dried and calcined above 450°C to produce STPP, according to Equation 2.1. Calculating the Na_2O/P_2O_5 molar ratio of the product gives a molar Na/P ratio of 5:3 that in solution is pH 7 before calcination and about 10 afterward:

$$NaH_2PO_4 + 2Na_2HPO_4 \xrightarrow{>450°C} Na_5P_3O_{10} + 2H_2O \quad (2.1)$$

To meet the product impurity specification and white appearance, STPP was first made with thermal acid when introduced in the 1940s. As demand for STPP began to rise in the 1950s, the industry discovered it was possible to make an acceptable product from WPA, thereby reducing manufacturing cost. WPA itself is generally much cheaper to make than thermal acid [2] as sulfur is normally cheaper than electricity, but in the middle of the twentieth century, many US producers had negotiated highly competitive electricity rates for their phosphorus plants. The *wet salts* process route was practiced around the world although there was only one corporation in the United States, Olin, which both used it and patented various developments.

In principle, all the commercial sodium phosphates can be made via this process by adjusting the Na/P ratio and precipitating out the impurities [3]. Further purification is possible through crystallization of the sodium ortholiquor [4,5]. An analogous process is the production of dicalcium phosphate (DCP) from WPA by the addition of a calcium source; whereas the sodium phosphate liquor is made at pH 7–9 and the impurities precipitate out, DCP itself precipitates at pH 4–5.5, leaving the impurities in the acidic liquor.

So far so simple; however, both major and minor impurities in the feed acid affect the STPP product purity, appearance, and function as well as the process to make it. At certain concentrations, some impurities interact with others to hinder processing such as filtration or concentration. Table 2.5 shows examples of impurities and their effect.

Chemical purification plants typically receive filter acid at 28% P_2O_5 with 2% SO_4, the WPA reaction operating with sulfuric acid excess to aid gypsum filterability. Other impurities depend on the phosphate rock, but a fluoride content of 1.5% is typical. STPP specifications limit sodium sulfate content to 2%; thus, the sulfate content of the acid must be reduced to less than 0.5%. Sulfate is reduced with a desulfation

TABLE 2.5
Impurities in STPP

Impurity	Effect	Action
SO_4	Forms sodium sulfate that reduces product purity. Higher sulfate lowers calcium levels that act as filter aid.	Desulfation to reduce concentration below 0.5%
F	May form colloidal compounds with other major impurities (Al, Mg, Fe) hindering filtration. Any sodium silicofluoride remaining will hydrolyze and form colloidal silica. High levels in ortholiquor may corrode downstream equipment producing STPP.	Defluorinate by precipitation and separation of the sodium silicofluoride
Cr	High levels in STPP give green color [2]. Was linked to interference with oxidizing agents. Chromium and nickel associated with dermatitis. Interacts with vanadium reducing precipitation efficiency [105].	Reduce acid and precipitate, Cr/Ni levels <1 ppm
Fe	High levels in STPP give yellow color [2].	Reduce acid and precipitate
Mg	High levels lead to turbidity in ortholiquor [106]. High levels may form fibrous material in STPP.	Various strategies including lowering levels with NaF as part of defluorination and precipitation during neutralization
Mn	High levels in STPP give pink color [2].	Reduce acid and precipitate
Si	High levels lead to turbidity in ortholiquor [106] and poor filtration of precipitates.	Lower levels with NaF as part of defluorination
V	High levels in STPP give yellow color [2].	Reduce acid and precipitate
Zn	High levels in STPP affect hydration [107].	Add sodium sulfide, reduce acid, and precipitate
Organics	High levels in STPP give a brown/gray color.	Remove by oxidation or adsorption with activated carbon

unit that is described in more detail in Section 2.3; however, to summarize, typically the filter acid is pumped to an agitated desulfation tank to which phosphate rock is added that introduces calcium, and so sulfate is precipitated as gypsum. Residence time in the desulfation tank is 2–4 h and the liquor is pumped to a belt filter to filter the gypsum. Some plants have used vacuum drum filters; others have used thickeners. The desulfation tank operates at 70°C–80°C; the filtered, desulfated acid is pumped to intermediate storage where it cools to ambient temperature and in doing so causes further precipitation.

The presence of excess fluoride can, either during cooling or concentration or as a result of neutralization, cause the formation of complex compounds that filter very poorly. Consequently, defluorination follows desulfation in this purification process. Different defluorination processes are described later; however, in this case, it is convenient to lower the fluoride levels by making sodium silicofluoride (Na_2SiF_6) that is only slightly soluble in phosphoric acid. The reaction is shown in the following equation:

$$2Na^+ + SiF_6^{2-} \rightarrow Na_2SiF_6 \qquad (2.2)$$

The sodium source can be either sodium hydroxide or soda ash, but utilizing sodium ortholiquor produced downstream simplifies the plant design and improves process control. Fully treated sodium ortholiquor ready for concentration before spray drying has an Na/P ratio of 5:3 and contains 19% P_2O_5. Enough sodium is added to the acid to raise the Na/P ratio to 0.15 (a significant excess over stoichiometry); on some plants, this was done in two stages to improve the silicofluoride crystal size and therefore the facility of its filtration. Overdosing of ortholiquor at this stage, beyond a Na/P ratio of 0.4, leads to the hydrolysis of the silicofluoride to sodium fluoride and silicon dioxide and is undesirable. The defluorinated acid flows to a thickener that allows the silicofluoride crystals to settle, forming a slurry containing 35%–40% solids. The slurry is then pumped to vacuum drum filters and the silicofluoride and some gypsum separated from the acid; filter cake washing reduces P_2O_5 losses at this point. Sodium silicofluoride is a saleable product and may be processed further for sale. Typically, this defluorination process reduces fluoride content to 0.1%–0.2%.

Some impurities form complexes with phosphoric acid and are difficult to remove by precipitation. Molybdenum and vanadium form molybdo- and vanadophosphoric acids as well as phosphates. Other elements such as iron, chromium, and uranium are normally present in an oxidation state that precipitates poorly. Therefore, the next stage of chemical purification is to alter the oxidation state of these elements by chemical reduction that conditions them for the following neutralization step. Some practitioners used $FeSO_4 \cdot 7H_2O$ [6] to reduce the acid, although others questioned its efficacy; however, the use of iron sulfide is effective and has the additional benefit of precipitating the heavy metals as sulfides (e.g., AsS, PbS, ZnS). The FeS addition rate is 0.5% by weight relative to P_2O_5 content. Sulfiding is discussed in more detail in the section(s) on dearsenication; however, two points are noteworthy at this stage. Firstly, sulfiding, whereby either a metal sulfide or hydrogen sulfide gas is added to the acid, inevitably leads to the evolution of some H_2S gas. H_2S is highly toxic. It causes nausea, vomiting, and breathing difficulties above about 10 ppm in air and coma, convulsions, and risk of death above 100 ppm. Consequently, great care is taken to control and scrub all vapor from equipment that is carrying out this process. Secondly, metal sulfide solubility varies by both metal and pH, as illustrated in Figure 2.2 that shows metal sulfide solubilities in water [7]. This property is utilized in the metal industry for selective metal extraction. As Figure 2.2 shows, above pH 8 some sulfide solubilities start to increase, that is, of course, unhelpful if the object is to precipitate out these impurities. Equally as the solution is more acidic so also are impurities more soluble; thus, precipitation of, say, As_2S_3 from pure acid is much harder than from a neutral or slightly alkali ortholiquor.

Some operators filter the sulfides that have precipitated at this stage or carry out a partial neutralization to pH 4–5 and then filter. The sulfide crystals are not particularly easy to filter at this stage; consequently, a filter aid is required. One option is to use activated carbon as both filter aid and to adsorb organic species; this technique is used on solvent extraction plants.

The defluorinated, sulfided, and carbon-treated acid is pumped to the neutralization section. The acid is co-fed with sodium carbonate slurry into the first agitated tank and overflows to the second neutralizer where sodium hydroxide is added to

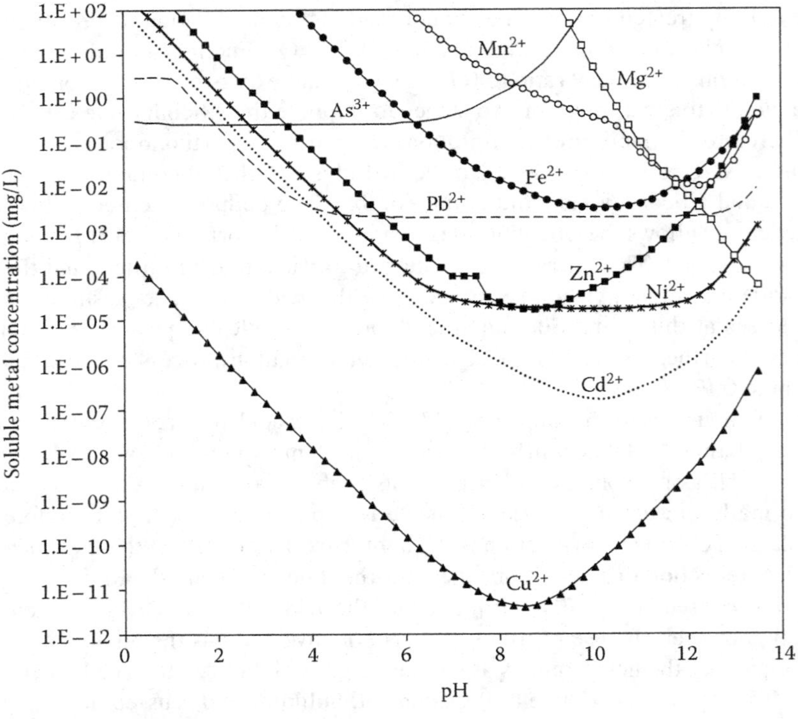

FIGURE 2.2 Metal sulfide solubilities in water. (Adapted from *Hydrometallurgy*, 104, Lewis, A.E., Review of metal sulfide precipitation, 222–234, Copyright 2010 with permission from Elsevier.)

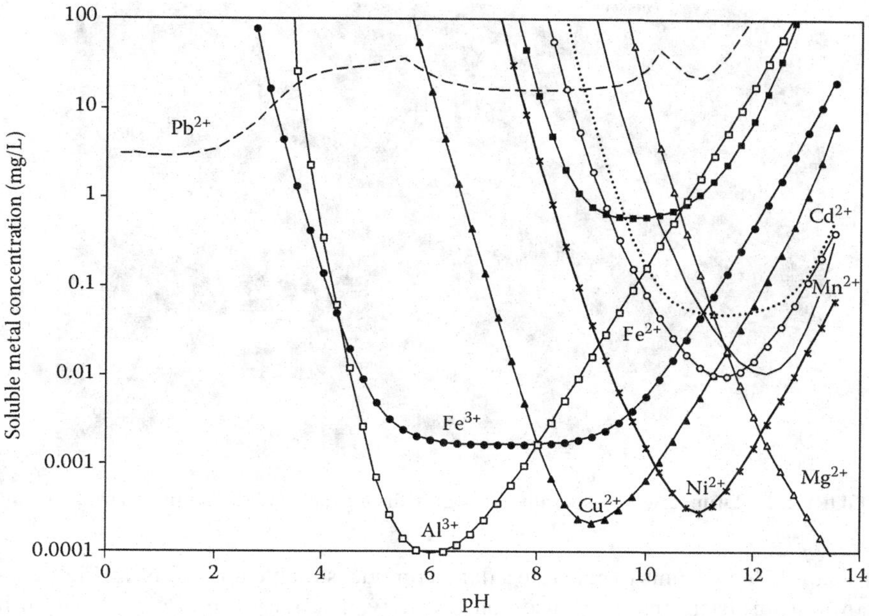

FIGURE 2.3 Metal hydroxide solubilities in water. (Adapted from *Hydrometallurgy*, 104, Lewis, A.E., Review of metal sulfide precipitation, 222–234, Copyright 2010 with permission from Elsevier.)

trim the pH to the desired level. Ortholiquor for STPP production is neutral; however, additional purification is possible by raising the liquor to pH 9, then filtering off the precipitated impurities, and then lowering the pH by adding clean acid. Figure 2.3 shows metal hydroxide solubilities in water [7], and at pH 9, many solubilities are at a minimum.

The solubility diagrams do not show precipitation times; however, for industrial practice, a residence time of 2–3 h is satisfactory. Solid content is typically 3%; P_2O_5 content of the solution is approximately 19%. The neutralizers are held at 95°C–98°C. The ortholiquor is pumped to filters; vacuum drum or belt filters permit cake washing and therefore reduce P_2O_5 loss. On some plants, the filter cakes are reslurried for secondary or tertiary filtration. The solids may be recovered and sold as fertilizer.

At this stage, purification is essentially complete although an oxidizing agent, such as hydrogen peroxide, sodium chlorate, or sodium nitrate, is sometimes added to ensure all organic species are destroyed. As well as destroying organics, the oxidizing agent may induce further precipitation of impurities so this addition is accompanied by a polishing filter.

The final stage of the chemical purification process is concentration when the ortholiquor is brought up to a suitable solid content to feed to a spray drier to make dry *orthobeads* that are fed to the calciner and converted to STPP. Prior to the practice of spray drying, the concentrated ortholiquor was dried on the surface

FIGURE 2.4 Drum driers on sodium ortholiquor drying duty. (With permission from Solvay.)

of steam-heated drums to generate a flaked product suitable for calcining; Figure 2.4 shows drum driers that were used on this duty. The target solid concentration in the liquor going to the spray drier atomizer is in the range 55%–60.5%, equivalent to 29.0%–31.9% P_2O_5 compared to 19% P_2O_5 in the neutralizer. Concentration is carried out in a variety of process units including pumped circulation vacuum evaporators, either single- or multieffect, and submerged flame evaporators. The advantage of the evaporators is that the heat source is usually heat-recovered steam from both within the STPP plant itself and on a large phosphate site from the sulfuric acid plant.

In this context, the net effect of the chemical purification process is the capability to produce technical grade STPP from WPA. The cost of this process is the loss of up to 6%–8% of the input P_2O_5 in precipitates.

2.3 SOLVENT EXTRACTION–BASED PROCESSES

2.3.1 Introduction

Solvent, or liquid–liquid, extraction is a standard technique for partitioning dissolved compounds between two immiscible or only partially miscible solvents, one of which is often water, with the object of purifying the liquid feed. This technique is used both in the laboratory and on an industrial scale. The range of application is wide and includes the manufacture of antibiotics and vitamins in the pharmaceutical field, agrochemicals, biofuels and biochemicals, food, fragrances, oil refining, and mineral extraction. There are many good textbooks on the subject, some that provide an excellent introduction [8–10] and others that are more advanced [11–14]. Space does permit a comprehensive treatment of the theory; thus, only the highlights are given as they pertain to its application in the purification of phosphoric acid. It is also worth noting that even 50 years after the first plants started to emerge, there are still no comprehensive and reliable mathematical models that will predict either the distribution of impurities or the exact number of stages of purification for any given feed acid analysis.

Solvent extraction has many similarities with distillation as a separation technique. In distillation, heat is used to facilitate the separation of materials with different volatilities. Solvent extraction exploits different solubilities in the introduced solvent. Both techniques can make use of McCabe–Thiele diagrams to estimate the theoretical number of stages required to effect separation.

Solvent extraction systems are affected by temperature and pressure (e.g., supercritical carbon dioxide is used as a solvent in the coffee and pharmaceutical industries). Across this industry, all the solvent extraction sections of phosphoric acid purification plants operate at atmospheric pressure but at differing temperatures, depending on the solvent system in place. Atmospheric pressure variation and what goes on in the vapor phase are of little consequence to the separation going on in the liquid phases but are of relevance to the plant design with regard to safety, the environment, and solvent cost. The operating temperature, however, is important as it affects solubilities. Partially miscible phases usually exhibit a critical solution temperature (CST), the temperature at which the phases merge into one. The purification plant in Coatzacoalcos, Mexico, operates with isopropyl ether (IPE) as the solvent. The extraction is performed at low temperature to maximize the extent to which phosphoric acid dissolves in the solvent as well as chilling the solvent to maintain the phases. Three immiscible phases may be formed in the IPE system due to the formation of a loose complex with phosphoric acid and water in the molar ratio of 1–2 mol ether for each mole of acid and water. The composition of this *complex* phase is temperature dependent with the ether content being lower at higher temperatures and exploits the temperature dependency of the solubility of the acid in IPE.

The three plants in North America, at Aurora, North Carolina; Geismar, Louisiana; and Coatzacoalcos, each use a single solvent (methyl isobutyl ketone, MIBK, TBP, and IPE, respectively). Some plants operate with two solvents; Prayon-designed plants (Belgium, Brazil, Morocco, and elsewhere) use a TBP/IPE mix as the solvent; these are treated as pseudoternary systems with two phases. TBP being the minor component (up to 15%) of the solvent system is present as a solubilizer that prevents the formation of the separate *complex* phase that is formed when IPE is used alone. Some may consider TBP as a co-solvent, but as in hydrometallurgical solvent processes, the action is best described as being that of a solubilizer.

The now closed FMC Foret plant in Spain used TBP plus a kerosene diluent. Kerosene has no solvent properties with respect to phosphoric acid.

Aliphatic alcohols, containing four or five carbon atoms, are or have been used as solvents. The Israeli Mining Institute (IMI) designed plants that operated in Israel and India using isoamyl alcohol (AmOH). Isopropanol (C3) was used in Germany for many years in the Budenheim phosphoric acid purification process. While isopropanol is fully miscible in all proportions with pure aqueous phosphoric acid, this is not the case with wet process phosphoric acid. In order to have a liquid aqueous phase, the system was modified by the addition of caustic soda. TVA patented a purification process based on methanol that caused a precipitate to be formed that had to be removed by filtration [15]. If methanol, ethanol, or isopropanol were to be used as the solvent, then separation of the purified acid could only be achieved by distilling the

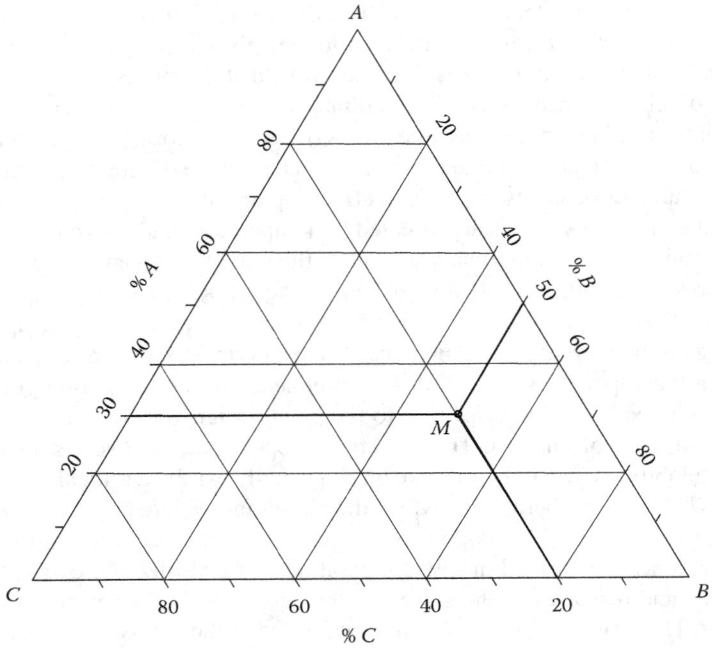

FIGURE 2.5 General ternary diagram.

alcohol from the acid. The phosphate could of course be recovered by neutralization to yield a sodium or ammonium phosphate solution.

Treybal [14] describes four different ternary systems; of these, the system of interest is type 1, the formation of one pair of partially immiscible phases.

Ternary systems are usefully described with triangular diagrams. Referring to Figure 2.5, each side of the equilateral triangle represents the weight percentage composition of one component with the component label at 100%. Viviani's theorem states that the sum of the perpendicular distances from each side to any point in the triangle adds up to the triangle's height. By setting the height to 100%, any composition is described; thus, in Figure 2.5, point *M* represents 30% *A*, 50% *B*, and 20% *C*. The curve *DPE* in Figure 2.6 is known as the *binodal curve* and represents the boundary of the two-phase region beyond which the mixture is homogenous. Point *P* is known as the *plait point* and is defined as the limit of immiscibility. In other words, two phases in equilibrium may start with compositions at *D* and *E*; as component *A* is added, the *C*-rich phase proceeds up the *DP* curve, and the other *B*-rich phase up the *EP* curve. Points on these curves where both phases are in equilibrium, for example, at *F* and *G*, are joined by the *tie line FG*. There are an infinite number of tie lines that are not parallel. At all points along the line *FG*, the two phases combined have the same composition. As more *A* is added, the tie lines get shorter until at point *P*, the plait point, the composition of both phases is identical, and beyond this point, only one phase exists. The composition at point *M* on tie line *FG* obeys

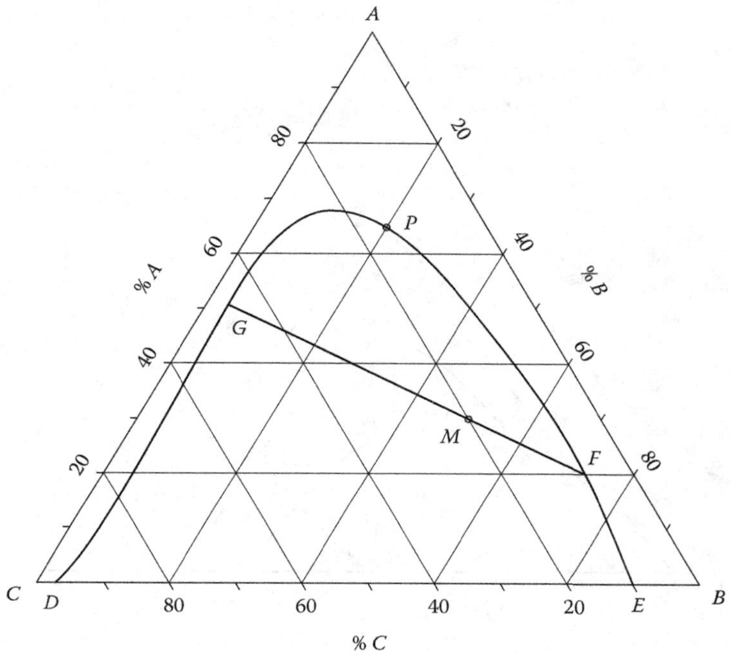

FIGURE 2.6 General ternary diagram with binodal curve.

the lever arm rule and is in proportion to the lengths of *FM* and *MG* as shown in the following equation:

$$\frac{F}{G} = \frac{\overline{GM}}{\overline{FM}} \tag{2.3}$$

Figure 2.7 shows the effect of differing temperature on solubility. As the temperature changes, so does the system solubility (in the case shown, there is a no CST for all three components). Consequently, when experiments are carried out to plot the binodal curve, the mixtures are maintained at the same temperature, and the temperature is always stated.

The binodal curve is determined simply with a stoppered separation funnel, Figure 2.8. For example, the solvent is carefully titrated into a quantity of phosphoric acid of known concentration and, therefore, water content. The mixture is shaken and allowed to stand for a few minutes and observed. More solvent is added until the point at which turbidity is observed, the *cloud point*, indicating the onset of two phases. This mixture is of known composition, is a point on the binodal curve, and is therefore plotted. The process is repeated until sufficient points are established to draw one side of the curve. For the other side of the curve, water is added to various solutions of concentrated phosphoric acid (90% H_3PO_4) in the solvent to determine the cloud point. At all times, the liquids are maintained at a chosen temperature.

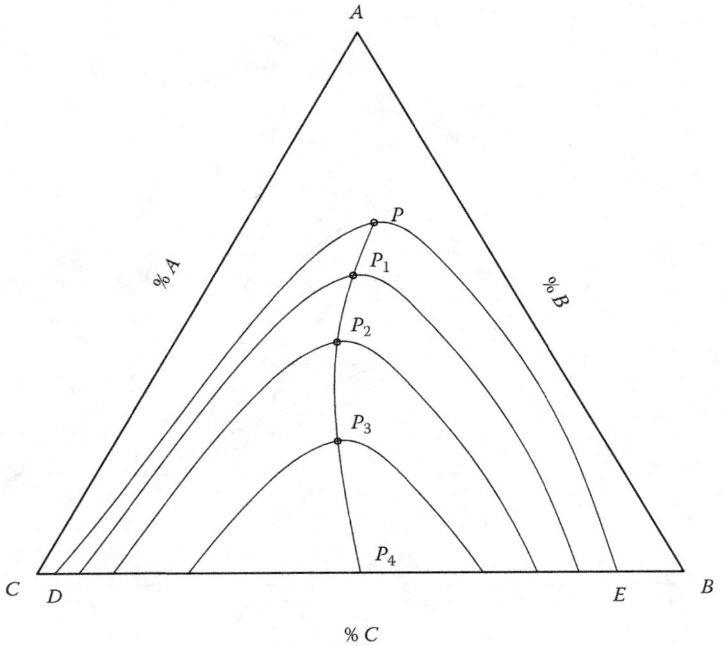

FIGURE 2.7 Effect of temperature on solubility.

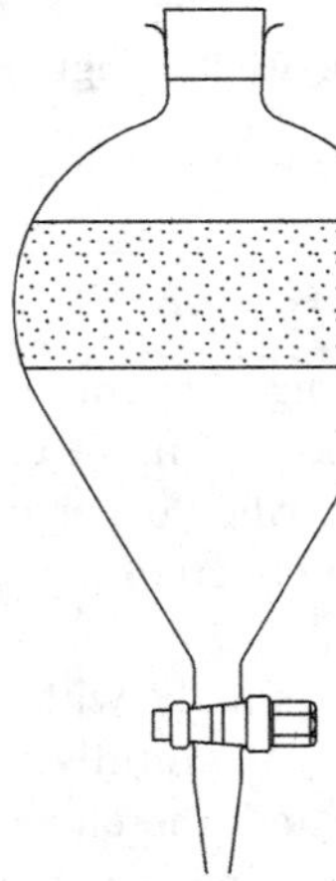

FIGURE 2.8 Separation funnel.

Once the binodal curve is established, tie lines are derived. Mixtures of known composition are shaken and allowed to stand and coalesce into two phases. The phases are weighed; the weight ratio is proportional to the distance from the starting mixture to the binodal curve, after Equation 2.3. Thus, a number of tie lines are constructed.

It is possible to construct more tie lines using the conjugate curve. Lines are drawn parallel to *AB*, *AC* through the tie line end points. The intersections of these

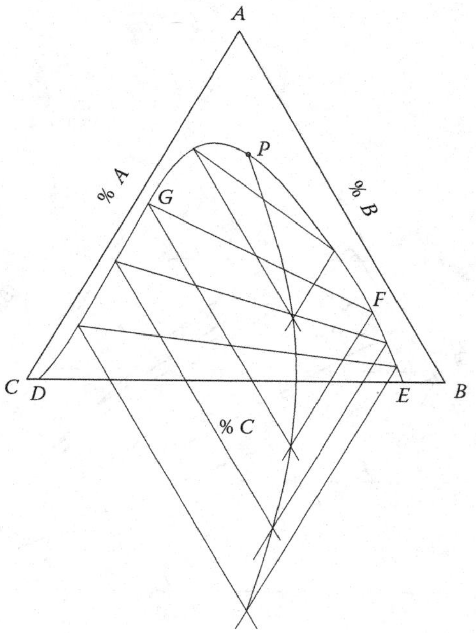

FIGURE 2.9 Conjugate curve construction.

lines are points on the conjugate curve from which new tie lines are constructed as shown in Figure 2.9.

Figure 2.10 shows the binodal curve and tie lines for the system of analytical grade phosphoric acid, MIBK, and water at 40°C plotted from Ananthanarayanan and Rao's data [16]. Other workers have also plotted the MIBK–H_3PO_4–H_2O system [17]. From Feki et al. [17] and using Hand's method [18], correlation curves representing the dependence of the logarithms of the ratios of the concentrations of phosphoric acid and MIBK in the aqueous phase, and phosphoric acid and water in the organic phase, were plotted for the tie line and binodal data. These lines cross at the plait point, giving a composition of 59.6% phosphoric acid, 15.0% MIBK, and 25.4% water. Using the graphical method and neglecting the three tie lines close to zero phosphoric acid, the conjugate curve crossed the solubility curve at 61.7% phosphoric acid, 10.0% MIBK, and 28.3% water (see Figure 2.11). Given the paper's declared solubility and mass balance accuracies of 1% and 1.5%, the two approaches to establishing the plait point are reasonably close in this case. In passing, it is worth noting from Figure 2.10 that from the MIBK–water line, where the acid content is 0%, the solubility of water in MIBK is only 2.0%, and MIBK in water 1.4%, which increases slightly at lower temperatures.

To begin to consider a process design, one must estimate the maximum and minimum amount of solvent required to effect separation. Following Treybal [14], an ideal stage is shown in Figure 2.12. *F*, *E*, *R*, and *S* designate feed, extract, raffinate (sometimes referred to as underflow), and solvent. With the system held at 40°C and taking *F* to be 1000 kg 70% H_3PO_4 and *S* as pure MIBK (never the case in a working process), both points are located on the ternary diagram, Figure 2.13.

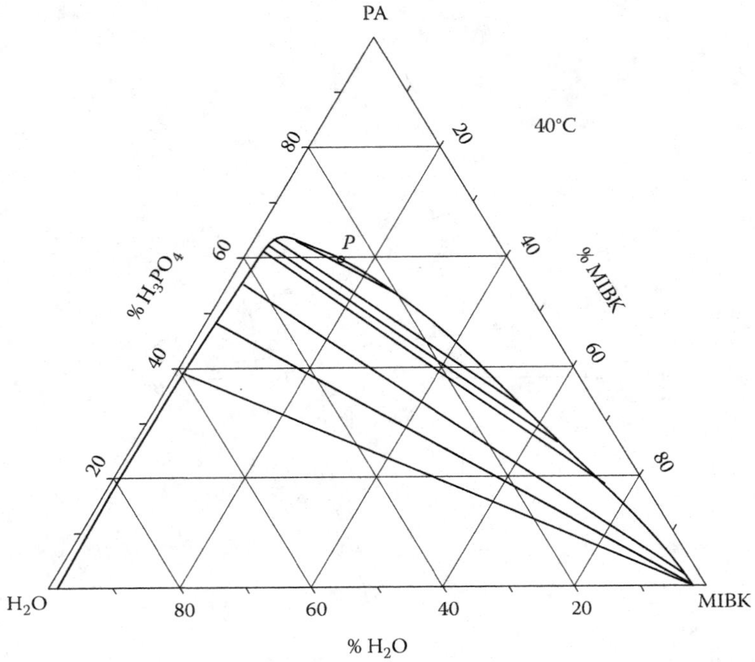

FIGURE 2.10 MIBK–H_3PO_4–H_2O ternary diagram. (Adapted with permission from Ananthanarayanan, P. and Rao, P.B., Ternary liquid equilibria of the water-phosphoric acid-isoamyl alcohol, cyclohexanol, or methyl isobutyl ketone systems at 35°C, *J. Chem. Eng. Data*, 13(2), 194–196. Copyright 1968 American Chemical Society.)

The feed F intersects the solubility curve at D, where $X_{BD} = 0.124$ (where X_{BD} represents the weight fraction of B in a D-rich solution). The minimum quantity of solvent, $S_{\min}$, is given by

$$S_{\min} = B_{\min} = \frac{F(X_{BD} - X_{BF})}{X_{BS} - X_{BD}} = \frac{1000\,(0.124 - 0)}{1 - 0.124} = 141.6\,\text{kg} \tag{2.4}$$

The maximum quantity of solvent, $S_{\max}$, is given by

$$S_{\max} = B_{\max} = \frac{F(X_{BG} - X_{BF})}{X_{BS} - X_{BG}} = \frac{1000\,(0.909 - 0)}{1 - 0.909} = 9989\,\text{kg} \tag{2.5}$$

Setting S at 1000 kg, by way of example, the solvent-free extract E and raffinate R are calculated as follows: firstly, locate M, the mixture point:

$$F + S = M = E + R = 1000 + 1000 = 2000\text{ kg} \tag{2.6}$$

$$X_{BM} = \frac{FX_{BF} + SX_{BS}}{M} = \frac{1000\,(0) + 1000\,(1)}{2000} = 0.50 \tag{2.7}$$

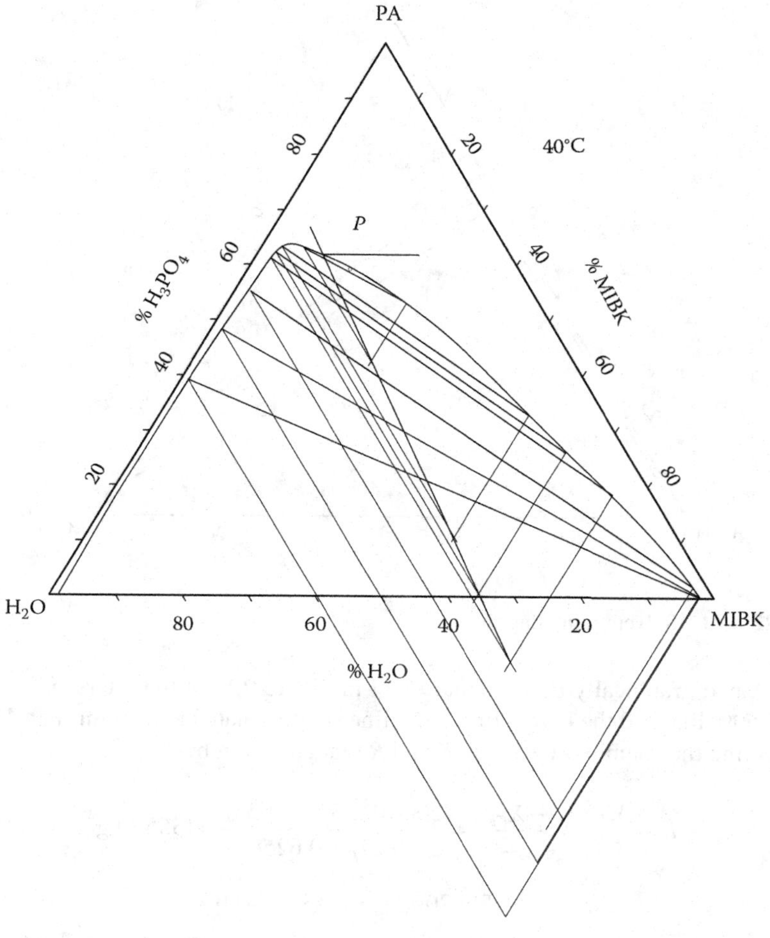

FIGURE 2.11 MIBK–H_3PO_4–H_2O ternary diagram plait point evaluation. (Adapted with permission from Ananthanarayanan, P. and Rao, P.B., Ternary liquid equilibria of the water-phosphoric acid-isoamyl alcohol, cyclohexanol, or methyl isobutyl ketone systems at 35°C, *J. Chem. Eng. Data*, 13(2), 194–196. Copyright 1968 American Chemical Society.)

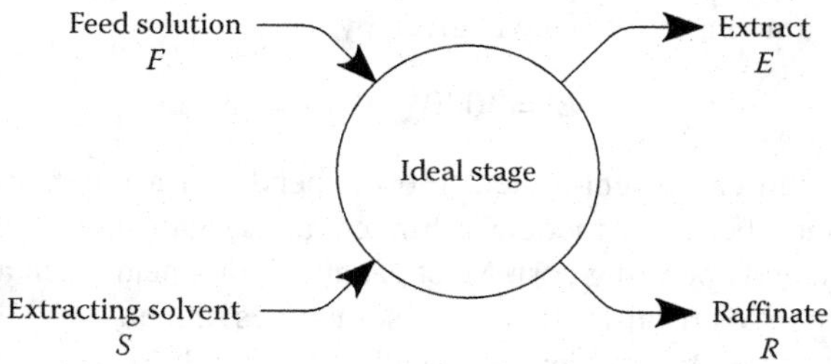

FIGURE 2.12 Ideal stage.

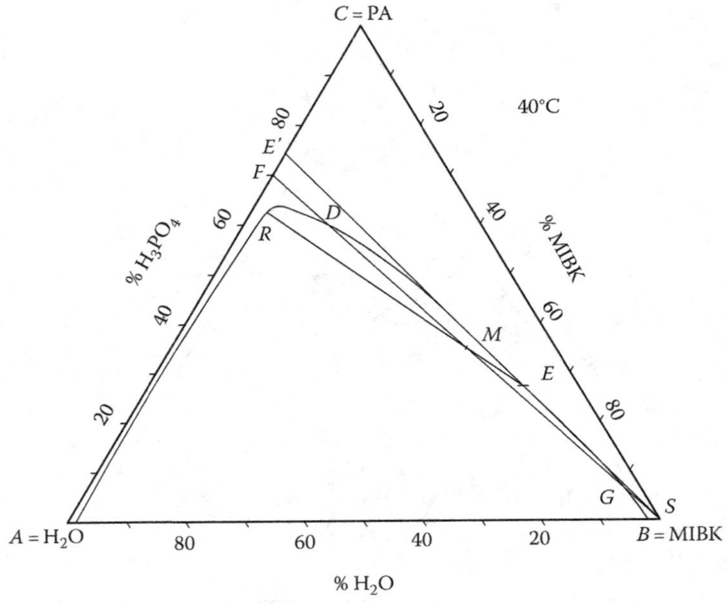

FIGURE 2.13 Solvent ratio diagram.

M is located graphically in the ternary plot in Figure 2.13 at the intersection of the MIBK 50% line and the FDGS line. A tie line is interpolated to pass through M, also intersecting the solubility curves at E and R that are given by

$$E = \frac{M(X_{CM} - X_{CR})}{X_{CE} - X_{CR}} = \frac{2000\,(0.350 - 0.625)}{0.272 - 0.625} = 1558.1\,\text{kg} \tag{2.8}$$

$$R = M - E = 2000 - 1558.1 = 441.9\ \text{kg} \tag{2.9}$$

E' represents the solvent-free composition of the extract and is located at the intersection of the water–phosphoric acid line and the extension of line SE. From the graph, $X_{CE'} = 0.744$. Thus, the solvent-free extract weight is given by

$$E' = E(1 - X_{BE}) = 1558.1(1 - 0.635) = 568.7\,\text{kg} \tag{2.10}$$

The solvent-free amount of raffinate is given by

$$R' = F - E' = 1000 - 568.7 = 431.3\ \text{kg} \tag{2.11}$$

In practice, the chosen solvent-to-acid ratio depends on a number of factors and often must balance better extraction at higher ratios against increased equipment size, capital cost, and operating cost. Most patents in this field teach a range of acid-to-solvent weight ratios from 1:1 to 5:1 and sometimes higher.

The binodal curves for wet process acids are slightly different to Figure 2.10 due to the presence of metal phosphates in the raffinate that act to increase the

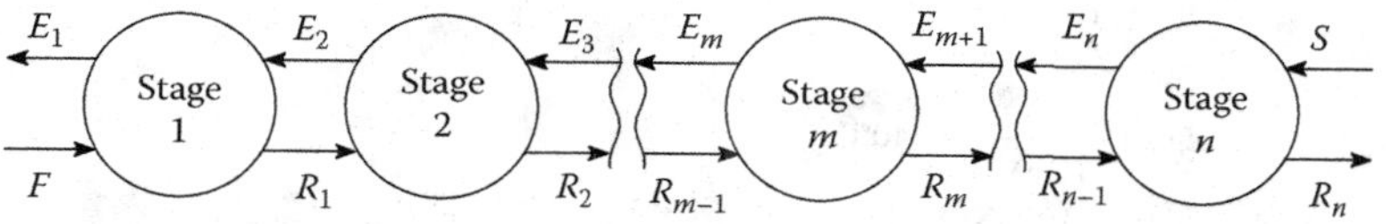

FIGURE 2.14 Countercurrent multiple-contact solvent extraction.

solvent-free phosphoric acid content of the extract. In the practical case of treating wet process phosphoric acid, the extract will normally be washed to further reduce the impurity content of the product that is released from the solvent. The wash solution will normally be recycled to the extraction stage(s) that further complicates the diagrammatic representation of the overall process.

All PWA plants have many stages that operate continuously in a countercurrent fashion. In theoretical terms, this is known as *countercurrent multiple-contact* solvent extraction. The feed and solvent enter at opposite ends of a cascade of stages as depicted in Figure 2.14. An overall mass balance gives

$$F + S = R_n + E_1 = M \tag{2.12}$$

where M represents the total quantity in the system. Equation 2.12 is rearranged to give O, the operating point:

$$F - E_1 = R_n - S = O \tag{2.13}$$

The operating point, O, is located by extending the lines FE_1 and R_nS.

A mass balance on the mth stage gives

$$R_{m-1} + E_{m+1} = R_m + E_m$$

$$R_{m-1} - E_m = R_m - E_{m+1} = O \tag{2.14}$$

As shown in Figure 2.15, lines FS and E_1R_n are easily drawn and point M plotted. Looking at the tie lines, shown dotted, it is clear that E_1 and R_n do not fall on the same tie line, therefore a number of extraction stages are required. The tie line running through E_1 intersects the solubility curve at R_1. To construct the second extract, E_2, a construction line is drawn from R_1 to O, the operating point; E_2 is located at the intersection of the solubility curve and the line R_1O. R_2 is located on a new tie line in the same way as R_1. This process is repeated until line R_nS is crossed. In this case, the desired extraction required approximately three and a half stages. The concentrations at each stage are obtained from the ternary diagram. The masses at any stage m may be calculated from Equations 2.15 and 2.16, and in the first instance, for $m = 1$, making $R_0 = F$:

$$R_m = \frac{R_{m-1}(X_{AR_{m-1}} - X_{AE_{m+1}}) + E_m(X_{AE_{m+1}} - X_{AE_m})}{(X_{AR_m} - X_{AE_{m+1}})} \tag{2.15}$$

$$E_{m+1} = R_m - R_{m-1} + E_m \tag{2.16}$$

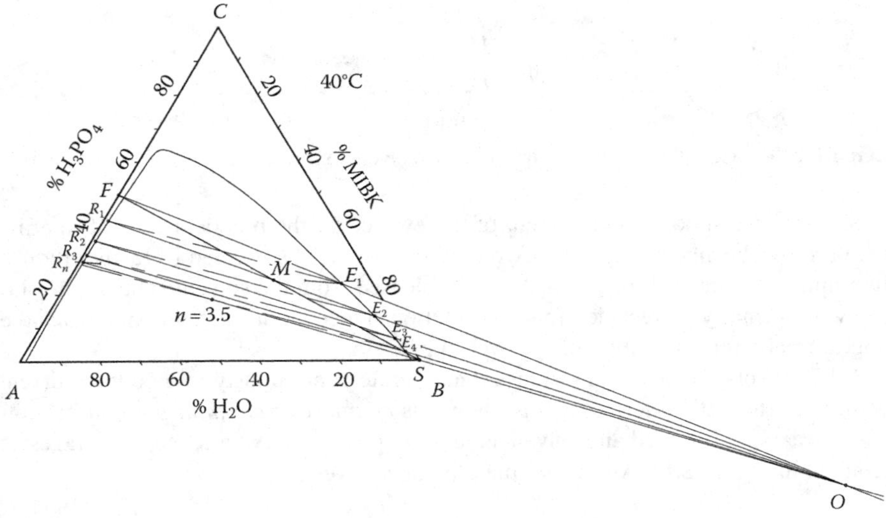

FIGURE 2.15 Countercurrent multiple-contact stage evaluation.

Another approach is to plot the weight fraction acid in the solvent phase (X_{CB}) against the weight fraction in the aqueous phase (X_{CA}) on Cartesian coordinates, as shown in Figure 2.16. To do this, the extract and raffinate points on the binodal curve are projected across to the new plot. The extract values are plotted on the y-axis; the raffinate values are projected from the 45° line onto the x-axis. From these values, which represent tie lines, the solvent-rich part of the equilibrium curve is drawn. An operating curve is also drawn. The first point is given by the coordinates (F, E), the last by the intersection of the desired raffinate concentration, R, and the equilibrium curve. A series of operating points, (R_m, E_m), are drawn on the triangular plot

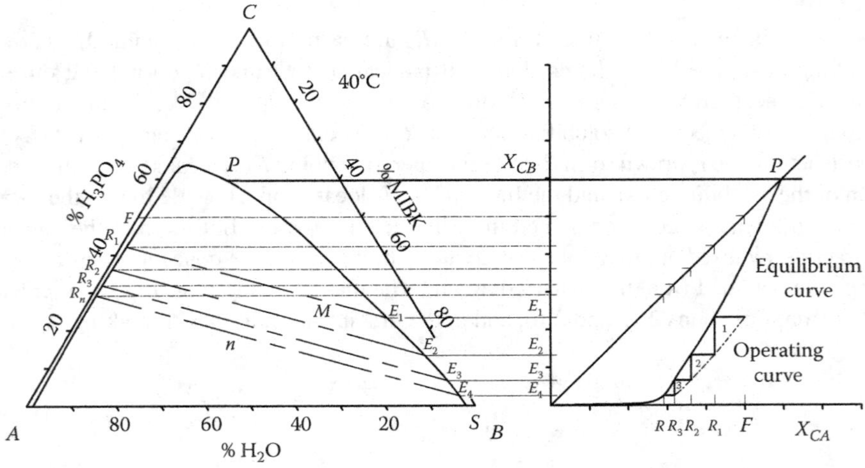

FIGURE 2.16 Countercurrent multiple-contact ternary diagram and distribution curve.

emanating from the operating point, *O*. These points are projected across to create the operating curve. Commencing from the feed concentration, lines are drawn between the operating curve and the equilibrium curve, creating stage steps to the desired raffinate concentration. The plot of X_{CB} against X_{CA} is known as the distribution diagram, and the ratio of these values is known as the distribution (or partition) factor (or coefficient, or ratio), *m*:

$$m = \frac{X_{CB}}{X_{CA}} \tag{2.17}$$

The effectiveness of the solvent, *B*, in separating the acid, *C*, from the water, *A*, is known as the selectivity, β (or, more generally, the solute *C* and the third component *A*). The selectivity is expressed by comparing the ratio of *C* to *A* in the solvent-rich phase to that in the aqueous phase, for any given tie line, and is given by the following equation:

$$\beta = \frac{X_{CE}/X_{AR}}{X_{CR}/X_{AR}} \tag{2.18}$$

The purification factor is of particular interest to the process designer and expresses the distribution of each impurity between the aqueous and organic phases and that of the acid and is given by the following equation:

$$PF(M) = \frac{\%M\,aqueous/\%M\,organic}{\%P_2O_5\,aqueous/\%P_2O_5\,organic} \tag{2.19}$$

Different metals have different purification factors in different solvents.

So far, we have seen the simple case of extracting a proportion of the phosphoric acid into the solvent, MIBK. The solvating properties of both MIBK and amyl alcohol lead to the requirement for only two stages of extraction for most PWA plants. Other solvents, like TBP, may require five or more extraction stages.

Some prediction is possible based on previous laboratory and plant experience, and some is possible from modeling; however, because of the complexity of the real system and differing interactions between some impurities, there is as yet no substitute for carrying out laboratory tests on a new feed acid to establish the distribution of impurities.

Once the acid is extracted, the next logical step is for it to be *stripped* (also *released* or *re-extracted*) from the solvent. This was the approach used in the earliest plants and is done by adding clean water to the separated solvent and re-extracting the acid back into the water. Although some purification has occurred, in practice, the solvent carrying the acid is *scrubbed* in a number of scrubbing stages by contacting the solvent with a proportion of clean acid from the final stripping stages. Scrubbing is essential to achieve the purification required to meet typical food grade standards. A typical PWA plant producing food grade acid, and using MIBK as the solvent, has 2 extraction stages, 5–15 scrubbing stages, and 2 or 3 release stages. A plant using TBP as the solvent may have 5 extraction, 5–15 scrubbing, and 5 stripping stages. In all cases, a stage may be thought of as a single mixer–settler unit or its equivalent.

Scrubbing is further enhanced by dosing with sodium hydroxide or ammonia; this increases the pH in the scrubbing section and improves purification. Although water is normally used for stripping, it is also possible with sodium or potassium hydroxide or ammonia. The use of sodium hydroxide in this context creates sodium phosphate liquor that is suitable for the manufacture of sodium phosphates such as STPP and was practiced on a number of plants.

As solvents are partially miscible, they must be stripped out of the product and raffinate streams; this is usually done by distillation or steam stripping. TBP has a very low solubility in water and is removed by other means outlined in the discussion about the Rhône–Poulenc (R–P) process.

In normal operation, a small amount of solvent is consumed for every ton of acid purified. Some solvent is lost in evaporation from piping joints, valve and pump seals, and storage vents. Pure solvent is often flammable to some degree and is therefore stored under a nitrogen blanket; nevertheless, there is still some diffusive loss. A small amount remains with a stripped acid. The largest normal loss is due to solvent *aging* or degradation, for example, MIBK degrades to trimethyl nonanone and TBP to mono- and dibutylphosphate. Most plants incorporate a solvent cleaning unit to remove the degradation products [19]. All losses, including degradation, are made up with fresh solvent.

Abnormal or maloperation can lead to significant loss of solvent and misdirection of product acid (product to raffinate stream, excess solvent in either raffinate or product stream sufficient to overwhelm solvent stripping). Plant failures or misdirected valves occasionally occur and are relatively easy to fix. Much harder to diagnose and rectify is an abnormal operation when the solvent extraction process just stops working. Symptoms of this situation might include the formation of an excessively deep third layer between the solvent and aqueous layers in the solvent extraction plant, an inversion of phases, formation of emulsions, and loss of any meaningful coalescence. The causes of these symptoms may be a simple loss of flow control of the aqueous or solvent stream due to a control valve or pump failure or loss of temperature control; far more serious is disruption of the solvent performance by excess organic species or colloidal silica or other contamination in the feed acid. The latter disrupt the fundamental mass transfer that is the solvent extraction process.

Referring back to Figure 2.8, the separating flask, the mass transfer of acid into the solvent is achieved by shaking the flask vigorously and then allowing it to stand and coalesce into two phases. If no shaking took place, the acid would slowly distribute between the two phases, but by shaking, irregular, misshaped droplets form, increasing the interfacial area and increasing the opportunity for mass transfer of acid from aqueous to solvent phase. On the plant, mixing is achieved by agitation in mixer–settlers or agitated columns or, in a controlled analogy of the laboratory shake-up, in pulsed columns. There is a practical limit to interfacial area that if exceeded leads to the formation of stable emulsions that hamper mass transfer and can stop the process. Because so many factors have a bearing on whether an emulsion may form, it is only possible to declare a rule of thumb that droplets must be larger than 100 μm and that in normal operation drop size is in the range 0.7–2.4 mm. Feed acid concentration influences the effectiveness of the extraction process, and in general, in the normal range of acid concentrations (between say 52%–60% P_2O_5), the higher the better. A rule of thumb is that for every additional percentage of feed acid concentration, the overall split improves by 2%–3% that brings the attendant operational savings.

A mass transfer system influenced by interfacial area and driven by concentration gradients can be modeled simplistically by Fick's second law, given by the following equation:

$$\frac{\partial \phi}{\partial t} = \nabla \cdot (D \nabla \phi) \tag{2.20}$$

where

ϕ is the concentration
t is the time
D is the variable diffusion coefficient with dimensions of area per unit time
∇ is the vector differential operator del

Equation 2.20 is the generalized form, and further simplifications are possible by fixing the interfacial area and setting desired boundary conditions for concentrations with reference to simple laboratory tests and reducing the problem to one dimension. With these constraints, it is possible to begin to get an idea of the size of the mixing section of, for example, a mixer–settler.

Even in a simple ternary system with three pure components, the behavior of the contents of a droplet is highly complex. Numerical simulation of droplet behavior is not straightforward [20], accounting for a solvent that both solvates and chelates the acid and for the many impurities, some of which behave simply, others form different complexes, making analysis *in silico* very difficult.

There are several chemical mechanisms that govern the distribution of the acid and its impurities and to what extent both migrate to the solvent or stay in the aqueous phase; Rydberg [11] classifies these mechanisms into four groups. One might simplify this further to two broad classifications: those where the solvent simply solvates the solute (the acid and its impurities) and those where a reaction takes place such as chelation as well as solvation.

2.3.1.1 Dispersion and Coalescence

Droplets are created to increase the surface area between the organic and aqueous phases and so maximize the rate of mass transfer of the acid into the organic phase. As the acid and solvent are mixed, whether in the separation funnel, the mixer–settler, or the column, two phases are present: the continuous phase and the dispersed phase. The droplets make up the dispersed phase and are the lighter density material. Practitioners refer to a process as *solvent or aqueous continuous*; in most PWA processes, the solvent is less dense than the acid, so most are aqueous continuous. The concentration of droplets in the continuous phase is related both to droplet size and to relative amounts of aqueous and solvent materials.

The process is completed by allowing the droplets to coalesce into two phases and then separating these phases by gravity. This principle is most easily observed in the laboratory with a separation funnel. After shaking the organic/aqueous mixture, the funnel is allowed to stand, and the time taken for the two phases to appear is noted. The coalescence time, τ, usually runs to minutes and is directly applicable to the sizing of separation equipment. For a mixer–settler, the specific volume, V_S, of the settler is given by the following equation:

$$V_S = \frac{\dot{V}}{\tau} \tag{2.21}$$

where $\dot{V}$ is the volumetric flow rate of the organic phase through the mixer–settler.

Figure 2.17 is an idealized representation of the birth and death of an individual droplet. Initially, a droplet of the organic (light) phase rises to the interface and pushes against it; as it rises, the dominating forces are density difference, driving it upwards, and viscosity, which is temperature dependent and linked to droplet velocity and surface area, arresting the rise. As the droplet pushes against the phase boundary, its velocity drops and surface tension effects become important. The droplet starts to deform from a flexible nominal sphere to a more ellipsoid shape; the aqueous phase drains around the droplet that continues to flatten; eventually the aqueous phase is drained and the droplet starts to incorporate into the organic phase. To the left of the drawing, small rigid droplets have formed that exhibit poor mass transfer and coalescence and start to form an emulsion. It is easy to imagine extraneous particles becoming trapped between these droplets and the development of what is known as crud that inhibits coalescence.

Figure 2.17 points toward the critical factors that affect coalescence. In a functioning system, the dominating factor is the density difference of the two phases that should be at least 200 kg/m^3. Two supporting factors are viscosity and surface tension that in a poor or nonfunctioning system will overwhelm density difference.

In the laboratory, using clean liquids, one usually observes a clear separation. If the liquids have a propensity to form an emulsion, this may happen with vigorous shaking of the separation funnel. In an agitated laboratory mixer–settler, one can cause an emulsion to form with overagitation. On an operating plant, or using plant samples in the laboratory, one may observe a *dispersion band* rather than a sharp interface. This phenomenon is referred to as a *rag* layer and is subtly different to

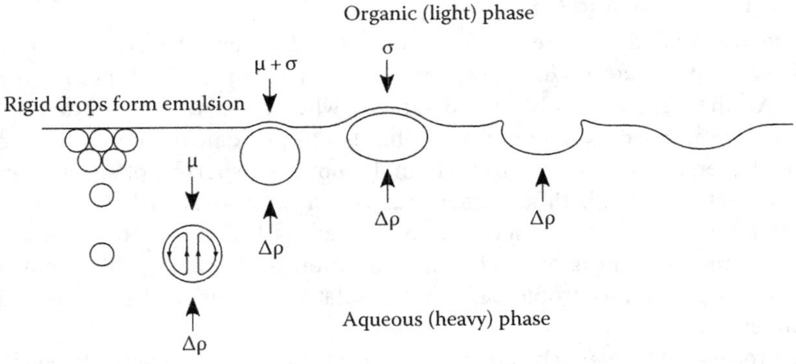

FIGURE 2.17 Droplet coalescence.

crud although the net effect may be similar. In the cleanest systems, it comprises an emulsion of small droplets that are slow to coalesce. A thin, stable dispersion band is both expected and allowed for in circumspect designs; however, if it gets out of hand, a thick band hampers separation and can lead to cross-contamination of the two phases. There are several possible causes:

- Overagitation causing small rigid droplets
- *Organics*, meaning either humic acids or other organic species originating from the phosphate rock that made the phosphoric acid, surfactants used in the wet process to knock down foaming caused mainly by carbonates in the rock, and solvent by-products, all of which affect surface tension
- Colloidal silica arising from the phosphate rock
- *Fine solids* that could include activated carbon (used in a pretreatment stage to remove *organics*!), filter aid such as diatomaceous earth (sometimes used in pretreatment), and complex phosphatic precipitates arising from the WPA as it cools in storage

As in the laboratory, so on the plant. In a mixer–settler, the two phases are mixed by an agitator in the mixer (either a separate vessel or integral with the settler), creating droplets, and then the droplets are allowed to coalesce in the settler. In columns, the same principle applies in that droplets are created (in different ways depending on the column design), then move up or down the column (depending on density) and, either within the column or at top and bottom, are allowed to settle and coalesce. Figure 2.18 shows an integral mixer–settler similar in concept to that originally developed by Davy McKee. Organic and aqueous feeds enter the mixer zone, travel through a draft tube, and are mixed by an impellor that both mixes, creating droplets, and pumps the mixture forward; the mixer is separated from the settler by a baffle that smoothes the pumping action in the settling zone; as the mixture enters the settler, it passes through a picket fence baffle, two rows of offset vertical tubes running the full width of the settler to smooth the flow; in the settler zone, droplets move to the dispersion band and coalesce into their homophase. The organic phase overflows

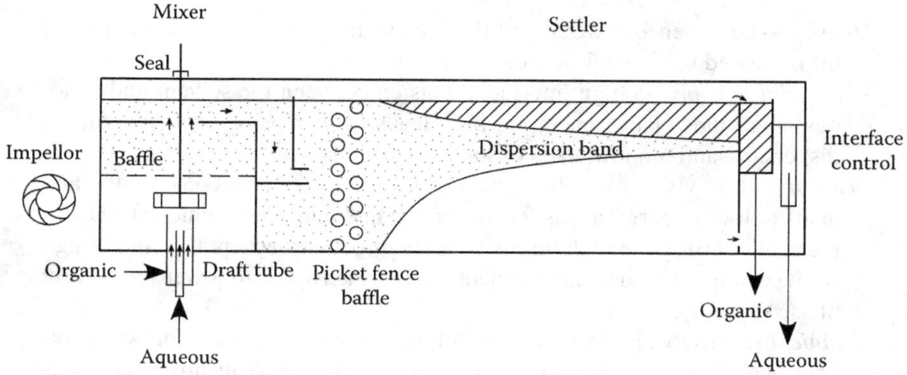

FIGURE 2.18 Mixer–settler.

a fixed weir and passes out of the mixer–settler for further processing. The aqueous phase passes into an overflow chamber, with an adjustable weir that sets the height of the interface, and also passes out for further processing. As long as the dispersion band is stable (and the emulsion in it unstable) and the settler sized to allow the band to become narrow, clean organic and aqueous streams are recovered.

2.3.1.2 Solvent Selection

Many of the criteria for the selection of a solvent for the purification of phosphoric acid are listed in the following text. Over the years, many solvents have been tested in the laboratory; the final choice for the industrial scale is always a compromise, a balancing of many factors that boil down to three:

1. Does it work?
2. How much is the total cost to the business?
3. Are the safety, health, and environmental consequences acceptable?

Solvent selection criteria are as follows:

Selectivity—the ability of the solvent to extract phosphoric acid in preference to water.

Distribution coefficient(s)—the ratio of concentrations of both the phosphoric acid and its impurities in the equilibrium liquid phases.

Capacity (loading)—the ability to dissolve large quantities of phosphoric acid; clearly, a high-capacity system requires less solvent, smaller equipment, and less energy input.

Low mutual solvent solubility—the solvent–water and water–solvent solubilities must be low to permit high selectivity and aid recoverability.

Recoverability—the presence above parts per million of organic species in most grades of phosphoric acid is unacceptable. It is essential, therefore, that the solvent can be removed from the acid as easily as possible. Solvent is removed from the acid leaving the solvent extraction section by steam stripping or distillation. The ease of this step has obvious implications for the volatility of the candidate solvents.

Density—a difference in density of the solvent and the acid of at least 200 kg/m^3 is needed to aid coalescence.

Interfacial tension—a high interfacial tension between the solvent and aqueous phase aids coalescence; too high leads to increasing energy input for dispersion, and too low leads to stable emulsions.

Viscosity—low viscosity aids mass and heat transfer and coalescence and leads to lower energy needs for dispersion and pumping. Some solvents, in particular TBP, have high viscosity; some processes accept this and design for it, while others use inert diluents such as kerosene to lower the solvent phase viscosity.

Chemical reactivity and stability—unhelpful chemical reactivity between solvent and phosphoric acid produces, irreversibly, wasteful products and is undesirable. A stable solvent does not produce excessive by-products that at best lead to the need for solvent regeneration or at worst to the formation of hazardous materials (e.g., ethers can break down to explosive peroxides). Resistance to heating, oxidation, and hydrolysis are all properties of a stable solvent.

Corrosiveness—a solvent that, of itself, leads to the need for more expensive materials of construction is less preferable than a noncorrosive solvent.

Vapor pressure—low vapor pressure is linked to lower solvent loss due to evaporation and simpler sealing technology.

Boiling point—a solvent is usually recovered from the acid product by steam stripping or distillation; a boiling point below that of the acid facilitates these processes.

Freezing point—once in the operating plant, most solvents operate well above their freezing point; however, fresh solvent is stored to replace losses, and this storage is simplified if the freezing point is below the coldest ambient conditions.

Flammability—many solvents are flammable, each with its own flash point; a solvent with a higher flash point presents a lower hazard than one with a low flash point. Only a few PWA plants have been built around the world, one of which burned down at Knapsack, Germany, in June 1987, when a storage tank containing amyl alcohol was ignited by a faulty pump. The plant was not rebuilt.

Toxicity—many solvents are toxic to a greater or lesser extent; solvent toxicity presents a challenge to both the designer and operator of the plant in containing this hazard within the plant and ensuring the final product is free of it. In the chemical industry in general, toxic solvents used successfully for many years are being replaced with less toxic alternatives, for example, benzene, which is carcinogenic, by toluene or xylene.

Cost and availability—even for the standard solvents used on PWA plants, the cost of filling a full-scale plant with fresh solvent is not insignificant nor is the annual cost of replacing lost solvent. If the solvent supply is local and secure, the on-site inventory, working capital and hazard are minimized.

Figure 2.19 shows the ternary diagram for IPE that has a three-phase zone. Figure 2.20 shows the superimposed ternary diagrams for MIBK, isoamyl alcohol (AmOH), and TBP. MIBK and AmOH were measured at 35°C and TBP at 50°C. Even a rudimentary inspection of these curves allows an initial estimation of the relative complexity of the solvent extraction section flowsheet depending on the solvent selected and given the same feed acid and final product specification.

2.3.2 Pretreatment Processes: Desulfation

Prior to solvent extraction, it is essential that the acid is conditioned. Deficiencies in pretreatment lead to excessive impurities in the final product, inefficient and costly operation, and consequential risk to either downstream processes such as fertilizers or the environment. The first stage of pretreatment is nearly always desulfation.

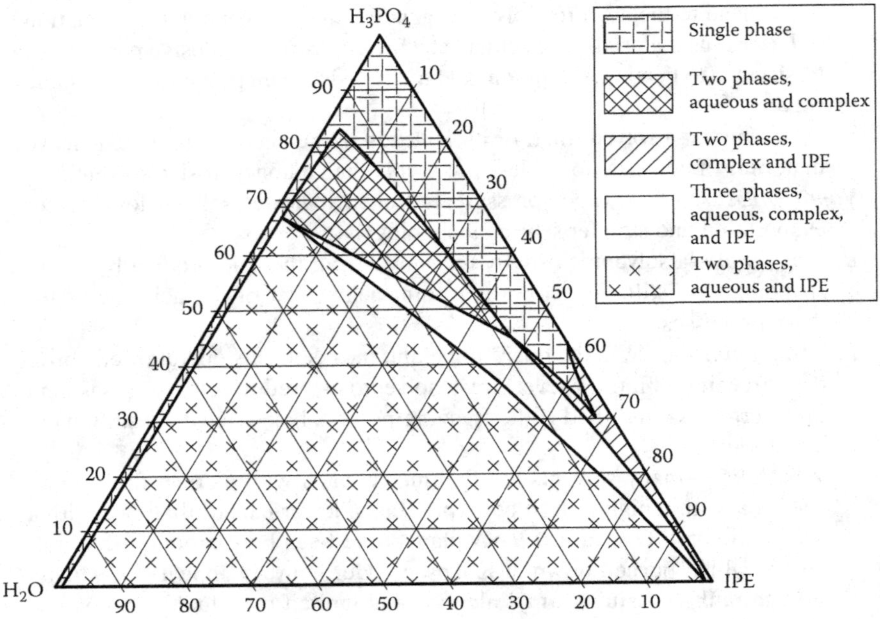

FIGURE 2.19 IPE–H_3PO_4–H_2O ternary diagram at 30°C. (Adapted from P.O. Schallert and C.C. Fite, Process for purifying phosphoric acid, US Patent 3318661, May 1967.)

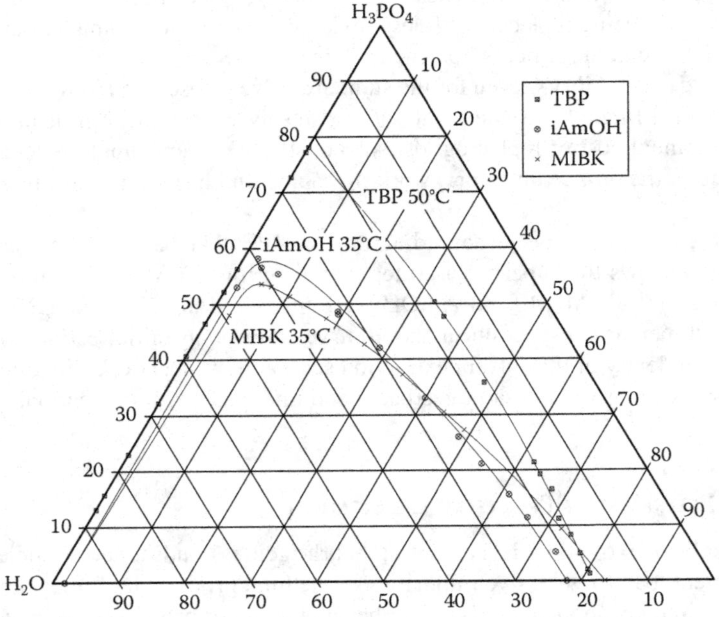

FIGURE 2.20 MIBK, iAmOH, and TBP–H_3PO_4–H_2O ternary diagram.

The production of phosphoric acid by the sulfuric acid route was discussed in Chapter 1. In order to avoid P_2O_5 losses as calcium phosphate, the sulfate level of the WPA is maintained at 2%–4% SO_4. Compared to most cations and for most solvents, the purification factor for SO_4 is poor. Consequently, without alternative treatment, the final product acid might have sulfate levels too high for downstream processing of derivative products, or as a product itself. The usual approach to lower sulfate levels is with precipitation through the addition of a calcium source. Calcium is most easily available in this context as limestone ($CaCO_3$), lime (CaO), or phosphate rock. Each has advantages and disadvantages:

1. Limestone—is likely to be the cheapest source of calcium; however, depending on the mine source, it may also contain magnesium. If the WPA already has relatively high levels of magnesium, adding more is unhelpful. The carbonate content might lead to excessive foaming and consequent control problem issues.
2. Lime—is likely the most pure form of calcium addition but also the most expensive. As with limestone, there are different lime purities, with attendant cost.
3. Phosphate rock—on a site that is already making WPA, this material is readily available. Rock is priced on its P_2O_5 content and is more expensive per ton than limestone; however, the P_2O_5 is not lost, and therefore, the calcium might be considered free of charge. On the other hand, phosphate rock brings the full range of impurities with it and, at the ppm level, may reintroduce impurities already dealt with upstream.

Common to all of these calcium donors is the need to control the addition: too much calcium leads to calcium phosphates, too little and the sulfate levels are undertreated. All three are supplied in solid form, and for each ensuring a consistent calcium feed is not easy.

The location of the desulfation step in the overall scheme requires careful consideration. Because its objective is to reduce SO_4 at percent levels, rather than ppm, it most obviously comes before the solvent extraction stage. Given that it is a precipitation process, ideally it should not be immediately before solvent extraction because fine solids affect that process and even the best separation step will pass some fines. Also at this stage, the feed acid is most concentrated, typically in the range 52%–59% P_2O_5, and at its highest viscosity and density, so less easy to filter. Therefore, desulfation should sit prior to the final concentration of the WPA. If the WPA is coming from a dihydrate plant at 28% P_2O_5, the choice is to desulfate either the 28% filter acid or a partially concentrated 42%–45% acid. If the WPA is coming from a hemihydrate plant (or its variants), there is no choice and the acid must be desulfated at 45%–48%. The case for desulfating at 28% is that it is slightly easier to integrate the desulfation unit with the main WPA reaction stage. The case for desulfating 42%–45% acid is that at this concentration, in addition to the advantages already set out, the solubilities of many impurities are at a low level; therefore, a filtration unit will be most productive at this stage.

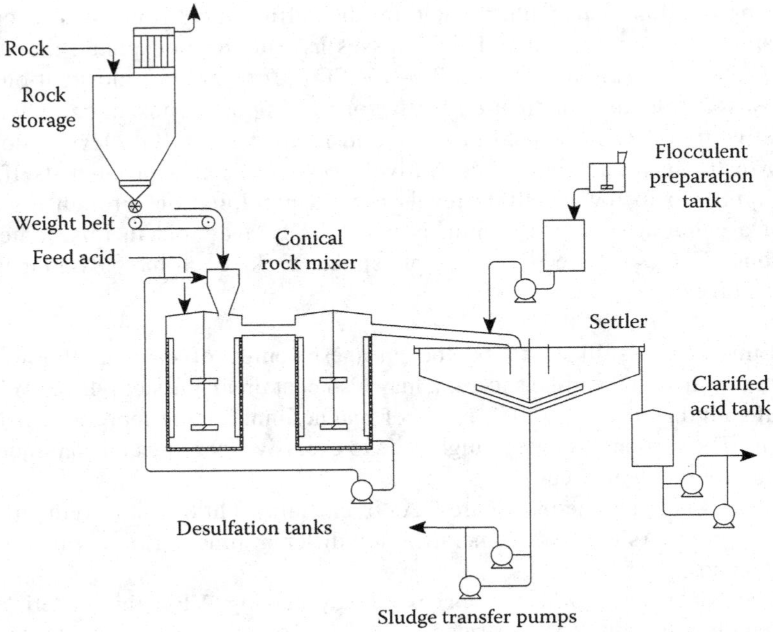

FIGURE 2.21 Desulfation flow sheet.

Figure 2.21 depicts a desulfation unit. In this case, phosphate rock is stored in a hopper and added to the process via a weight belt. The rate of addition of the rock is calculated and input to the control system with reference to regular sulfate analysis of the acid coming forward and the calcium content of the rock. Some operators target desulfated acid at 0.5% SO_4; however, variations in the feed acid and rock analysis, and in the reaction efficiency in the desulfation tanks, can lead to overshoot and the formation of small DCP crystals that hamper separation of the precipitate. A target of 1% SO_4 at this stage is usually sufficiently low for most final products out of the overall plant and is a little more controllable. The rock contacts a proportion of the feed acid and a recycle of partially desulfated acid and gypsum precipitate in a static, conical mixer. Some plants add the rock straight into the reaction vessel. The purpose of the mixer is to ensure good mixing, avoiding localized concentrations, and to promote growth of the gypsum crystal. The mixture drops into the first desulfation tank and is mixed with the bulk that is held at 80°C–90°C. Both desulfation tanks are baffled and agitated; agitators are usually simple pitch blade turbines rotating at a suitable speed to ensure good mixing. Total residence time over the two tanks is between 2 and 4 h. Materials of construction are usually rubber-lined carbon steel with an additional lining of graphite bricks in the agitated zone. The desulfated acid together with calcium sulfate crystals overflows to a clarifier. Clarified acid is pumped forward to the next stage and the acidic calcium sulfate sludge is pumped either to the WPA plant or to a filter to recover more P_2O_5.

2.3.3 Crude Defluorination

High levels of fluoride in acid are corrosive and can form complexes with the major impurities present prior to the solvent extraction stage, in particular silica and aluminum that are difficult to filter. It is usual therefore to carry out crude defluorination prior to solvent extraction or ensure that the levels are 0.5% or less. One approach to defluorination is described in Section 2.2, with the use of a sodium source to make Na_2SiF_6. Another approach is to combine crude defluorination with concentration, there are several patents in the field [21–23] and more recently [24]. Figure 2.22 shows a combined single-stage defluorination and concentration. Activated silica may be added to the crude acid (depending on the Si/F ratio) that is then pumped to the pumped circulation vacuum concentrator. Silica tetrafluoride, SiF_4, comes off as a vapor with water and is then condensed to fluorosilicic acid, FSA, as shown in the following equation:

$$HF + Si + H_2O \rightarrow SiF_4 + H_2O \rightarrow H_2SiF_6 \quad (2.22)$$

In this case, following [24], the vapors from the concentrator vapor head are contacted by sprays of circulating FSA in the duct to the entrainment vessel. FSA flows from the entrainment vessel down the barometric leg to the FSA sump. Some FSA is recirculated, and the rest is exported either for sale as the acid or precipitation with sodium hydroxide, separation, and sale.

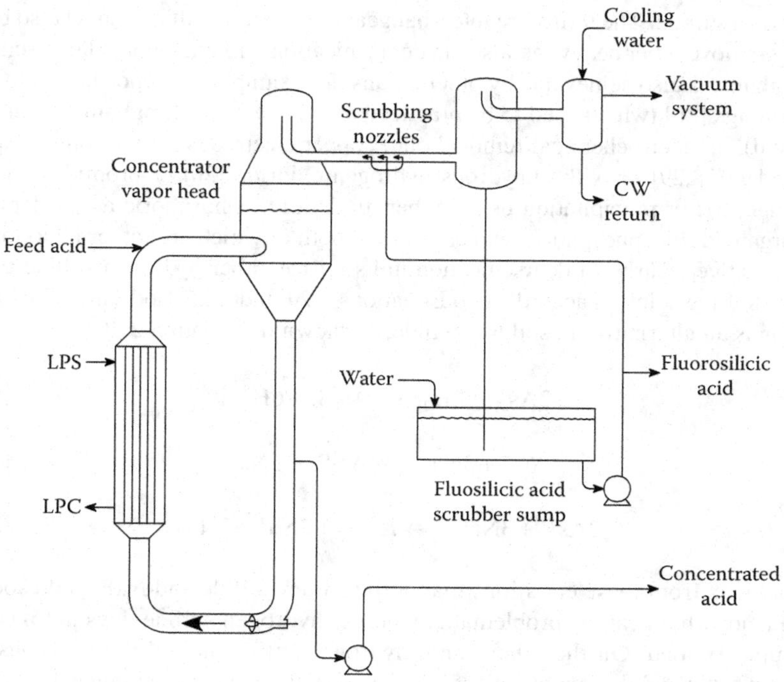

FIGURE 2.22 Combined crude defluorination and concentration flow sheet.

2.3.4 Crude Dearsenication

Most PWA plants remove arsenic before the solvent extraction section. There are two main reasons for this positioning: firstly, that the heavy metal profile of the feed acid, and the partitioning properties of these heavy metals in the subsequent solvent extraction stage, may lead to excessive levels in the underflow acid often exported for use in fertilizers; secondly, prior to solvent extraction, there may be several steps where solids are precipitated and separated and it is sometimes convenient to combine sulfide and other precipitates in one separation. Care is taken to present a clean acid for extraction–dearsenication after extraction reintroduces solids to a clean and now purified acid, and even greater care is essential to ensure the final product meets specification on solid content. The Albright & Wilson (A&W) plant at Whitehaven placed the dearsenication unit after solvent extraction as at this stage less acid required treatment so the capital and operating costs were slightly lower; it was permitted to do so because in the early days of operation, the feed acid impurity levels allowed it, and in the latter days, the plant was a total exhaustion process that did not make an underflow acid for export. The PotashCorp (PCS) plant at Aurora does not have a dearsenication unit as the WPA feed is already sufficiently low in arsenic not to require further treatment. Dearsenication before and after solvent extraction is slightly different, and impurity concentrations are lower after than before; however, the principles are identical. The mainstream process technology is well proven and was first developed for use on thermal acid [25].

In some corporations, dearsenication is referred to as sulfiding because the process uses a sulfide. The terms are interchangeable; however, sulfiding may also beneficially remove other heavy metals, and dearsenication, although not at large scale on PWA plants, is also achievable by other means, for example, by deposition onto copper turnings [26] (which used to be practiced at the A&W Buckingham site on thermal acid), or electrochemical removal onto copper electrodes [27], or onto graphite electrodes [28,29] or with anhydrous hydrogen chloride [30] or bromides and ion exchange [31] or precipitation using either dialkyldithiophosphoric acid esters [32] or diorganyldithiosphosphoric acid esters [33], both of which are rather malodorous.

In practice, nearly all dearsenication units utilize either hydrogen sulfide gas or sodium sulfide solution according to Equations 2.24 and 2.25. Sodium hydrosulfide solution is an alternative to sodium sulfide, as shown in Equation 2.26:

$$2As^{3+} + 3H_2S \rightarrow As_2S_3 + 6H^+ \quad (2.23)$$

$$2As^{3+} + 3Na_2S \rightarrow As_2S_3 + 6Na^+ \quad (2.24)$$

$$2As^{3+} + 3NaHS \rightarrow As_2S_3 + 3Na^+ + 3H^+ \quad (2.25)$$

It is obvious from these equations that both sodium sulfide and NaHS add sodium to the acid, which can be problematic, whereas hydrogen sulfide does not increase the impurity load. On the other hand, hydrogen sulfide is highly hazardous; not only is it toxic and flammable, it is also explosive in the air in the range 4.3%–45%

by volume. When used at full scale, hydrogen sulfide is supplied and stored in drums in 600 kg lots and can easily represent the most significant hazard on-site. Within the dearsenication plant itself, regardless of the sulfide chosen, care in both design and operation is required to avoid release of H_2S gas or the creation of explosive conditions.

The effectiveness of dearsenication is affected by the acid concentration. In general, higher acid concentrations lead to better sulfide precipitation: this statement summarizes a complex and not fully understood balance; on the one hand, increased acid concentration has increased availability of H^+ ions that mop up S^- ions, therefore reducing the rate of sulfide precipitation; on the other hand, the acid is more viscous, thus slowing down movement of hydrogen sulfide gas bubbles, thus increasing the time for mass transfer into the acid and the arsenic reaction. The higher viscosity also makes filtration more difficult, slowing the filtration rate; on the other hand, the higher concentration reduces the acid volume and therefore the filter size. One patent [34] teaches the addition of liquid carbon disulfide after the sulfide reaction at a higher (54%) acid concentration, to extract all the sulfide species, followed by liquid–liquid separation. An R–P patent [35] describes dearsenication of lower concentration acid, less than 40% P_2O_5, typically 20%–30%, after solvent extraction although the plant at Geismar dearsenicates before extraction. At atmospheric pressure, sulfide precipitation decreases with increasing acid temperature, which reflects reduced H_2S solubility in the acid and less sulfide availability, such that there is little sulfide precipitation above 100°C. Conversely, sulfide precipitation increases at lower temperatures, as does viscosity. A reasonable balance of these factors locates dearsenication where the acid concentration is approximately 40% P_2O_5 at an operating temperature of 50°C with viscosity about 3.6 cP.

Different operators carry out the dearsenication reaction at different pressures. Reaction at higher pressure, one patent [36] teaches reaction at 1.5–10 bar, aims to keep more hydrogen sulfide in the acid for longer.

Most operators have found that the separation of the arsenic and other heavy metal sulfides by filtration is not straightforward. There are two reasons for this: firstly, because the sulfide crystals are small and not particularly well formed, sometimes to the point of colloidality; secondly, that some heavy metals are in oxidation states that must first be reduced in order to form the metal sulfide precipitate during which process solid sulfur is also precipitated. One cause of poor crystal morphology is supersaturation. This may occur throughout the reaction medium because of the high stoichiometric excess needed to achieve dearsenication, or it might be a local effect caused by poor distribution of the hydrogen sulfide. Operators have found stoichiometric excess of 300%–500% molar satisfactory for atmospheric or near atmospheric pressure reactions or 200%–300% for the higher pressure reactions [36]. Hydrogen sulfide is produced in the reaction of sodium sulfide with phosphoric acid as per the following equation:

$$Na_2S + H_3PO_4 \rightarrow H_2S + 2Na^+ + HPO_4^{2-} \tag{2.26}$$

Because of the influence of impurities in the acid, hydrogen sulfide solubility in phosphoric acid is determined by experimentation and is temperature, pressure, and acid concentration related. The following equation relates hydrogen sulfide solubility, $[H_2S]$ weight percent to acid concentration, C, (40%–90%), at 50°C and atmospheric pressure:

$$[H_2S] = 1.6382C^{-0.753} \quad (2.27)$$

Thus, for 40% P_2O_5 (55.2% H_3PO_4) acid, H_2S solubility is 0.08%.

Assuming an arsenic content of 25 ppm, equivalent to 25 g As per ton of acid, or 0.33 mol, then 0.5 mol of Na_2S is required, equivalent to 39 g to precipitate the arsenic as As_2S_3. Na_2S of 39 g reacts initially to form 34 g H_2S or 0.0034% in the acid and well within the solubility limit. A molar excess is essential to drive the reaction; however, even a 500% molar excess is still within the solubility limit.

Operators are frequently concerned to remove other heavy metals as well as arsenic. For example, lead is proscribed in all food grade products and the heavy metal test is described as *heavy metals as Pb*; cadmium levels are restricted in fertilizers and usually concentrate in the solvent extraction underflow; high levels of molybdenum and antimony are problematic in the higher-purity grades for electronic applications. The obvious approach to precipitating these heavy metals is simply to increase the sulfide addition; however, this is not always effective. One reason is the oxidation state issue; for example, see Equation 2.28 for the case of As^{5+}. Clearly, hydrogen sulfide is consumed by just altering the oxidation state of the arsenic (and deposits sulfur):

$$2As^{5+} + 4H_2S \rightarrow As_2S_3 + S + 8H^+ + 2e^- \quad (2.28)$$

A further challenge is the variation of sulfide solubility with temperature. For example, antimony sulfide is soluble up to 4 ppm at 75°C but only 1 ppm at 25°C; a clear, warm, post-filtration acid may have an unacceptable cloudy precipitate as a cool product acid. FMC taught the addition of hydrogen peroxide to address this issue [37].

As hydrogen sulfide is formed in the acid, according to Equation 2.26, or is introduced directly, it starts to dissociate according to Equations 2.29 and 2.30, with dissociation constants 1.985×10^{-7} and 1.335×10^{-13}, respectively:

$$H_2S \rightleftharpoons HS^- + H^+ \quad (2.29)$$

$$HS^- \rightleftharpoons S^{2-} + H^+ \quad (2.30)$$

The concentration of H^+ ions is calculated from the first dissociation constant of phosphoric acid and by assuming H^+ and $H_2PO_4^-$ concentrations are similar, as shown in the following equation:

$$7.5\times10^{-3} = \frac{[H^+][H_2PO_4^-]}{[H_3PO_4]} \approx \frac{[H^+]^2}{[H_3PO_4]} \quad (2.31)$$

For 40% P_2O_5 (55.2% H_3PO_4) acid at 50°C with density 1364 kg/m^3, [H_3PO_4] is 7.69 mol/L. Therefore, by substitution, [H^+] concentration is 0.67 mol/L. Hydrogen sulfide at 0.0034% concentration is equivalent to 6.2×10^{-4} mol/L.

From Equation 2.29, and using the first dissociation constant for hydrogen sulfide, the [HS^-] concentration is 1.8×10^{-10} mol/L. It is clear that the S^{2-} concentration is tiny. Consequently, we may deduce that the sulfide reaction is between hydrogen sulfide molecules rather than ionic sulfur. Without being specific about the reaction mechanism, a system that raises the hydrogen sulfide concentration in the acid will drive the reaction. Therefore, a well-mixed, pressurized reaction is likely to be more effective than one at atmospheric pressure.

After reaction, precipitated sulfides are separated by filtration. The presence of small sulfide particles necessitates filter aid. Diatomaceous earth, cellulose, and activated carbon are all used. Diatomaceous earth has the disadvantage that it is essentially reactive silica and that gross fluoride content in the acid will attack it, reducing its effectiveness and adding impurities to the acid. Activated carbon has the advantage that as well as capturing sulfides it will also capture undesirable organic species. Different types of filter presses and pressurized leaf filters are used. Some plants have a single filter, others multiple filters. The use of a filter and the need to precoat it and discharge the filter cake when the pressure drop across it becomes too great usually mean that the operation is either batch or semibatch. One operator claims to use a continuous process; however, batch operation is satisfactory.

Hydrogen sulfide management is important for safety, health, and product quality reasons. Once separation is complete, the acid is air blown to remove residual hydrogen sulfide. Gas from air blowing and ventilation from all plant items is directed to a liquid scrubber where the hydrogen sulfide is absorbed in NaHS solution. Sodium hydroxide is added periodically to maintain pH.

Figure 2.23 shows the flow sheet of a crude dearsenication unit. The core of this unit is the sulfide reactor and filters. As drawn, this is a semibatch operation, hence two filters; as one is filtering, the other is either discharging the filter cake or undergoing precoating. Both reactor and filters are operating at pressure, nominally at 2 bar pressure and, to minimize the explosion hazard, under nitrogen blanket.

Desulfated, defluorinated acid is pumped and held in an intermediate storage tank. The plant imports sodium sulfide solution. Typically, this is sold as a 2% solution; however, because it is made with the hydrated salt, the effective concentration is 1.2%. The storage tank is under slight vacuum and vents to the unit scrubber; thus, H_2S vapor concentration above the liquor is minimized and held outside its explosive range. The sodium sulfide liquor is pumped to the sulfide feed tank, where it is blended with NaHS solution from the scrubber. Both the acid and the sulfide solution are pumped to a static mixer where two streams mix intimately at pressure and the precipitation reaction gets underway. The acid enters the reactor through a dip pipe.

Activated carbon is charged to the agitated carbon slurry tank where it is mixed with dearsenicated acid from the dearsenicated acid tank. The slurry solid content is around 25%, the actual level a balance of space available in the reactor and the quantity determined in the laboratory to ensure satisfactory filtration of all sulfides and sulfur. An aliquot of carbon slurry is pumped to the reactor and mixes with the acid.

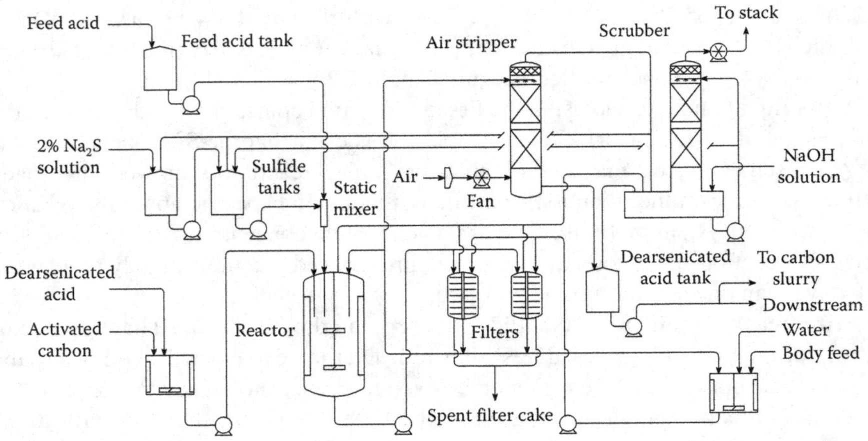

FIGURE 2.23 Dearsenication flow sheet.

When the reactor is full and the filter precoated, the filtration circuit is established. Solid body feed is charged to the precoat tank and mixed with water. The precoat mix is pumped around a clean, empty filter to establish a filter bed on the filter leaves; as the body feed becomes established, the liquor returning to the precoat tank becomes clear water. Once precoated, the filter is ready to receive acid. The reactor pump circulates the acid slurry to the body of the selected filter, and acid passes through the filter leaves and out of the filter. Pumping continues with regular sampling of the circulating acid. When the acid is found to be clear, it is routed to the stripping column. As the filter cake becomes more clogged with sulfides, the pressure drop across the filter rises. At a predetermined level, valves are opened and closed to route the acid to the second filter. A small water wash is sometimes used to maximize acid recovery before the filter cake is discharged. The discharge method depends on the filter design; in this flow sheet, water from the precoat tank fills the filter, the filter leaves rotate, and the filter cake is discharged through an outlet valve to be pumped away for further treatment.

Acid enters the column above two layers of randomly packed polypropylene rings and below a demister pad. Filtered air is blown into the base of the column and flows upward and contacts the acid, stripping out any remaining hydrogen sulfide that has not reacted or flashed on entry to the column. Stripped acid flows by gravity to the dearsenicated acid storage. Some operators install a small guard filter at this stage, through which the acid is pumped prior to transfer forward or for use slurrying the active carbon. Occasionally, the acid is overreduced by the hydrogen sulfide that leads to corrosion problems downstream. To overcome this, a small amount of hydrogen peroxide is added to the dearsenicated acid.

Vapor from the top of the stripping column flows to the scrubber entering the scrubber body below the scrubbing column. Atmospheric vents from the sulfide tanks and the dearsenicated acid tank are also connected to the scrubber body. The reactor and filters have both emergency overpressure relief vents and normal pressure letdown connections to the scrubber body. An exhaust fan ensures that the scrubber operates at slight vacuum so that those vessels operating at atmospheric pressure that have the potential for hazardous hydrogen sulfide content in the vapor space are constantly swept with fresh air. Consequently, operation of the scrubber is critical and would normally be interlocked with operation of the acid and sulfide feed pumps to ensure there is no possibility of generating hydrogen sulfide without the scrubber being in operation. On start-up, the scrubber is filled with weak sodium hydroxide solution that reacts with the hydrogen sulfide to form NaHS that is then recycled back to the sulfide feed system. Fresh hydroxide is added periodically under pH control. The scrubber liquor circulates via the scrubber pump from the base to a distribution point above the packing. A demister pad minimizes droplets escaping the system. To ensure any remaining hydrogen sulfide is well dispersed, the scrubber fan often discharges to a stack that vents at high level. This also ensures that if there was an upstream malfunction, any breakthrough of hydrogen sulfide would be dissipated.

The principal effluent from this unit is the solid filter cake comprising As_2S_3, S, other heavy metal sulfides, activated carbon, and body feed. Together it constitutes highly hazardous waste. For plants close to the sea, and given the quantities are small, gross dilution and disposal at sea have been carried out in compliance with the local environment authority permissions. Otherwise the cake is placed in secure drums for hazardous waste disposal.

In general, equipment operating at atmospheric pressure is made from polypropylene-lined glass reinforced plastic (GRP). Rubber-lined carbon steel is a cheaper option but potentially slightly less robust. Pressure vessels and pumps are made from 904 L, a high-alloy austenitic stainless steel, or similar metals with high corrosion resistance.

2.4 SOLVENT EXTRACTION PROCESSES

This section deals with the core of the solvent extraction–based processes, the solvent extraction section itself. Most of the corporations operating in the industrial phosphates business in the 1960s developed solvent extraction processes as evidenced by the patent literature. Table 2.6 captures many of those that were granted patents in the field, and is not an exhaustive list. The motivation to do so was partly defensive, partly offensive. If a process could be developed to produce an acid of sufficient purity for the manufacture of STPP, and at significantly lower cost, then here was a weapon to increase market share. Equally, corporations without this process were potentially at a disadvantage. The application of solvent extraction itself to the purification of phosphoric acid was not patentable, being too broad and too well known as a standard laboratory technique; therefore, the principal differentiator was the specific solvent or solvent group. Thus, a named process tends to be associated with a specific solvent and the original, and now often defunct, corporation that developed it. The following sections describe several of the different solvent extraction processes that were commercialized including those that have so far survived the corporate rationalizations of the last 15 years. The detailed history of the development of all the processes is now thought lost (and space would not permit one here anyway); however, for illustration, that of the A&W process and parts of the Prayon and R–P process are included.

TABLE 2.6
Summary of Patents Issued in the 1960s and 1970s for the Purification of Phosphoric Acid by Solvent Extraction

Company	Patent	Issue
Albright & Wilson	US3947499	1976
	US3914382	1975
	US3912803	1975
Allied Chemical Corp.	US3723606	1973
Arad Chemical Industries	IS32320	1968
Armour Agricultural Chemical	US3408161	1968
Azote et Produits Chimiques	GB1357614	1971
Chemische Fabrik Budenheim	GB1344651	1974
Canadian Industries	US3298782	1967
Central Glass Co	GB1296668	1972
Cominco	US3388967	1968
Dow Chemical	US3449074	1969
FMC	US3684438	1972
	US3410656	1968
GIULINI	GB1350293	1974
Goulding	GB1342344	1974
Israeli Mining Institute	US3304157	1967
	US3311450	1967
	GB1199041	1970
	US3573005	1971
Produits Chimiques Ugine Kuhlmann	US4108963	1978
Monsanto	US3684439	1972
	US3479139	1969
Montedison	GB1172293	1969
Nippon Kokan KK	JP7215458	1971
Occidental Petroleum	US3694153	1972
Office National Industriel de l'Azote	US3497329	1970
Produits Chimiques Pechiney-Saint-Gobain	US3397955	1968
	US3366448	1968
	US3607029	1971
Prayon	US3970741	1974
Produits Chimiques et Metallurgiques du Rupel	GB1323743	1973
St. Paul Ammonia Products	US3375068	1968
Stamicarbon NV	US3363978	1968
Susquehanna Western Inc	US3359067	1967
Tennessee Corporation	US3318661	1967
	US3367738	1968
Toyo Soda Manufacturing	US3920797	1975
	US4154805	1977
	US3529932	1970
Typpi Oy	GB1129793	1968

2.4.1 Albright & Wilson Process

The development of the A&W process is interesting in the way that it continued to change and grow and develop in response to changing and often very challenging business conditions. Other solvent extraction processes saw much less change.

The A&W process is based on the solvent MIBK. Many other solvents were evaluated; however, the choice of MIBK was made initially because it was different from the solvent isopropanol, IPE, patented by IMI [38] and commercialized by Haifa Chemicals and what is now ICL in Israel, and because a Dutch company Stamicarbon had published patents [39] teaching that MIBK worked in separating nitric acid from phosphoric acid.

A proposal to develop a solvent extraction process was presented to the A&W board of directors during its consideration of the decision to invest in the huge new phosphorus plant in Newfoundland in 1967. By 1971, A&W was both bleeding cash and experiencing great difficulty achieving stable and productive operation of the Newfoundland plant. In this context, the A&W board authorized the modest expenditure of £24,000 on a demonstration plant to carry out the continuous purification of phosphoric acid by solvent extraction with MIBK and to provide data for a potential commercial plant. The plant was erected at the A&W Marchon division site at Whitehaven. An early configuration is shown in the photograph of Figure 2.24.

The operation of the pilot plant is best described with reference to Figure 2.25. Fifty-four percent P_2O_5 acid was pumped into Extractor 1. This acid was previously treated with activated carbon and clarified and made from Moroccan rock. Also entering the first extractor was recycled solvent from Extractor 2. Leaving Extractor 1 was a stream of acid at approximately 45% P_2O_5 concentration, loaded

FIGURE 2.24 A&W solvent extraction demonstration plant. (Courtesy of Solvay, Hugh Podger, and Brewin Books Limited, Studley, U.K.)

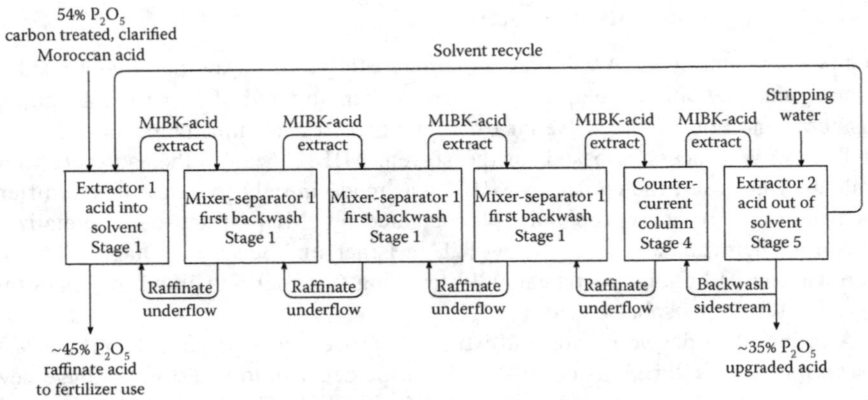

FIGURE 2.25 A&W pilot plant diagram.

with impurities, going to a holding tank for acid to be used for fertilizer manufacture. Also leaving Extractor 1 was the solvent stream loaded with acid. Acid-loaded solvent then countercontacted underflow acid in three mixer separators and a countercurrent backwash column. This column was fed a proportion of the purified acid from Extractor 2 to remove more of the impurities extracted with the acid in the solvent coming forward. In Extractor 2, distilled water contacted the acid-loaded solvent, stripping the acid into the aqueous layer. Purified acid at approximately 35% P_2O_5 concentration containing around 2% MIBK was then collected for further treatment and analysis. The stripped solvent was then recycled to Extractor 1.

The demonstration plant ran for 6 months. In parallel, chemists at A&W's Oldbury site carried out complimentary laboratory work. A system of five separating funnels was used. Different solvent/acid volume ratios were tried in the laboratory that in turn allowed the demonstration plant to start up with the right ratio. As often happens, once running, the plant overtook the laboratory in optimizing running conditions. Table 2.7 shows the metal analysis of the solvent and aqueous phases throughout the system and shows how the impurity levels are reduced. The results are normalized to either 54% or 67% P_2O_5.

The laboratory work was conducted in three phases: the first, to establish a starting point for the solvent/acid volume ratio; the second, to identify process conditions for the concentration and decolorization of the acid from Extractor 2; and the third, to evaluate the application of a Japanese ion exchange process [40] to reduce iron and arsenic levels. The second phase was successful in identifying conditions to concentrate the acid from solvent extraction at 35% P_2O_5 up to 43%, then achieve color specification through the addition of 0.4% activated carbon, and then achieve a sales concentration of 60%–66% P_2O_5. The acid color was measured against Haifa Chemicals commercial bulk acid and was satisfactory. The third phase verified the patent teaching but was found to require a more pure acid than could be provided at this stage of development of A&W's process.

Table 2.8 shows the analysis of five acids: the feed acid to the pilot plant; a Haifa bulk shipment; a typical, decolorized, and concentrated acid from the pilot plant; the pilot product acid after polishing; and a crystallized acid from the plant. At this stage

TABLE 2.7
Analyses of Stream around the A&W Pilot Plant

Sample Phase/Stage	%P_2O_5	Fe	Al	Mg	Ca	Cr	V	Mn	Cu	Zn	Cd	Pb
Solvent stage 1	67	880	490	800	<2	180	155	10	20	180	7.5	<4
Solvent stage 2	67	560	180	240	7	62	54	3	16	38	<2	<4
Solvent stage 3	67	280	65	51	<2	105	35	<2	5.5	32	<2	<4
Solvent stage 4	67	45	5	10	<2	8	8	<2	2.5	5.5	<2	<4
Underflow stage 1	54	3940	1190	2545	660	900	290	22	115	825	31	7
Underflow stage 1	54	1170	470	665	30	280	120	8.5	43	240	10	6
Underflow stage 1	54	810	150	185	10	90	75	1	17.5	130	4	5
Underflow stage 1	54	210	48	53	<1	40	30	<1	9	37	1	5
Underflow stage 5 product acid	67	42	<2	5	<2	9	7	<2	16	6	<2	<4

TABLE 2.8
Analysis of Feed Acid to and Product Acids from the A&W Pilot Plant Together with Analysis of a Bulk Supply from Haifa Chemical

	Feed Acid	Haifa Bulk	Pilot Decolorized	Pilot Decolorized, Polished	Pilot Crystallized
P_2O_5	54%	58.5%	61%	67%	66%
Fe	2100	52	50	<5	<1
Al	750	<1	9	9	<0.1
Mg	1630	2	1	<0.5	—
Ca	35	22	—	—	—
Sr	<1	<2	<0.5	<0.5	<0.5
Na	130	—	—	10	47
V	200	<0.5	4	5	<0.5
Cr	200	<1	3	5	<0.5
Mn	15	—	<4	<4	<1
Ni	—	1	<0.3	1	0.3
Cu	75	55	<5	<5	<1
Pb	5	<5	0.2	0.8	1.2
Cd	20	—	<0.5	<0.5	<0.5
Zn	335	205	3	3	0.5
C	160	40	280	260	5
As	10	0.5	11	1	0.5
F	1800	85	90	35	1
Cl	<10	<20	5	15	<10
SO_4	1.41%	<200	2.6%	2.8%	600

of development, A&W envisaged that for the purest applications, such as food grade products, a crystallization step would be necessary after solvent extraction. A&W carried out laboratory studies at this time, using a simple cooled glassware agitated crystallizer and laboratory-scale basket centrifuge. The results from this work were included in one of A&W's patents [41] primarily covering acid purification by solvent extraction.

In July 1972, authorization was given for the expenditure of £ 60,000 to build a pilot plant capable of 1 ton P_2O_5/day. This unit operated from December 1972 through August 1973. The plant utilized pilot-scale mixer–settlers supplied by Power Gas (later Davy Powergas). Operating conditions were verified and samples of underflow acid supplied to A&W Barton works and ICI for processing to NPK fertilizers.

A combination of laboratory and pilot plant study was used to optimize the number of mixer–settlers (stages) of extraction, scrubbing, and stripping. The results are presented in Table 2.9.

The number of mixer–settlers chosen at each stage was a reasonable compromise between the capital cost of a mixer–settler and operating cost.

In 1974, the A&W board authorized £2 m expenditure to build a 40,000 tpa P_2O_5 plant that was commissioned on the spring of 1976. The project was code-named *MO* for **M**archon **O**ldbury, reflecting the development input from both the Whitehaven

TABLE 2.9
Evaluation of the Effect of Different Numbers of Extraction, Scrubbing and Stripping Stages

Number of extraction stages	1	2	3
Percentage of P_2O_5 feed appearing as product acid	60	65	67
$\%P_2O_5$ concentration of raffinate (fertilizer acid)	49	42	41
Number of scrubbing stages	4	5	6
Percentage of P_2O_5 feed appearing as product acid	60	65	68
Number of stripping stages	1	2	3
$\%P_2O_5$ concentration of product acid	33	42	44

(Marchon) site and the Oldbury laboratories. The core of the plant consisted of two Davy Powergas mixer–settlers for extraction, five for water scrubbing, and two for stripping. The feed acid came from A&W's F3 and F4 wet acid plants at Whitehaven that made 28% P_2O_5 acid from Moroccan Khourigba rock. The product acid, 40,000 tpa P_2O_5, was *salts* grade, in other words, sufficiently pure for use on the STPP plants on-site. The raffinate acid, 20,000 tpa P_2O_5, was exported for NPK fertilizer production.

Davy Powergas mixer–settlers, as shown in cross section and elevation in Figure 2.26 and as installed in Figure 2.27, were chosen both because of successful scale-up from laboratory to pilot scale and because they were already operating successfully in the mining industry at a larger scale than needed at Whitehaven. Two sizes of mixer–settler were used, reflecting different volume flow rates at different stages. The mixing section of all nine was identical being 1.47 m by 1.47 m with a working depth of 1.35 m. The settling section of the two extraction and two stripping mixer–settlers was 6.91 m long by 2.29 m wide with a working depth of 0.69 m. The settling section of the five scrubbing mixer–settlers was 8.38 m long by 2.74 m wide with a working depth of 0.84 m. The first three mixer–settlers were fabricated from 316 L stainless steel, the rest

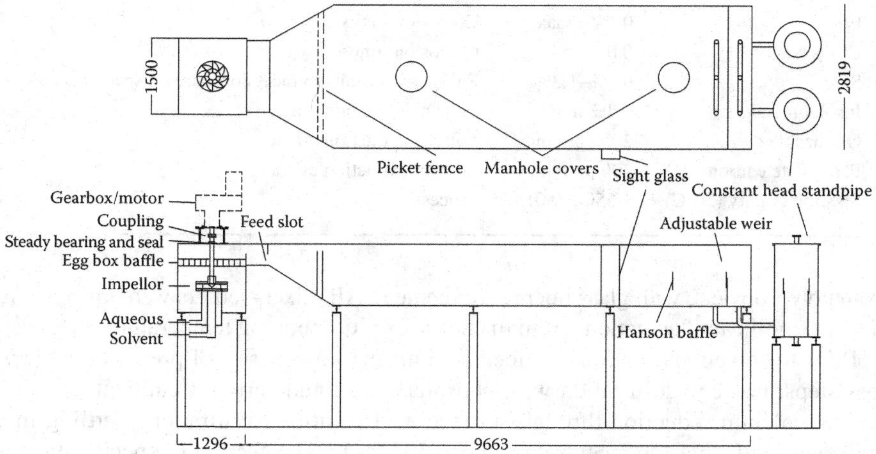

FIGURE 2.26 Davy McKee mixer–settler.

FIGURE 2.27 Mixer–settlers at A&W Whitehaven. (Courtesy of Solvay, Hugh Podger, and Brewin Books Limited, Studley, U.K.)

TABLE 2.10
MO Plant Feed Acid Specification

Constituent	Limits	Reason
P_2O_5	56.5% ± 1.0%	Maximize product split
CaO	0.2% max	Solids build up leading to blockages
Fe	0.3% max	Purification limits to meet downstream specifications
As_2O_3	40 ppm max	Downstream specifications
F	0.14% max	Corrosion limits
Cl (as NaCl)	0.04% max	Corrosion limits
SO_3	0.7%–1.3%	Purification limits to meet downstream specifications
Insoluble matter	0.4% max	Solvent extraction efficacy
Oxidizable carbon	200 ppm max	Solvent extraction efficacy
Particulate carbon	200 ppm max	Solvent extraction efficacy
Specific gravity (20°C)	1.650 ± 0.015	Process control

from polypropylene with glass fiber reinforcement. All mixer–settlers were insulated to prevent significant fluctuations from the normal 45°C operating temperature.

Prior to solvent extraction, the feed acid underwent a series of pretreatment process steps, namely, desulfation with phosphate rock and carbon treatment, concentration, calcium reduction through sulfuric acid addition (resulfation), settling in a thickener, and finally concentration to 56.5% P_2O_5. The feed acid specification is shown in Table 2.10.

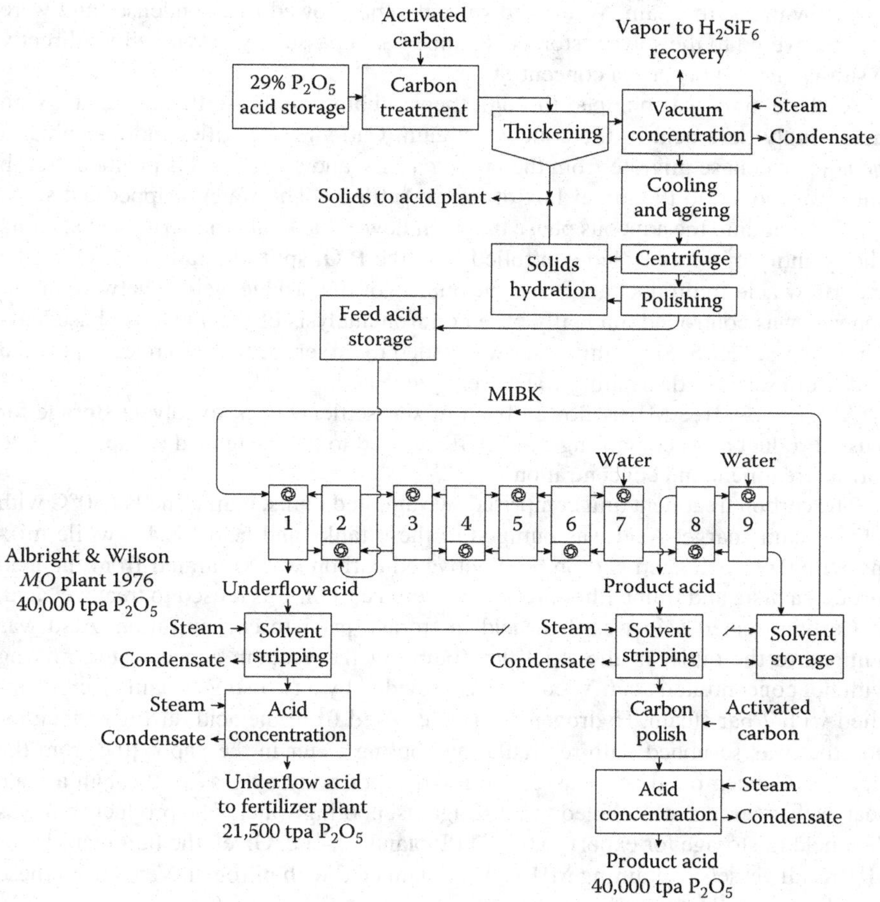

FIGURE 2.28 A&W *MO* plant block diagram.

Figure 2.28 shows a block diagram of the 40,000 tpa P_2O_5 *MO* plant. The plant achieved design rates within 2 months of start-up. Concentrated, filtered acid was pumped from storage through a steam-heated graphite tube heat exchanger raising the acid temperature to 45°C and on to mixer–settler 2 at 8.7 m³/h. MIBK was pumped from storage to mixer–settler 1 at 24.4 m³/h. Residence time in the mixer section was 60–90 s.

In mixer–settlers 1 and 2, acid is extracted into the fresh organic layer from both the acid feed and the aqueous stream pumped back from mixer–settler 3. The loaded organic stream is pumped forward to mixer–settler 3.

The aqueous stream from mixer–settler 1 (the raffinate or underflow) was 42% P_2O_5 and 1% MIBK and carried most of the impurities away. It was pumped to the top of the underflow stripping column via a preheater that raised the acid temperature to 124°C. The column was PTFE lined, with a 4.5 m high packed bed of graphite Raschig rings. The column was fitted with a thermosyphon reboiler supplied with

3 bar low-pressure steam. Water and solvent vapor flowed to a condenser and were then recovered to the solvent storage. Stripped acid passed by gravity either directly to storage for export or via concentration.

As the extract was pumped through the scrubbing mixer–settlers, it mixed with increasingly pure acid in the aqueous stream. Cationic impurities more soluble in the aqueous phase migrate from the organic phase and so the acid is purified. Scrub water was added to the mixer box of mixer–settler 7. The water stripped out some purified acid into the aqueous phase that then flowed back into the scrubbing section. The quantity of water added controlled both the P_2O_5 split (the ratio of P_2O_5 in the underflow acid to product acid) and the quality of the product acid. The water addition rate was controlled manually based on iron analysis of the aqueous phase leaving mixer–settler 8. Stripping water was added to mixer–settler 9 to release purified acid. Both scrub and stripping water were preheated.

Almost acid-free MIBK flowed from mixer–settler 9 back to solvent storage for reuse. Product acid, containing 2% MIBK, passed to the clean acid stripper, then to carbon treatment and concentration.

The carbon treatment unit comprised two agitated tanks, maintained at 80°C with a live steam sparge. Acid was pumped to these tanks and held for 4 h while mixing with 0.6% activated carbon. The activated carbon was separated from the acid through a plate and frame filter press. Recovered carbon was reused to treat wet acid.

Carbon-treated acid was then held in storage prior to concentration. Acid was pumped to the outlet of the axial flow pump of the vacuum concentrator, mixing with hot concentrated acid. Vacuum was provided by a two-stage steam ejector supplied with 7 bar steam. Hydrogen fluoride evolved from the acid during concentration that was scrubbed with recirculating cooling water in the vapor line from the ejectors. Concentrated acid passed from the vapor body by gravity through a plate heat exchanger and was cooled to 50°C against cooling water. The product acid was then held in storage for export to the STPP plants on-site. Given the flammability of MIBK, all vessels containing MIBK were blanketed with nitrogen. Vents from these vessels were collected and scrubbed.

Building on the success of the *MO* project, in 1977, A&W authorized a £20 m project named *MMO* (Modified Marchon Oldbury). This was a new plant with a 95,000 tpa P_2O_5 capacity. Furthermore, as well as a *salts* grade for STPP manufacture, it could also export a higher-purity technical grade suitable for derivative phosphates. This new plant came on stream in the spring of 1979 at a time of highest demand for STPP and allowed the Whitehaven site to shut down the old chemical purification plant (*wet salts*).

In concept, the new plant was very similar to the MO plant; there were two mixer–settlers for extraction, five for scrubbing, but now three for stripping. The mixer boxes remained the same size as on the MO plant, but the settler sizes increased to cope with higher throughput: extraction settlers 1 and 2, scrubbing settler 3, and stripping settlers 8–10 were 8.38 m long by 2.74 m wide with a working depth of 0.84 m; and the scrubbing settlers 4–7 were 9.86 m long by 3.2 m wide with a working depth of 0.84 m. Flow rates through the plant increased to 17 m^3/h for the feed acid and 51 m^3/h for the MIBK.

The principal change from the MO plant was the operation of the stripping mixer–settlers. As before, approximately 33% of the phosphoric acid in the organic phase in mixer–settler 7 was contacted with scrub water, stripping acid into the aqueous phase to pass back through the scrubbing mixer–settlers. The organic phase passing to mixer–settler 8 was contacted with more water, stripping another 33% of the acid. This product was known as *medium* grade 42% P_2O_5 acid. The remaining acid was released in a two-stage countercurrent sequence. Water was pumped to mixer box 10, stripping the acid from the organic phase into the aqueous phase; this aqueous phase was then transferred to mixer–settler 9, stripping more acid from the organic phase. By doing this, the acid that was 34% P_2O_5 in mixer–settler 10 was concentrated to 42% P_2O_5 in mixer–settler 9.

The impurity level in the product acid from mixer–settler 8 was about 10 times greater than that in the acid from mixer–settler 9. By varying water addition rates, impurity levels could be altered between the different products. The more impure acid from mixer–settler 8 was directed to STPP production, the more pure to technical phosphate manufacture, or both streams could be blended.

Other than a change in scale, the posttreatment steps remained almost the same. Blockages and tube fouling in the solvent stripper reboilers led to a change to direct steam injection to supply heat to the stripping columns.

Figure 2.29 shows a block diagram of the MO and MMO plants at this stage of development. Table 2.11 shows a comprehensive analysis of all the aqueous phases sampled during a trial using feed acid made from calcined Moroccan Yousouffia rock.

During 1980–1981, a new process was developed in the A&W laboratories to recover the P_2O_5 content of the underflow stream. This process, named UFEX, for *U*nder*F*low *EX*traction, contacted the underflow acid with MIBK loaded with 77% sulfuric acid. As the sulfuric-loaded MIBK contacted the weak phosphoric acid underflow, the sulfate and phosphate ions redistributed between the organic and aqueous phases, allowing recovery of 92% P_2O_5 from the underflow. Overall, this permitted 97% P_2O_5 recovery from the feed acid. The driver for this development work was the difficulty of disposing of the underflow acid into the fertilizer market at that time and the better margin from selling a more pure acid. The difficulty was part technical, part commercial: technical in that the magnesium content of the rock supplying the P_2O_5 had increased, leading ultimately to high magnesium levels in the underflow acid that are unhelpful in fertilizer manufacture, and commercial in that A&W were trying to place a relatively poor-quality acid into a declining fertilizer manufacturing base in the United Kingdom.

The initial experimental work was carried out batchwise in laboratory separating funnels. The development question was "how many mixer–settler stages are needed to separate a P_2O_5 stream from the underflow?" The hypothesis went as follows:

1. Unstripped underflow acid P_2O_5 concentration was too low to extract further P_2O_5 from it into MIBK.
2. Seventy-seven percent sulfuric acid is extracted into MIBK in a similar way to 74% H_3PO_4 and could be introduced into a clean MIBK stream.

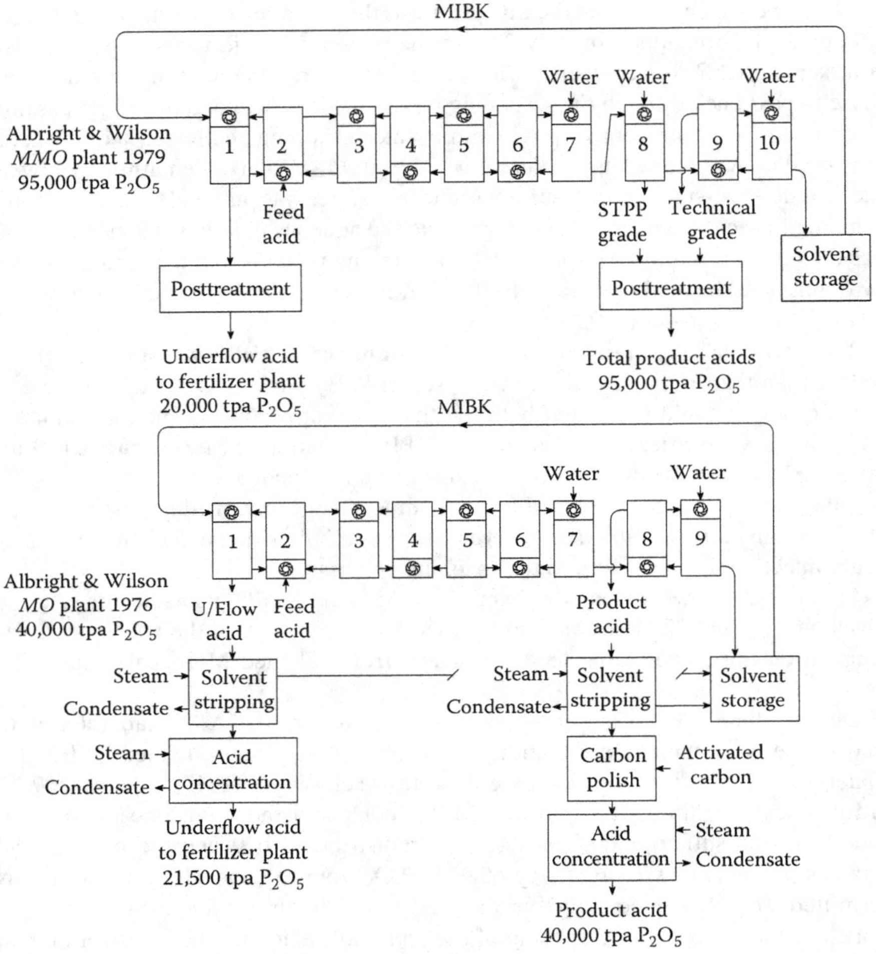

FIGURE 2.29 A&W *MO* and *MMO* plants block diagram.

3. Contacting the weak underflow acid stream with a sulfate-loaded organic stream leads to a redistribution of phosphate and sulfate ions. The net effect is the recovery of P_2O_5 into the organic stream.

Assuming a 1.6 purification factor between P_2O_5 and SO_3, it was calculated that nine stages would suffice. The purification factor in this case is defined as in the following equation:

$$\text{Purification factor} = \frac{\text{Mass of } P_2O_5 \text{ in organic phase}}{\text{Mass of } P_2O_5 \text{ in aqueous phase}} \times \frac{\text{Mass of } SO_3 \text{ in aqueous phase}}{\text{Mass of } SO_3 \text{ in organic phase}} \tag{2.32}$$

TABLE 2.11
MMO Plant Aqueous Phase Analyses

			Aqueous Phase Analyses								
	Feed Acid	Underflow	M-S 2	M-S 3	M-S 4	M-S 5	M-S 6	M-S 7	M-S 8	M-S 9	M-S 10
P_2O_5%	58.3	43.2	49.2	48.4	47.9	46.9	46.6	45.6	44.3	43.6	36.7
SO_4%	1.7	3.1	2.6	1.7	1.3	1.2	1.1	1.1	1.0	0.7	0.3
ppm											
Fe	1900	7,300	5,100	4500	2500	1200	650	310	95	7	<1
Mg	4700	19,600	13,500	6100	1450	345	100	30	15	1	6
Al	275	1,100	965	200	70	40	30	15	15	10	20
Na	195	680	600	325	155	60	40	30	7	11	<10
K	465	1,910	1,495	615	185	60	25	<25	<25	<25	<25
U	105	330	260	315	275	200	135	90	40	7.5	2
V	340	1,500	1,100	1000	530	250	130	60	16	1	<1
Cr	510	2,300	1,700	1200	570	235	115	60	13	2	<1
Zn	1100	4,000	3,300	4300	2800	1200	630	270	60	4	<1
Ca	260	590	1,500	100	50	60	5	5	5	5	15
Sr	2	9	5	3	2	<1	1	<1	<1	<1	<1
Ni	205	760	550	310	115	80	25	10	7	<5	<5
Cu	40	185	130	80	30	11	2	2	<1	<1	<1
Ti	175	605	430	310	195	145	115	85	55	20	5
Mn	15	50	35	25	12	5	<1	<1	<1	<1	<1
Pb	0.2	0.5	0.2	0.5	0.3	0.1	0.1	0.2	0.2	0.2	0.1
Cd	60	230	215	135	55	20	8	3	1	0.9	0.6
As	5.3	2.8	3.1	3.6	4.4	4.1	4.6	3.8	3.7	4.0	5.5
B	120	123	90	60	60	60	105	55	130	60	275
Si	390	430	320	300	145	160	340	130	220	70	80
F	3400	3,500	5,000	1800	1500	1400	1300	1400	1400	1700	4100

TABLE 2.12
Evaluation of the Effect of Different H_2SO_4/P_2O_5 Ratios on Plant Split

H_2SO_4/P_2O_5 Ratio	Stage to Which H_2SO_4 Added	% Clean P_2O_5 Split	Reject Acid Analysis % P_2O_5	Reject Acid Analysis % SO_3	Reject Acid Analysis % Fe	Product Acid Analysis % P_2O_5	Product Acid Analysis % SO_3	Product Acid Analysis ppm Fe
0.50:1	1	82.1	12.6	32.3	1.02	43.2	0.43	550
0.61:1	1	88.6	8.8	35.6	1.05	43.7	0.75	725
0.76:1	1	94.2	4.4	40.4	0.89	43.2	1.74	855
0.61:1	2	93.9	5.0	37.9	0.90	42.4	1.87	830

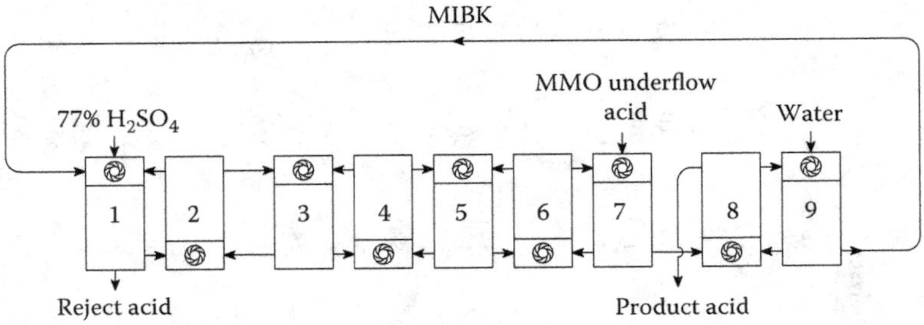

FIGURE 2.30 A&W *UFEX* mixer–settler scheme.

The initial batch experiments that were carried out at 21°C and 50°C gave purification factors in the range 1.8–3.0. Further work was carried out simulating a nine-stage countercurrent process during which different P_2O_5/SO_3 ratios were tested at different operating temperatures. It was found that as the H_2SO_4/P_2O_5 ratio increased, so also did the P_2O_5 split between product and reject acid and the iron and sulfate content of the product acid. The results of tests at 40°C are shown in Table 2.12. Further P_2O_5 split improvement was possible; however, eventually the sulfate content of the product acid would become too high, yet again demonstrating that impurities do not disappear but are merely redistributed.

The broad aim of the development work at a nine-stage solution reflected the idea that a near copy of the MO plant might suffice for the UFEX process. The envisaged configuration is shown in Figure 2.30.

The experimental work also revealed that complex solids precipitated as shown in the following equations:

$$\begin{gathered} K_2Fe_5^{II}Fe_4^{III}(SO_4)_{12} \cdot 18H_2O \\ Na_2SiF_6 \\ (Fe,Al)_3(K,Na)H_{14}(PO_4)_8 \times 4H_2O \end{gathered} \tag{2.33}$$

In early 1981, A&W carried out laboratory-scale trials of the UFEX process on a 60 mm diameter agitated column at Kühni AG, Switzerland (now Sulzer Chemtech). Several different column technologies are used in solvent extraction processes, and several companies in the phosphoric acid field had assessed various columns in the

1960s and 1970s. The Kühni column comprises a column with a central vertical shaft (the largest diameter units, 3 m and above, have three) that carries a number of shrouded impellors. The shaft drive is located at the top of the column and is always variable speed. Perforated baffle plates are located above and below each impellor, and two plates and one impellor comprise a compartment. Analogous to theoretical stages in distillation, a number of compartments are equivalent to a single mixer–settler.

The trials proved the UFEX process was viable on a Kühni column and allowed an estimation of the number of compartments. Kühni was also able to assess the distribution of perforations in the column plates and different turbine diameters at different levels in the column. At the conclusion of the trials, A&W ordered a 150 mm diameter column that was set up on-site, integrated into the *MO* plant, and over the years, used for many different trials.

In 1984, A&W commissioned the UFEX plant and mothballed the MO plant. The process arrangement is shown in the block diagram of Figure 2.31. The Kühni column, 1.6 m diameter with 48 compartments, was installed together with a lime silo, neutralization vessel, and sulfuric acid day tank. MIBK from the site storage

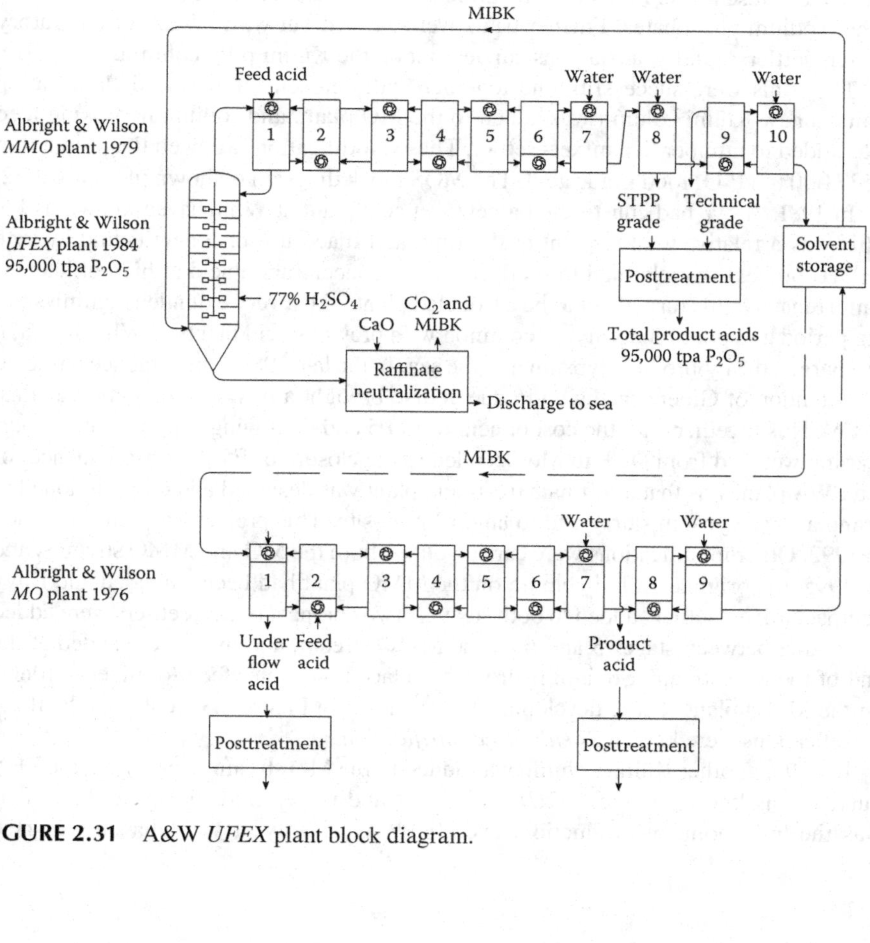

FIGURE 2.31 A&W *UFEX* plant block diagram.

was pumped into the lower part of the column, as was 77% sulfuric acid; the loaded solvent flowed up the column and contacted the underflow acid passing down the column. The aqueous stream containing 46% SO_4 and 4.4% P_2O_5 passed out of the column to the neutralization vessel where it was mixed with lime. The resultant calcium sulfate slurry was then pumped to drain and discharged to the Irish Sea.

In 1987, A&W commissioned the *MOS* process (designated from **M**archon **O**ldbury **S**odium). The MO and MMO plants were originally designed to produce technical grades of solvent-extracted phosphoric acid suitable for STPP, other industrial phosphates, and liquid fertilizers. These plants did not carry out dearsenication or defluorination, and the sulfate levels in the product acid were also too high for food grade products. During the conception of the MO process, it was envisaged that food grade acid sufficiently pure to displace thermal acid would be made via crystallization in the so-called *MOX* process. Unfortunately, this process would have distributed impurities into the technical acid stream and was a relatively high-cost process. During development work in the 1960s, it was noted that the scrubbing stage could be improved by the addition of an alkali to the scrubbing water. Sodium, ammonia, and potassium hydroxides were contemplated. Laboratory trials with separating funnels were carried out with sodium hydroxide. Sodium hydroxide was chosen because the sodium could be directed to product acid going to make STPP or other sodium phosphates. Further trials were carried out with a bank of laboratory mixer–settlers; finally, a trial was carried out on the Kühni pilot column.

The trials were successful, and together with dearsenication and defluorination units, a new Kühni column was added to the MO plant, and sodium hydroxide feed was added to number five mixer–settler. These modifications allowed the production of 37,000 tpa P_2O_5 food grade acid. The MOS block diagram is shown in Figure 2.32.

In 1981, A&W had constructed a new wet acid plant at Whitehaven known as F5. This was a relatively large plant at the time and made use of a new hemidihydrate (HDH) process that claimed to produce a high-concentration acid at high efficiency. Unfortunately, this turned out to be a troubled plant with a very extended commissioning period and poor reliability. In common with previous acid plants at Whitehaven, it discharged the by-product gypsum into the sea. In the late 1980s, this practice attracted the attention of Greenpeace who in due course brought a private prosecution against A&W. This together with the cost of acid from F5 and the savings opportunity of purchasing wet acid from OCP in Morocco led to the closure of F5. The consequence for the PWA plant was that a raffinate treatment plant was designed and commissioned to granulate the gypsum slurry and to landfill it on-site. This project was commissioned in 1992. Other modifications were carried out on both the MO and MMO streams, and by 1995, the original Kühni column on the MMO plant had been extended and Davy compact mixer–settlers added to both steams. Two compact mixer–settlers were added in parallel between stages 5 and 6 on the MMO stream, and three were added at the end of the MO stream. Sodium hydroxide replaced the water feed to mixer–settler 7 on the MMO plant. These developments are shown in Figure 2.33; collectively, these modifications were known as *enhanced purification*.

In 1996, another Kühni column was added to the MO stream to carry out a UFEX duty; it was known as the *MOUFEX* project and is depicted in Figure 2.34. This was the high point of production; overall PWA capacity at Whitehaven was now

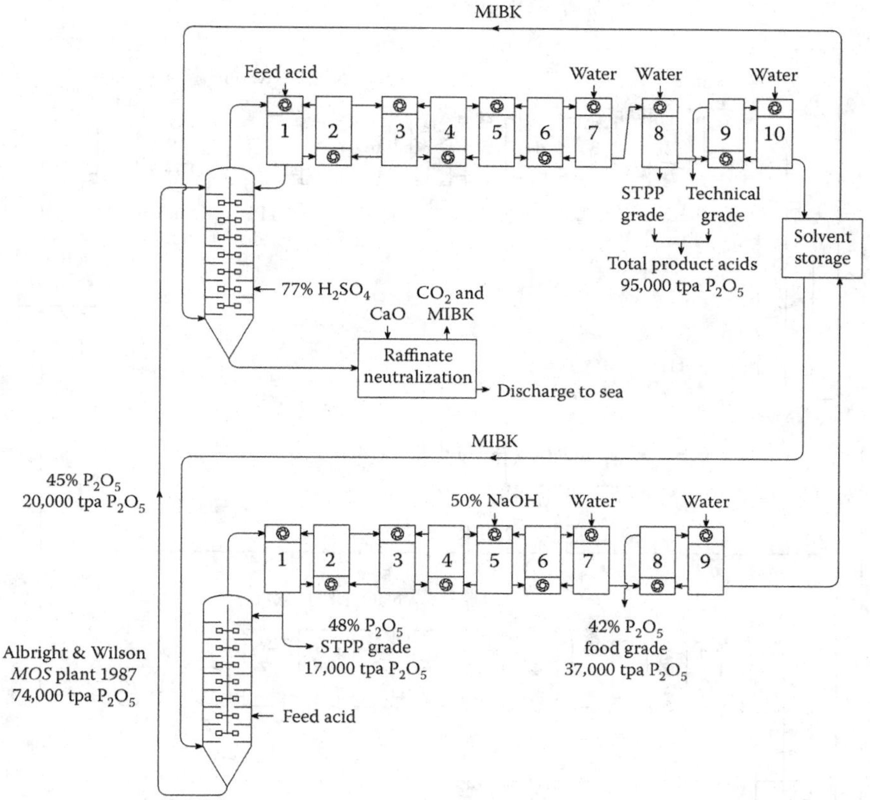

FIGURE 2.32 A&W *MOS* plant block diagram.

179,000 tpa P_2O_5, including 60,000 tpa P_2O_5 food grade. Within 5 years, the plant was closing down due to a combination of market forces, a decline in product demand (particularly the collapse of STPP demand), and consolidation and takeovers in the industry.

The PWA complex at Whitehaven had seen near constant development and modification since 1974. The plant was therefore complicated and not the cheapest to operate; nevertheless, during the decline after 1996, many modifications were made that allowed it to adapt to market demands in both volume and purity requirements. Figure 2.35 shows a mass balance of the PWA operation in 1999; unit capacities are given within the boxes and flows in tons per year P_2O_5 shown above the arrows. Rates are based on an operating time of 7660 h/year.

While A&W considered entering into several PWA joint ventures to which it would supply the technology, in fact it only concluded one. The timing of this joint venture in the PWA development cycle meant that the plant designed for the Purified Acid Partnership (*PAP*) at Aurora incorporated all the important A&W learning. The design is almost optimal, the capital cost is minimized, and the plant produces arguably the cheapest purified acid in the world.

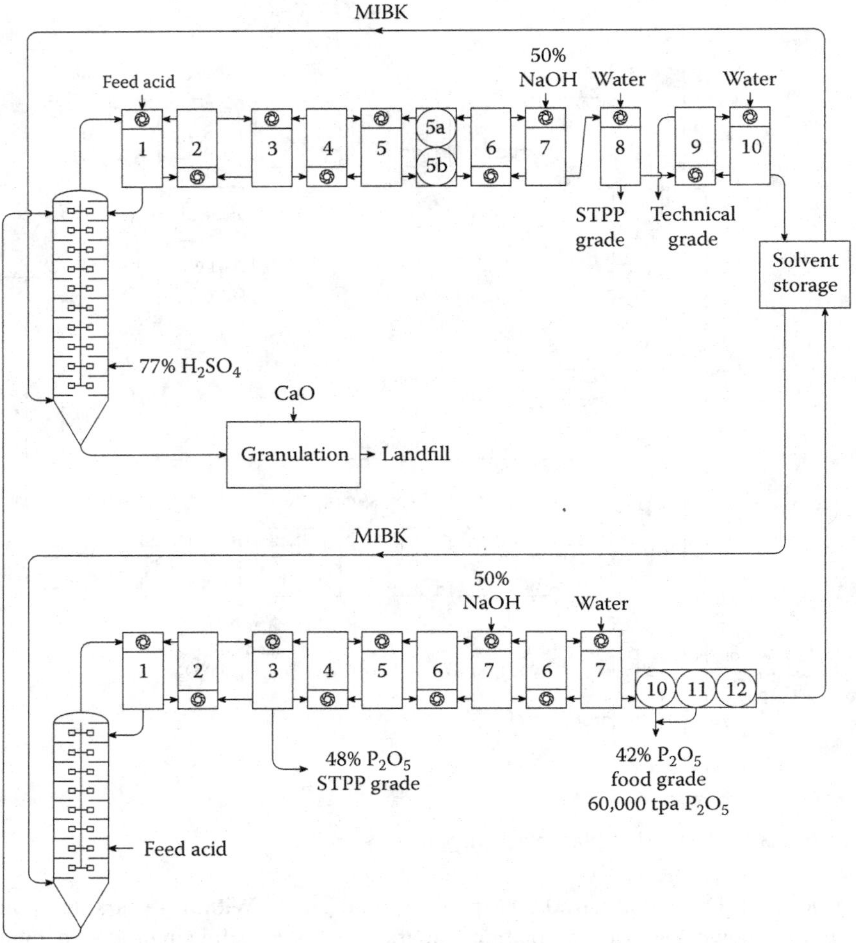

FIGURE 2.33 A&W *enhanced purification* plant block diagram.

The plant was designed in 1989 and commissioned in 1991 with an initial capacity of 110,000 tpa P_2O_5. During commissioning, the operating team was supported both on-site and from Whitehaven by the A&W engineers and chemists who designed the plant. Commissioning was for the most part straightforward; however, an unexpected change in the feed acid impurity profile caused several months of head scratching. In the years following commissioning, the output was increased through optimization and the switch from sodium hydroxide to ammonia scrubbing. Ten years after initial commissioning, capacity was up to 165,000 tpa P_2O_5. The design proved sufficiently robust that when the current owners, PCS, installed additional trains 3 and 4 in 2003 ($70 m) and 2006 ($73 m), no major changes were made to the basic technology. Aurora capacity is now 327,000 tpa P_2O_5.

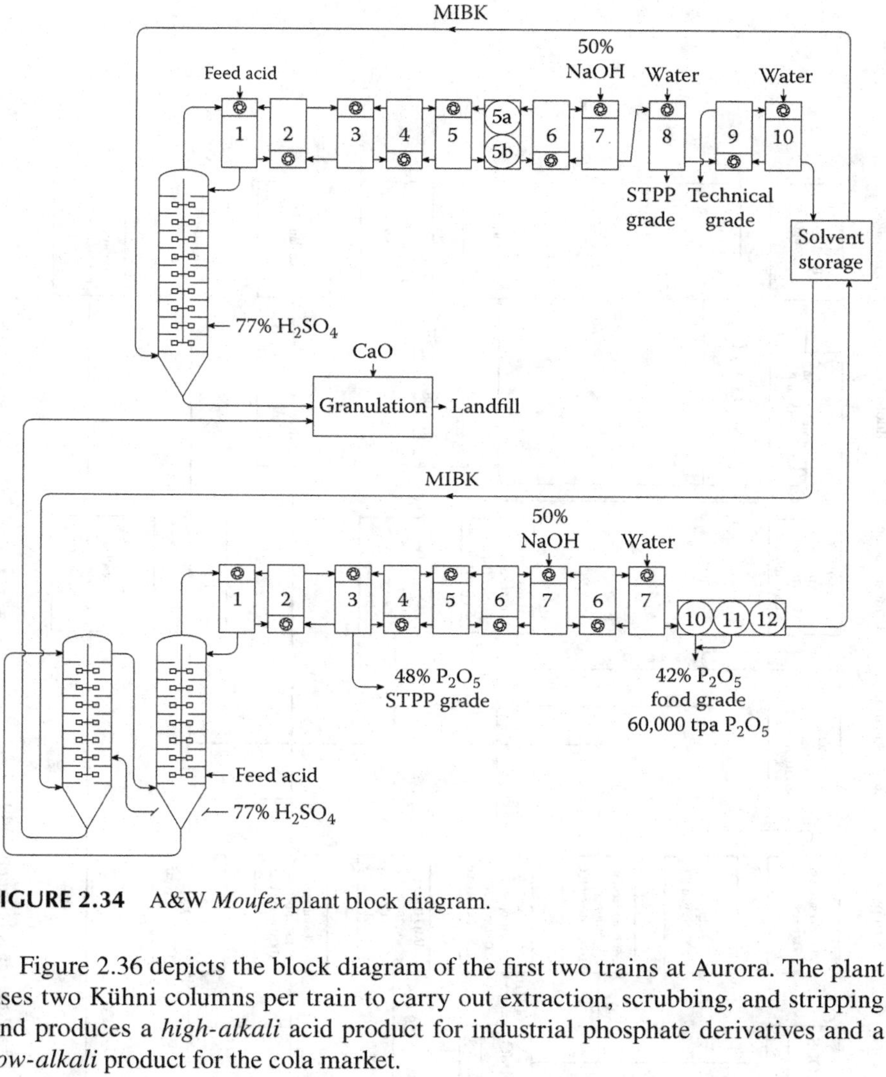

FIGURE 2.34 A&W *Moufex* plant block diagram.

Figure 2.36 depicts the block diagram of the first two trains at Aurora. The plant uses two Kühni columns per train to carry out extraction, scrubbing, and stripping and produces a *high-alkali* acid product for industrial phosphate derivatives and a *low-alkali* product for the cola market.

PCS makes wet acid at Aurora from its own locally mined phosphate rock. The rock has a low arsenic content that obviates the need for a dearsenication section.

Figure 2.37 shows the flows around the solvent extraction section of one train. For each train, the wet acid is transferred to a 700 m^3 feed tank at about 50°C. The acid is already pretreated, with a low sulfate content (0.7% SO_4), at 56%–57% P_2O_5 concentration. The acid is pumped to the top of the extraction column via an acid cooler. The cooler lowers the feed temperature to 35°C so that, with the temperature rise that accompanies the acid solubilization, the column is maintained at 50°C. The extraction column is 2 m diameter with 48 compartments. MIBK is pumped from the solvent storage tank to the base of the extraction column, also via a cooler, and rises through the column contacting the wet acid as it passes downward. A solvent/acid

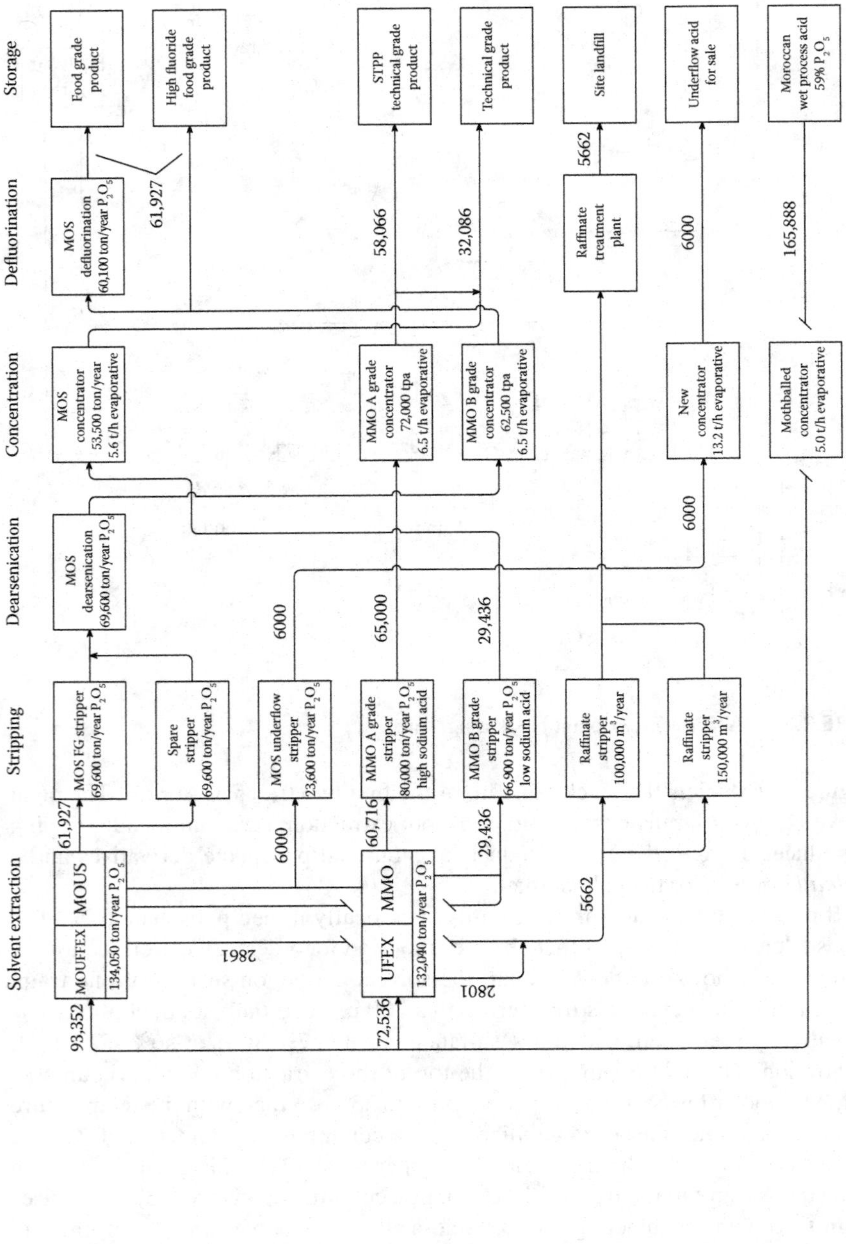

FIGURE 2.35 A&W Whitehaven site P_2O_5 mass balance 1999.

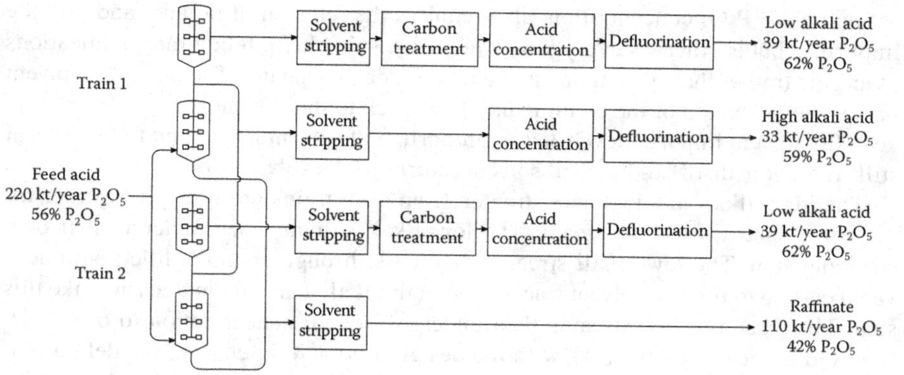

FIGURE 2.36 A&W Aurora plant block diagram (first two trains).

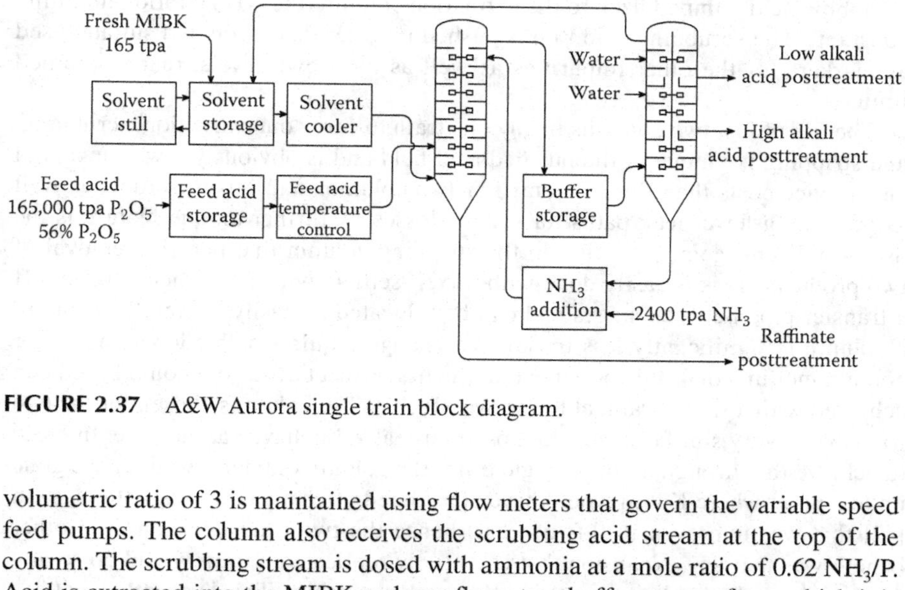

FIGURE 2.37 A&W Aurora single train block diagram.

volumetric ratio of 3 is maintained using flow meters that govern the variable speed feed pumps. The column also receives the scrubbing acid stream at the top of the column. The scrubbing stream is dosed with ammonia at a mole ratio of 0.62 NH_3/P. Acid is extracted into the MIBK and overflows to a buffer storage from which it is pumped forward into the base of the stripping column. Raffinate acid, carrying most of the impurities in the feed acid, coalesces in the base of the extraction column and is pumped away under level control, for solvent stripping and then to the ammonium phosphate fertilizer plant. As the loaded solvent rises through the stripping column, it contacts the falling aqueous stream of purified acid, and more impurities are transferred into the aqueous phase. The aqueous stream exiting the base of the column is the *high-alkali* acid at about 47% P_2O_5 concentration. Apart from fluoride and organic levels, this acid has a food grade impurity profile. The now partially loaded solvent continues up the column into the final water addition zone where the remaining acid is stripped from the solvent and passes out of the column. This stream, at about 42% P_2O_5 concentration, also requires the removal of organic and fluoride impurities but is otherwise very low in impurity, suitable for food grade applications as an acid (rather than a component of a food grade phosphate). The unloaded solvent passes out of the top of the column and flows back to the solvent storage. In order to keep the solvent impurity levels low, a proportion of solvent is pumped to the solvent still where it is distilled. Purified solvent returns to the solvent storage.

The high-alkali and raffinate streams from both trains are brought together for solvent stripping with live steam. The low-alkali streams each undergo their own stripping step. The low-alkali streams then pass through columns filled with activated carbon to remove solvent traces. The high-alkali acid stream does not take this step. The three purified streams then undergo vacuum concentration to 62% P_2O_5 followed by defluorination. As with the design used at Whitehaven, the defluorination column is fed with a combination of hot air and steam to remove fluoride as an HF vapor.

The split, or ratio, of acid going forward to the stripping column to that passing out of the extraction column as raffinate is maximized by control of the flow of scrubbing acid pumped back to the extraction column (the NH_3/P ratio remaining constant). The scrubbing acid rate is pushed until levels of iron and sulfate, used as markers for the other impurities as well as their own levels, reach specified limits.

The use of just two columns in any one train to carry out extraction, scrubbing, and stripping is unique in the purified acid field and is obviously lower in capital and service costs than three columns or two columns and mixer–settlers as well as presenting fewer leak paths for solvent losses. A further unique feature is the use of a disengagement section in the stripping column that permits removal of two products. This is easily done with mixer–settlers where a branch is taken off a transfer pipe between stages and can be relocated as easily. Doing the same in a column is significantly less trivial as a change requires a shutdown and major column modification. It is possible that the first project at Aurora could have been achieved with a single train, at lower overall cost. The columns chosen for the two trains were very similar in size to those in use at Whitehaven at the time. In order to achieve the throughput on a single train, the column diameter would have gone up to a point where Kühni would have considered offering a three-agitator system [42] that was unproven in this process, hence the two-train decision. The other benefit was that any expansion that essentially replicated a train would do so at a production scale that better matched the market capacity. This is exactly what has happened.

Figure 2.38 is a photograph of the four trains at Aurora. The plant on the right-hand side of the photograph comprises trains 1 and 2. The sheeted in columns on the right-hand side of each plant are access elevators. To the left of the elevators are the two Kühni columns, most clearly seen on trains 3 and 4. The stripping section is most clearly seen in the center of the photograph to the left of the Kühni columns. The acid concentrator bodies stand out on the extreme left of the photograph and in a line of three in the center of trains 1 and 2 on the right. Figure 2.39 shows the outlines of the major equipment sections from the photograph of Figure 2.38.

FIGURE 2.38 Aurora trains 1–4. (Courtesy of PotashCorp, Saskatoon, Saskatchewan, Canada.)

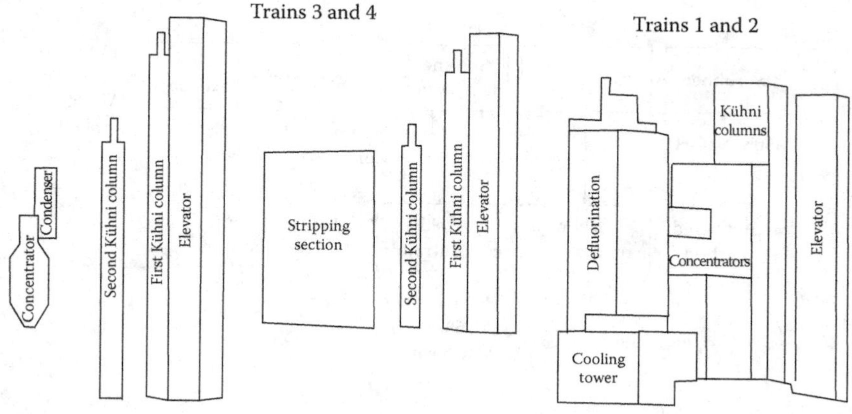

FIGURE 2.39 Aurora plant outlines.

2.4.2 Budenheim Process

Chemische Fabrik Budenheim (CFB) developed a purification process in the 1970s and was granted a number of patents in this field, for example [43,44]. The Budenheim process was seriously considered during negotiations for the joint venture at Aurora that is now owned and operated by PCS. CFB operated a plant in Germany for a number of years but in 1998 entered into a joint venture with OCP and Prayon to build a Prayon technology PWA plant in Morocco. The Moroccan plant now supplies CFB requirements for purified acid.

Figure 2.40 shows a block diagram of the Budenheim process. Twenty-eight to thirty percent P_2O_5 wet process acid is received into a day tank that also serves to age the acid and allow some impurities to precipitate out and settle. The acid is pumped forward to the pretreatment stage where the acid is mixed with sodium hydrogen sulfide, filter aid, and activated carbon and allowed to mix and hold in an agitated tank.

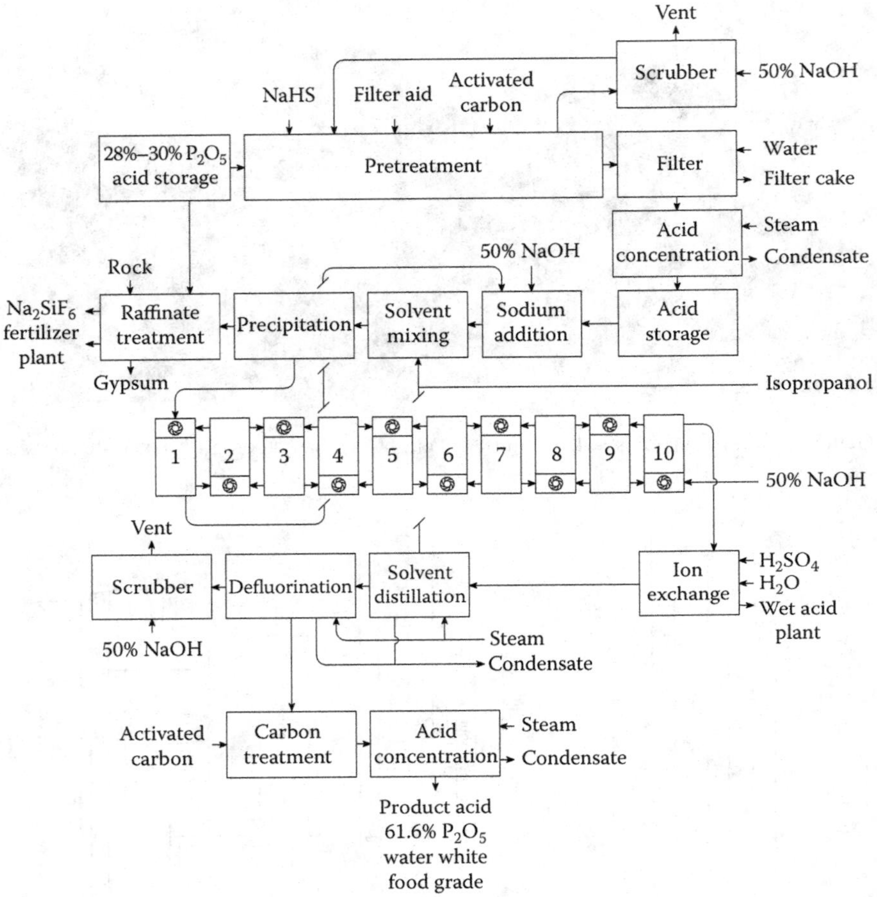

FIGURE 2.40 Budenheim plant block diagram.

Vent gases from the tank include hydrogen sulfide, and these are directed to a sodium hydroxide scrubber that converts any hydrogen sulfide gas to sodium hydrogen sulfide that is then recycled back to the feed acid pretreatment. The sulfide precipitates arsenic and the carbon removes organic species; both are filtered on plate-and-frame filters.

The filtered acid is concentrated to 55% P_2O_5 in a vacuum concentrator and held in buffer storage. The concentrated acid is then mixed with fresh 50% sodium hydroxide and monosodium phosphate solution from the first mixer–settler. Isopropanol is added to this mixture in a 5:1 ratio; this stream then flows to the precipitation tank. Most of the impurities either precipitate in, or distribute to, the aqueous phase that is approximately 30% P_2O_5. This raffinate stream is directed to fertilizer manufacture and is first mixed with WPA. As the raffinate has both high sodium and sulfate content, mixing with WPA allows sodium fluosilicate to precipitate. Phosphate rock is added to reduce sulfate levels through calcium sulfate precipitation. Finally, the acid is concentrated to about 40% P_2O_5, suitable for fertilizer manufacture.

TABLE 2.13
Budenheim Process Stream Composition

	Premixer–Settler Stream	Postmixer–Settler Stream	Final Product
% P_2O_5	9.0	8.9	54.0
% isopropanol	72.3	72.3	
ppm SO_4	20	50	30
ppm F	10	50	20
ppm Na		400	10
ppm Fe	10	1	5
ppm Al	90	1	5
ppm Ca	10	1	5

The precise solvent stream composition arising from the precipitation tank depends on the feed acid composition, but an indicative analysis is given in Table 2.13 [43]. Comparison with wet acid analyses in Table 2.4 gives an indication of the level of purification achieved at different stages.

The organic phase mixes with an aqueous phase of 46% monosodium phosphate in a number of mixer–settlers (10 are shown in the diagram). As the organic stream progresses through the mixer–settlers, impurities are transferred to the aqueous stream and most impurities are reduced to low levels. The three notable exceptions are sulfate, fluoride, and sodium. The washed organic stream is transferred from the last settler to a bank of fluid bed ion exchangers where sodium ions are removed by cation exchange. The exchangers are regenerated with sulfuric acid and the regenerate is passed back to the wet acid plant.

Following ion exchange, the isopropanol is separated from the acid by atmospheric distillation against a chilled condenser, followed by vacuum distillation. Both steps require steam heating. The product from the second column is 50% P_2O_5 together with trace solvent, fluoride, and other organics. This stream is steam stripped to remove the fluoride and trace solvent, is cooled, and is fed forward to a bank of active carbon beds. Finally, the acid is concentrated, against vacuum, to achieve sales grade concentration of 54%–62% P_2O_5 and food grade purity. The third column in Table 2.13 shows some relevant analysis.

2.4.3 FMC Process

FMC was a major player in phosphates in the United States and through its Spanish operation, FMC Foret in Huelva, in Europe. In the United States, its business was founded on phosphorus operations in Idaho. Foret, however, relied on wet acid for its processes and installed a small purification plant based on IMI technology in the early 1970s. FMC had undertaken a process development program in the 1960s in the United States and was granted a patent [45] utilizing a 3:1 by volume TBP/kerosene mixture as solvent, columns for extraction and stripping, and solvent regeneration. The process described was capable of producing a water white acid to meet food grade standards. In the 1970s, FMC evaluated different solvent extraction equipment including Karr, Scheibel, and rotating disk contractor (RDC) columns and Podbielniak centrifugal contactors. In the early 1980s, further extensive laboratory and pilot-scale study was carried out to facilitate the design of the new purification plant at Huelva. During the 1980s, FMC was granted a number of patents. Of note were improvements to the scrubbing (washing) stage [46] by the use of metal phosphates (one example was the addition of sodium ions, thus creating an aqueous stream with a Na/P ratio of 0.05–0.14) and solvent regeneration [19]. Further patents were filed, extending the use of sodium ion addition creating a sodium phosphate stream suitable for conversion to STPP [47] and modifying the solvent for improved efficacy by replacing kerosene with diisobutyl ketone (DIBK). None of these later patents were implemented on an industrial scale.

In 1983, a 12,000 tpa P_2O_5 plant was installed at Foret, principally to supply the STPP plant. Figure 2.41 shows a block diagram of the process; Figure 2.42 is a photograph of the plant itself. Other views are currently available via a well-known search engine. Feed acid came from the existing wet acid plant on-site and was pretreated to

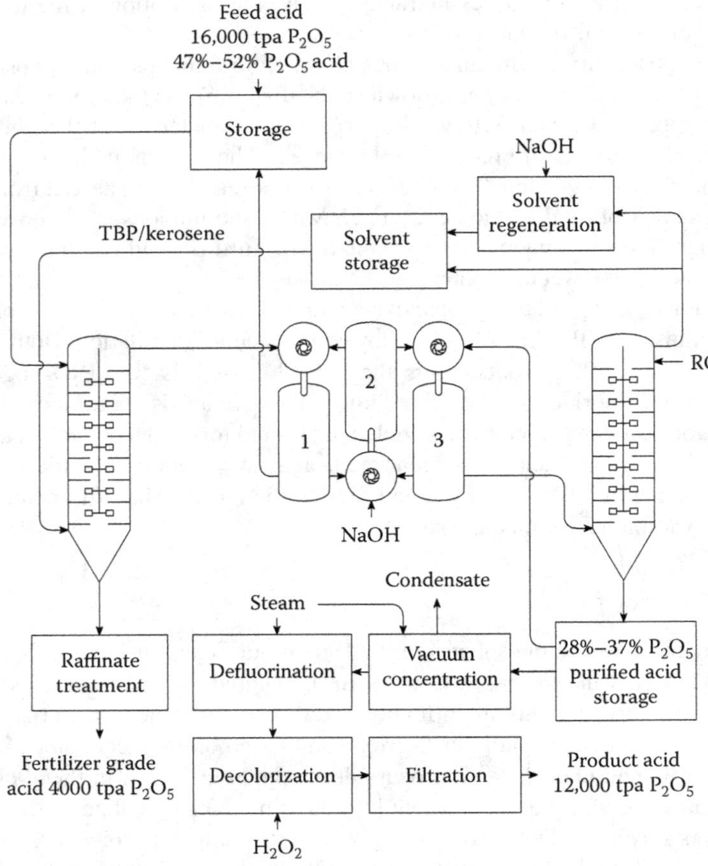

FIGURE 2.41 FMC Foret Huelva plant block diagram.

FIGURE 2.42 FMC Foret Huelva PWA 2004 (author's photograph).

reduce sulfate, arsenic and other heavy metals, and fluoride levels to those compatible with the final product specifications. Finally, the feed acid was concentrated to 47%–52% P_2O_5.

Feed acid at 50°C entered the top of the extraction column, a Kühni column with 30 compartments equivalent to 5 mixer–settler stages, and descended the column contacting the TBP/kerosene solvent. The solvent to feed acid ratio was approximately 4:1 by weight. The organic stream leaving the top of the column was 12% P_2O_5 with Fe and SO_4 levels less than 50 ppm. The aqueous raffinate stream leaving the column base was 20% P_2O_5 with Fe and SO_4 levels close to 1%.

The organic stream was pumped via intermediate storage to a cascade of three countercurrent mixer–settlers and scrubbed with a sodium-enriched aqueous stream. The aqueous stream flowed back to the extraction column via the feed storage, the organic stream forward to stripping in a second Kühni column entering at the column base.

The scrubbed organic stream contacted water countercurrently in the column and phosphoric acid was transferred from the organic to aqueous phase. The ratio of water and organic flow was controlled to give an acid concentration in the range 28%–37% P_2O_5, depending on the desired purity of the final product acid.

During production, a side stream of solvent passed continually to the solvent regeneration unit as the TBP became steadily contaminated with silica, organic species, sulfides, etc., and itself started to hydrate to DBP. The solvent regeneration unit comprised three mixer–settlers where firstly, the solvent was contacted with water to remove phosphoric acid and soluble impurities; secondly, the solvent was contacted with sodium hydroxide to remove any remaining contaminants into the aqueous phase—this phase then passed to effluent treatment; and thirdly, the treated solvent contacted the aqueous phase from the first mixer–settler neutralizing any sodium hydroxide from the second mixer–settler. The regenerated solvent is pumped to the solvent storage.

The weak product acid passed to a double-effect evaporator for concentration to 54% then to a single-effect evaporator to achieve 62% P_2O_5.

To achieve fluoride and color specification, food grade acid passed first to a defluorination column. The defluorination column was PTFE lined and fitted with graphite plates operating under vacuum. The acid was steam stripped to reduce fluoride levels to <10 ppm.

Finally, the defluorinated acid was mixed in a static mixer with hydrogen peroxide to remove color bodies leaving a water white food grade acid.

The Foret plant operated satisfactorily until its closure in 2010 due largely to the decline of the STPP market in Europe.

Toward the end of the 1990s, FMC's phosphorus operation in Idaho was faced with expensive environmental upgrades. Consequently, given most of the phosphorus was subsequently burned to make phosphoric acid for STPP and other phosphates, a purified acid plant project was evaluated. FMC chose to apply its own technology, developed at its technical center in Princeton and implemented in Huelva, in Idaho at the site of a wet acid and fertilizer producer, Agrium.

The original plan was to build an 80,000 tpa P_2O_5 single-train plant that would take pretreated acid at 52% P_2O_5, purify it to food grade, and return the raffinate to site for ammonia fertilizer (MAP/DAP) production. During discussions with the site owners, FMC agreed to design and build the pretreatment stages and utilities. The feed acid was now an untreated 28% P_2O_5 made from a Western States ore. Broadly speaking, the standout characteristics of this rock are high organics and vanadium content. Prior to the date of implementation, the rock was calcined to destroy organics as part of the rock treatment processes prior to acidulation to wet acid; however, a combination of environmental and cost drivers led to the decision to use uncalcined rock and manage the organics in the wet process chemically.

The block diagram of FMC's process at Idaho is shown in Figure 2.43.

The overall pretreatment process goals were to reduce sulfate levels to 0.2%, fluoride to 0.3%, arsenic to less than 1 ppm, and cadmium to less than 15 ppm. 120,000 tpa 28% P_2O_5 was pumped via acid coolers to the desulfation tank where it was mixed with 4000 tpa P_2O_5 phosphate rock slurry. Other sulfate-reducing agents were considered including soda ash but rock was chosen on price and availability criteria. The site received the rock as slurry via pipeline; consequently, it was only available in this form. The quantity required on the purified acid plant for desulfation was a small percentage of the total site demand, and in practice the control of the rock addition, and therefore sulfate and other impurity levels, proved difficult. Following desulfation, the acid was clarified and the slurry from the clarifier filtered on a belt filter with condensate washing. The solid waste from belt filtration was collected with other filter waste for disposal; in total P_2O_5 losses through filtration were

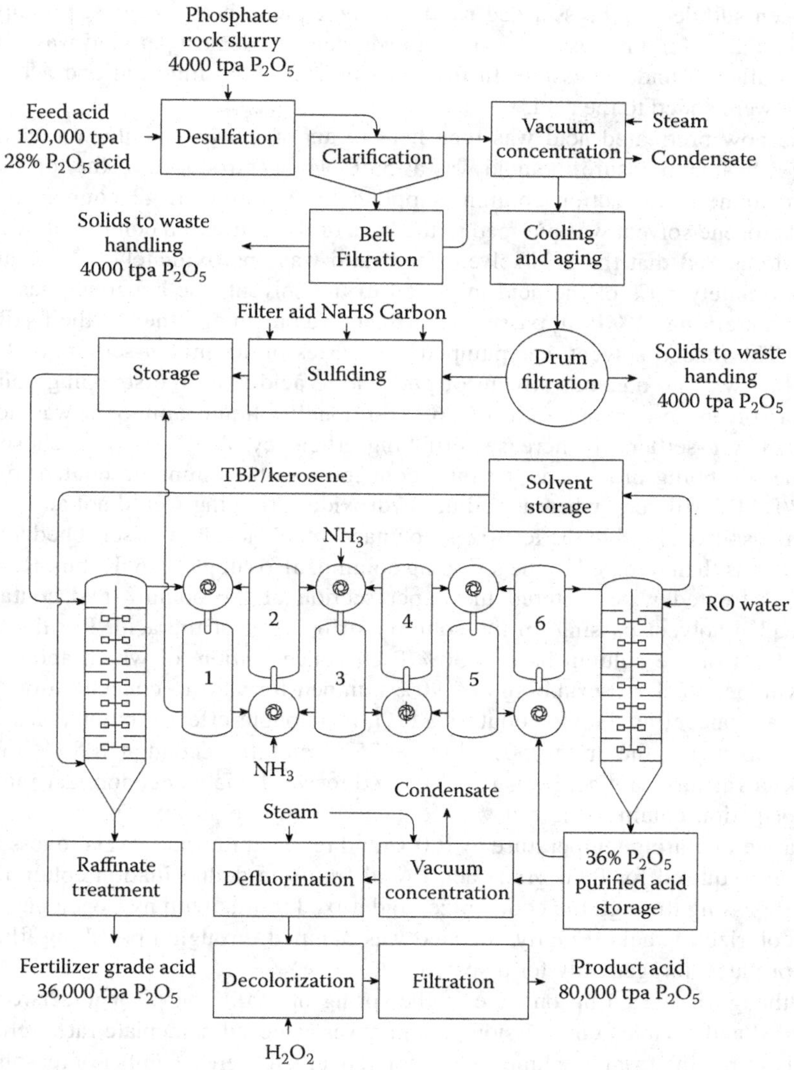

FIGURE 2.43 FMC Idaho plant block diagram.

significant; the diagram simplifies total P_2O_5 loss, ascribing it to filtration alone. The clarified acid was then pumped to the 52% P_2O_5 vacuum-pumped circulation concentrator with double-effect steam ejectors creating the system vacuum. Following concentration, the acid was cooled and held in intermediate storage.

The cooling of 52% P_2O_5 wet acid always causes the precipitation of complex phosphates. Therefore, following storage, the acid passed to a large vacuum drum filter that was precoated with diatomaceous earth. Filter cake, loaded with complex phosphates and some free acid, was sent for disposal with cake from the belt filter. The filtered acid was pumped to the sulfiding unit where it contacted sodium hydrogen sulfide in a pressurized reactor to precipitate heavy metals, principally arsenic and cadmium, passed through a hydrogen sulfide stripper, and was subsequently filtered under pressure. In the filtration step, both filter aid and activated carbon were added to the acid.

The now pretreated acid was then held in another intermediate storage with integral heater to ensure the acid was at 55°C when introduced to the top of the 2.6 m diameter extraction column, supplied by Kühni with 42 compartments. TBP/kerosene solvent was pumped to the base of the extraction column; flow ratio control ensured that the acid/solvent feed ratio was approximately 5 by volume. Approximately 80% of the acid migrated to the solvent; the aqueous phase raffinate containing 21% P_2O_5 passed to raffinate treatment and then to the fertilizer plant. The loaded solvent was pumped to a series of six mixer–settlers and was scrubbed with an aqueous stream of phosphoric acid from the stripping column with a solvent-to-aqueous ratio of 26. Additionally, liquid ammonia was added to the mixer–settlers to increase scrubbing efficiency. Ammonia was chosen to aid the scrubbing process as it would contribute to the ammonia content of the MAP/DAP fertilizers, whereas sodium hydroxide scrubbing would not and would raise the sodium level in the fertilizers to unacceptable levels. The scrubbed loaded solvent was then pumped to the stripping column, also supplied by Kühni. Reverse osmosis–treated water entered the upper section of the column and contacted the loaded solvent passing up the column, stripping it of the acid. Purified acid passed out of the column base at 36% P_2O_5 concentration, of which about 10% was pumped back to scrubbing and 90% pumped forward to concentration. The acid was concentrated up to about 63% P_2O_5 in a double-effect vacuum concentrator in two stages, the first to 50% P_2O_5, at 75°C, and the second at 105°C, both at 14.3 kPa vacuum. Hot acid was then pumped forward, via an economizer into the defluorination column where it was contacted with live steam. The column was held at an operating temperature to 160°C and reduced fluoride levels to less than 10 ppm, required for food grade acid. Acid from the defluorination column was cooled passing through the economizer and mixed in line with hydrogen peroxide to decolorize the acid. Finally, the acid was pumped through a polishing filter to final product storage ready for dispatch.

Although it started up on time in the spring of 2000, the plant underwent an extended and troubled commissioning and never achieved nameplate rates reliably. This is surprising from a technical standpoint, but there were a number of reasons for the lack of progress. The basic FMC process, as demonstrated in Spain, was good; the Idaho flow sheet was perfectly satisfactory if somewhat pinched in places, and much of the equipment was identical to that used on the other purified acid plants at Aurora and Geismar and operated well. Nevertheless, the commissioning team faced challenges on all fronts:

1. Firstly, the project scope included the provision of most utilities (steam, air, RO water, nitrogen, cooling water), all of which had to be commissioned and brought their own challenges.
2. Secondly, the pretreatment section was difficult to operate because the phosphate rock slurry system was difficult to control; in turn, this led to

poor control of the clarification unit that in part led to more frequent blockages in the concentration system than would normally be the case; furthermore, the drum filter proved problematic which may have been due both to aspects of its mechanical design and/or the duty required of it. These problems built on each other as well as hampering the sulfiding operation so that in the end the *treated* acid ready to feed the solvent extraction section was often of poor quality; indeed, at times so poor, it was remarkable that any purified acid was made.

3. The third challenge was that the host company was expanding and commissioning its own wet acid production at the same time and making changes to its rock feed. Consequently, at least in the early days of commissioning, the purified acid plant did not receive a stable wet acid feed of consistent composition.

The net effects of these challenges were very high P_2O_5 and solvent losses and low output. Despite these challenges, the plant operation did steadily improve; however, during this period, the whole industrial phosphate industry was undergoing major change. FMC and Solutia (itself spun out of Monsanto) formed a joint venture, Astaris, which comprised most of the phosphate operations of the two companies. The Idaho plant was included in the joint venture. At the same time, overcapacity in the STPP market led to a loss in sales for Astaris and consequent drop in internal demand for acid to make STPP. The final nail in the coffin for the Idaho plant was the expansion by PCS of its purified acid plant, with the lowest costs in the United States. Taking all these factors into account, Astaris decided to close the plant in 2003.

2.4.4 IMI Processes

The drive for the IMI processes came initially from the need to find a use for coproduced hydrochloric acid [48]. During the 1950s, Israeli scientists developed and patented the Aman process [49] to produce magnesium oxide from the chloride salts of the Dead Sea. The coproduct from this process was hydrochloric acid that was known to be as effective as sulfuric and nitric acid in dissolving phosphate rock. Unlike the sulfuric acid process where the liquid phosphoric acid is separable from the solid calcium sulfate, in the hydrochloric acid process, as shown in the following equation, both phosphoric acid and the coproduced calcium chloride remain in solution:

$$Ca_{10}(PO_4)_6F_2 + 20HCl \rightarrow 10CaCl_2 + 6H_3PO_4 + 2HF \quad (2.34)$$

The breakthrough came when Baniel et al. developed the solvent extraction process that allowed the separation of calcium chloride from phosphoric acid [50]. Also in the 1950s, geologists undertook surveys for phosphate rock formations in Israel to allow the superphosphate plant at Haifa [51] to run with alternative rock sources. Several sources were discovered of which the first to be mined, in 1952, was the Oron deposit, located 25 miles southwest of the Dead Sea. Subsequently, the Zin and

Arad deposits were mined. Clearly, the economics of a relatively pure phosphoric acid, made from a local rock with a *free* acid, looked very attractive. In 1966, Arad Chemical Industries was formed and built a purified acid plant, based on the IMI process and designed to produce 165,000 tpa P_2O_5. The process is highly corrosive and required the use of plastics (PVC and polyethylene) for materials of construction that were still in a relatively early stage of development. In turn, these materials limited maximum process temperatures to 60°C and affected the morphology of some equipment, for example, the cylindrical shape of the IMI settlers. A block diagram of the Arad process is shown in Figure 2.44.

Given most wet process acid is derived from sulfuric acid, IMI went on to develop other processes: firstly, one where a stream arising from hydrochloric acid dissolution of rock is combined with wet acid and then processed as per the standard route, and secondly, the *closed* route where the calcium chloride is recycled around the plant, a process that could run after start-up without hydrochloric acid [25]. Subsequently, these processes have developed in terms of both the solvent used and the extraction equipment.

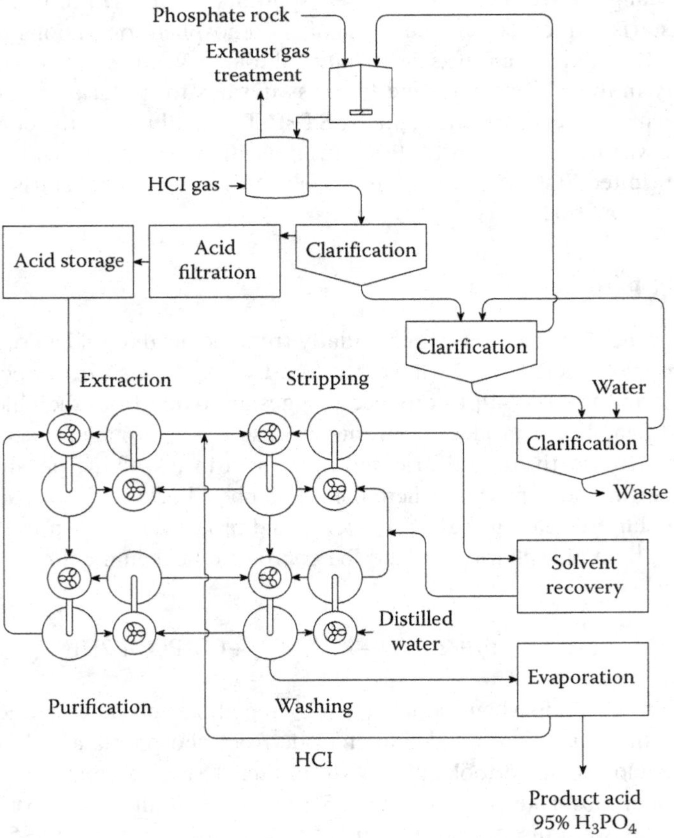

FIGURE 2.44 Arad plant block diagram.

In 1966, IMI announced it had given an exclusive license to the Litwin Corporation of Wichita, Kansas covering the United States and Canada. The license included the process package, engineering, and design, for IMI's *cleaning* process to purify wet process phosphoric acid. In the announcement, IMI gave some details of a 2 tpd pilot plant, including a flow sheet, and analysis of the *clean* and *residual* acids resulting from the solvent extraction of a 53.5% P_2O_5 Florida wet acid. The results were not particularly impressive, the plant being only a single-stage extraction and strip with no scrubbing; nevertheless, the principle was demonstrated, and it was clear that a range of solvents were screened for the process. One distinguishing feature of this process is the use of different solubilities at different temperatures. Figure 2.45 shows the IPE–H_3PO_4–H_2O ternary diagrams at 5°C and 55°C and can be compared to Figure 2.19 that shows the same system at 30°C.

Extraction is carried out at 5°C as it is more effective at that temperature. As the temperature is raised, the three-phase zone moves toward the plait point, and as a result, the extract separates into a solvent-rich (low phosphoric acid content) layer and a heavier more phosphoric acid–rich extract. When the acid is stripped from this solution at the higher temperature, it is at a higher concentration than if it were stripped at 5°C. It is also the case that the volume of extract to be handled at the higher temperature is significantly lower than if the temperature had not been raised; therefore, smaller equipment may be used.

At this time, IMI was working with A&W, and with input from A&W, Litwin Frenkel prepared a scheme for a potential plant at Whitehaven. The selected

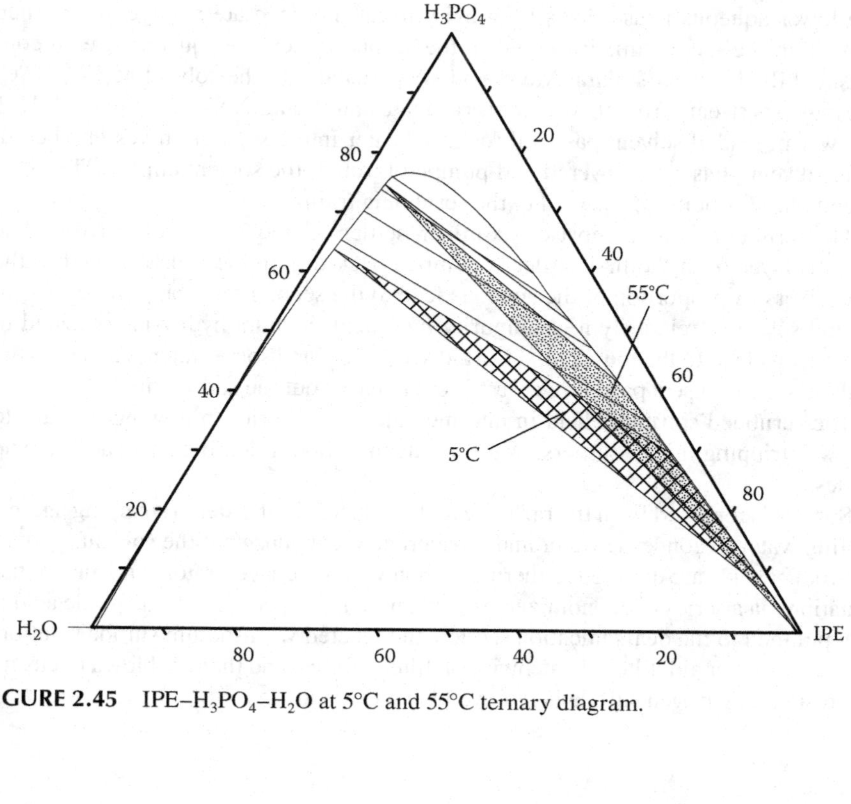

FIGURE 2.45 IPE–H_3PO_4–H_2O at 5°C and 55°C ternary diagram.

solvent was IPE, with extraction taking place at low temperature in a single mixer with two parallel settlers. The loaded solvent was then scrubbed in three mixer–settlers and then stripped in a single mixer–settler at 50°C. Both the raffinate from the extraction settlers and the product acid from the stripping settler were transferred to solvent stripping columns to boil off the IPE that was recycled back to the stripping section. Finally, the product acid was concentrated in a conventional vacuum concentrator.

In 1971, a large purified acid plant was built in Coatzacoalcos, Mexico. This plant is still in operation and is the largest example of the IMI process outside Israel. The plant is rated at 200,000 tpa P_2O_5 and has been the subject of considerable development since the mid-1990s when A&W assumed full ownership.

The IMI Coatzacoalcos process is shown in Figure 2.46. The production site lies on the Gulf of Mexico and has direct access to ship, rail, and road transport. Sulfur and phosphate rock (for many years Moroccan Khourigba K-10 but more recently from other sources) are imported. Sulfur is burnt to sulfuric acid on a Monsanto unit and wet acid made on a twin-stream Prayon dihydrate plant. Steam from the sulfuric acid plant is used to concentrate the wet acid. The PWA plant receives 52% P_2O_5 wet acid into two 180 m^3 day tanks. The wet acid, at about 50°C, is pumped to the first IMI mixer–settlers that are in parallel at 30 m^3/h. The mixers have a volume of 4 m^3 and the settlers, 66 m^3; both are made from polypropylene-lined glass-reinforced plastic. These mixers also receive an aqueous stream from the scrubbing train and a partially loaded cold solvent from the second extraction mixer–settler. As the acid and solvent mix, the extraction reaction raises the liquid temperature to 35°C–40°C. The lower aqueous phase flows by gravity to an intermediate storage and is then pumped to the second mixer–settler. In the second mixer, the aqueous stream contacts a chilled solvent stream. More acid is extracted into the solvent at 5°C–10°C. The aqueous stream from this settler is raffinate and is pumped away to make GTSP. The warm, loaded solvent passes through a heater into a separation vessel where a clean solvent phase is recovered and pumped through the solvent chiller. The acid-loaded phase is pumped forward to the scrubbing train.

The scrubbing train comprises six mixer–settlers, 3 and 27 m^3, respectively. The aqueous layer from the first settler is pumped back to join the wet acid feed to the first mixers. A proportion of the aqueous feed to the second scrubbing mixer is split off and sold as a relatively high-impurity *pure* acid. Sodium hydroxide is added to the aqueous feed to the fourth mixer to aid scrubbing, and some water is added to the sixth mixer to cause a partial release of acid to carry out the scrubbing.

The scrubbed solvent is held in intermediate storage prior to flowing forward to the two stripping mixer–settlers. Water is added to both releasing two purified acid grades.

Solvent is removed from the raffinate and product acids by steam stripping against cooling water to condense water and refrigerant to condense out the solvent.

Product acids are directed to thermosyphon vacuum concentrators and onto defluorination via a series of economizers in a steam stripping column. Food grade acid is then pumped to the dearsenication section and reacted with sodium sulfide solution, mixed with filter aid, filtered batchwise on filter presses, and then air blown to ensure that residual hydrogen sulfide is removed from the acid.

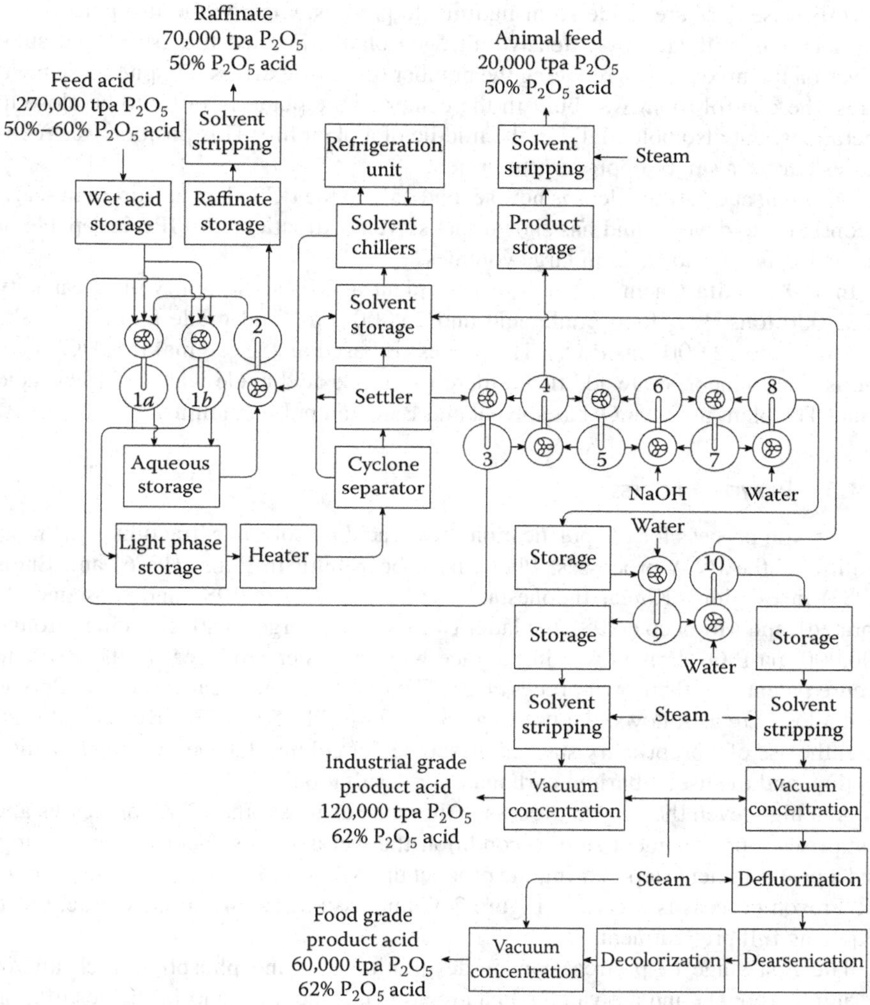

FIGURE 2.46 IMI Coatzacoalcos plant block diagram.

The final steps are decolorization utilizing carbon-filled columns, final concentration, and passage through a guard filter prior to final product storage.

The plant continues to undergo development. Under A&W ownership, two major additions were made to the plant, the dearsenication unit and the decolorization unit, as well as many smaller incremental improvements to increase consistency, output, and product quality. Development continued under Rhodia and now Innophos and also covered safety issues. The original plant was built to the standards of its time, the late 1960s and early 1970s, just before the accidents that led to the much needed development of hazard and operability studies, safer designs, and so on. The solvent, IPE, is flammable; the process requires that the solvent is chilled, which requires a refrigeration plant, which was originally a propane unit. Furthermore, the mixer–settlers are made from flammable plastics. Clearly, the fire potential is high. This is mitigated by extensive nitrogen blanketing and the use of hydraulic drives on the mixers, thus reducing the number of electric drives. Despite these measures, the control room was built in the center of the plant, in order to be close to operations, but also potentially in the middle of a plant fire. This position was rectified as part of a safety improvement project.

The plant at Coatzacoalcos is now second only to the PCS plant at Aurora in scale; it continues to develop and has shown that solvent extraction with IPE is capable of producing food grade acid in large volumes.

In 1993, Haifa Chemicals built a new plant at Mishor Rotem with a capacity of 32,000 tons P_2O_5 food grade acid and 25,000 tons food grade phosphate salts (equivalent to 12,000 tons P_2O_5). The process is based on the original IMI HCl route and is able to incorporate additional merchant grade acid made via the sulfuric acid route. The plant uses AmOH as solvent and Bateman pulse columns.

2.4.5 Prayon Process

The Prayon process for the purification of wet acid by solvent extraction is the most prolific of the PWA processes. Plants have been built in Puurs (1976) and Engis (1983) in Belgium, Korea, Indonesia, Brazil (originally in 1987 and subsequently uprated), and Morocco (1998), the latter two being the largest with capacities around 100,000 tpa P_2O_5. Prayon was in the race with the other producers in the 1970s to secure patents for their own process [52]. The distinguishing features of the Prayon technology are as follows: the use of an 85%–95% IPE, 5%–15% TBP solvent mixture; the use of a proprietary stacked mixer–settler column for the solvent extraction section; and the use of barium carbonate for desulfation.

At a high level, the Prayon process is the same as most other PWA processes and comprises a pretreatment step to condition the wet acid, a solvent extraction step, and a posttreatment step to bring the product up to food grade. The block diagram of the Prayon process is shown in Figure 2.47 and assumes a 54% P_2O_5 wet acid feed requiring full pretreatment.

The first stage of pretreatment is desulfation. Ground phosphate rock and/or barium carbonate and activated silica are added to the feed acid in the desulfation reactor. The calcium in the ground rock or the barium reacts to form a solid sulfate (barium being the less soluble of the two), and silica is added to mop up excess fluoride. The exact amounts of these additives depend upon the impurity levels in the feed acid; however, the goal is to reduce sulfate to around 0.3% and fluoride levels to less than 0.1%. The acid is pumped from the desulfation reactor to the concentration section where the acid is brought up to 62% P_2O_5 in a standard pumped circulation vacuum concentrator. Vapor from the concentrator body enters a separator; the gas phase, including both HF and SiF_4, passes to a fluorine recovery system where the gases are condensed, forming fluosilicic acid that is then exported for sale in the water fluoridation market. The amount of silica added is controlled to ensure the formation of fluosilicic acid. The acid from the concentrator is held in intermediate storage then pumped to filtration on filter presses. The filter cake is pumped back to the wet acid plant and the filtrate goes forward to settling, aided

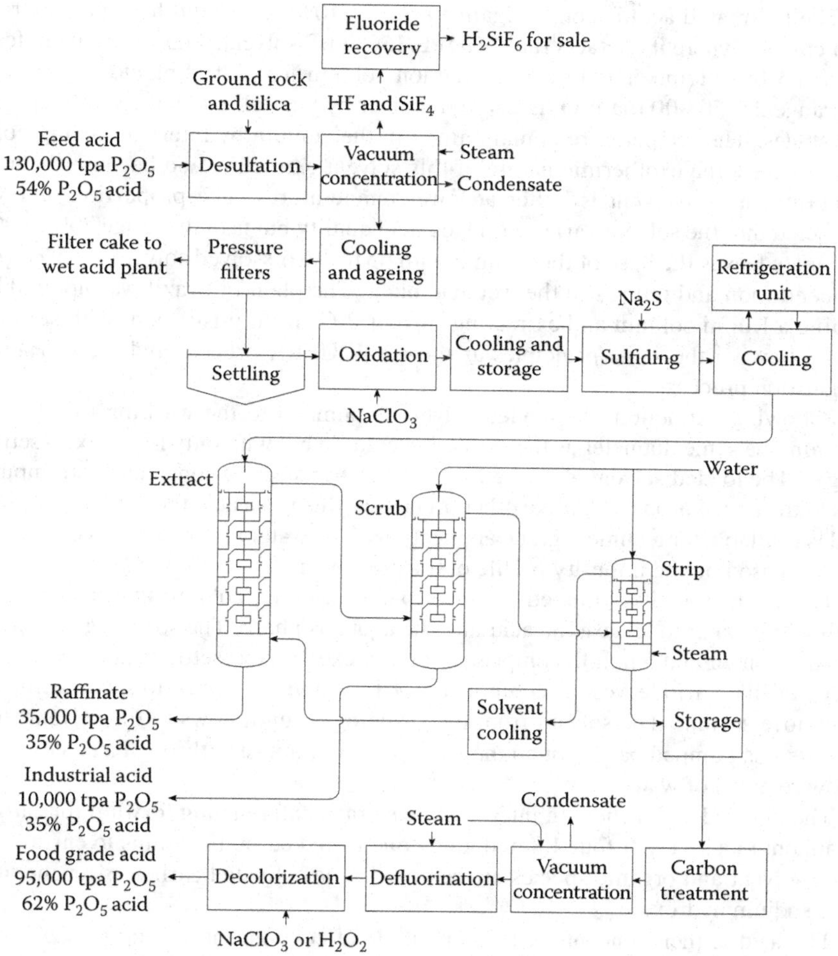

FIGURE 2.47 Prayon plant block diagram.

by a flocculating agent. Following settling, the acid is heated to 150°C with steam in a graphite tube heat exchanger, mixed with 5 kg/ton P_2O_5 sodium chlorate, and allowed to oxidize in the PTFE-lined oxidation reactor. Oxidation destroys organic compounds that both hinder solvent extraction and impart color to the acid; it also alters the oxidation state of some elements and may make the acid more corrosive to some equipment. The hot, oxidized acid passes out of the reactor and is flash cooled and is transferred to the dearsenication reactor that is also fed with sodium sulfide. The sulfide reacts with arsenic that is now at +5 oxidation state and therefore does not precipitate. Unlike other dearsenication/sulfiding processes, the plant in Brazil does not have either carbon addition or filtration; the acid is pumped forward to the solvent extraction section. In Morocco, sulfiding is carried out as part of the processing of wet acid prior to feeding the PWA plant.

The pretreated acid is cooled against brine to 10°C–15°C and fed to the extraction column where it contacts the 85% IPE/15% TBP solvent. The acid/solvent feed ratio is 5 by volume. The Prayon extraction column for overall plant capacities in the range 25–50,000 tpa P_2O_5 is 2.4 m diameter by 8 m tall and has five mixer–settler stages. The temperature is maintained in the column by internal cooling coils to counteract the exothermic nature of this solvent extraction and because the acid solubility in the solvent is better at lower temperatures. The proportion of P_2O_5 extracted into the solvent varies with feed acid quality but is in the range 70%–80%. Raffinate leaves the base of the column with up to 1% dissolved solvent at 35% P_2O_5 concentration and returns to the wet acid plant. The plant in Brazil was uprated by adding a Kühni column and extracting further P_2O_5 from this stream with sulfuric acid following the same principles as the A&W UFEX process and the R–P total exhaustion process.

Following extraction, the loaded solvent is pumped to the washing (scrubbing) column, the same diameter as the extraction column but with only four mixer–settler stages. The loaded solvent is washed with water, transferring some acid and impurities to the aqueous phase. Unlike other processes, this wash acid then leaves the plant and is available for technical grade applications. The water addition rate is controlled and set based on the impurity profile of the purified acid.

The washed acid is pumped forward to the stripping column that is contacted with warm water to release the acid into the aqueous phase. The stripping column is 1.5 m diameter and 5 m tall, comprises three stacked mixer–settlers, and operates at 50°C. Purified acid leaves the column at 45% P_2O_5 and is directed to steam stripping to remove solvent. The solvent from the stripping column flows to the solvent storage. As it is pumped back around the solvent circuit, it passes through a coalescer to allow removal of water.

The purified acid is held in intermediate storage before going forward for carbon treatment in a series of four 1.8 m diameter activated carbon columns to ensure any trace solvent and organic species are removed. The activated carbon is regenerated with sodium hydroxide.

The acid is then concentrated in a multiple effect concentrator up to 62%–63% P_2O_5. At this stage, the acid contains approximately 300 ppm F. Acid intended for food grade is then sent for defluorination. The original Prayon design was for batch defluorination; however, in Brazil and Morocco, this is done on a continuous basis with steam in a PTFE-lined column.

The final stage of treatment is decolorization that is carried out either with $NaClO_3$ at 170°C or with hydrogen peroxide.

If the sulfate level is too high, another posttreatment step is included where barium hydroxide is added to precipitate barium sulfate.

The Prayon process has proved to be flexible. The plants built in Indonesia, Korea, and Brazil were very similar to the first full-scale plant in Puurs, Belgium (which was initially 40,000 tpa P_2O_5, compared to 25,000 tpa P_2O_5 for the other three). Brazil in particular has undergone many developments, including the conversion from batch to continuous defluorination, extraction from the raffinate stream, and the use of barium in desulfation. Furthermore, the Brazilian plant has demonstrated a flexibility to deal with quite significant changes in feed acid impurity levels. The plant was initially commissioned with Cajati acid, a relatively pure wet acid from an igneous rock but has also managed with Goiásfertil acid, also from an igneous rock but with high iron levels. The plant also utilizes acid from Moroccan Yousouffia and Ben Guerir sedimentary rock with different levels of organics and heavy metals. A further strength of the Brazilian operation is the integrated pilot plant that allows the technical team to evaluate and plan the introduction of different acids.

2.4.6 Rhône–Poulenc Process

The R–P process is so named to cover the purification process, based on 100% TBP solvent, which arose from French chemical producers that amalgamated in the 1960s to form the R–P group. Relevant patents, with consistent authors, are assigned to different companies [53–55] that ended up under the R–P umbrella. The thinking behind the original decision to choose TBP is not known to the author but likely arises from the French nuclear industry. The R–P process was piloted at the Les Roches de Condrieu site with mixer–settlers and was first implemented at industrial scale at the Rouen site. This plant is no longer running but is noteworthy as the only example of a purification plant that produced purified sodium orthophosphate liquor that was pumped across to the adjacent STPP plant for direct conversion to STPP. Although this was a single-product plant, it was arguably the most efficient when compared to other plants that produce a concentrated acid that is then neutralized with a sodium source to form the sodium orthophosphate for STPP production. R–P also built and operated a rare earth recovery plant based on hundreds of mixer–settlers in La Rochelle, France. The second implementation of the R–P process was built in Geismar in 1991 and is still in operation with a nameplate capacity of 95,000 tpa P_2O_5.

Figure 2.48 is a block flow diagram of the integrated PWA/STPP plant at Rouen. Fifty-three percent P_2O_5 wet acid was received from the R–P wet acid plant in Rieme, Belgium. In the final years before it was shutdown, this acid was made from different igneous rocks such as Kola and Kovdor that produced good-quality acid, low in organics but high in titanium. The acid was exported having undergone a full pretreatment at Rieme; however, cooling during transportation led to precipitation of solids that were filtered before entering the extraction column. This was a 36-compartment, 2 m diameter Kühni column equivalent to 10 mixer–settler stages. Fresh wet acid feed was 8 m^3/h, and TBP solvent feed, having passed through a cooler in order to maintain the column temperature in the range 50°C–55°C, was 100 m^3/h. A small flow of 98% sulfuric acid was introduced to the column at a point equivalent to the fifth mixer–settler to improve extraction; as a result, the raffinate leaving the base of the column carried only 1%–2.5% of the P_2O_5 from the feed acid and flowed at 0.5 m^3/h. This stream was mixed with kerosene to extract the TBP from the raffinate. The TBP/kerosene mixture was then mixed with sulfuric acid to re-extract the TBP from the kerosene, and the TBP returned to the top of the column. The kerosene was recirculated. The raffinate now had a TBP content less than 25 ppb and was sent to the sulfo-phosphate d'ammonium (SPA) plant where it was reacted with ammonia and granulated to produce an ammonia-based fertilizer.

Loaded solvent passed from the top of the column at 95 m^3/h to the first of four mixer–settlers that comprised the scrubbing section (washing). A small amount of

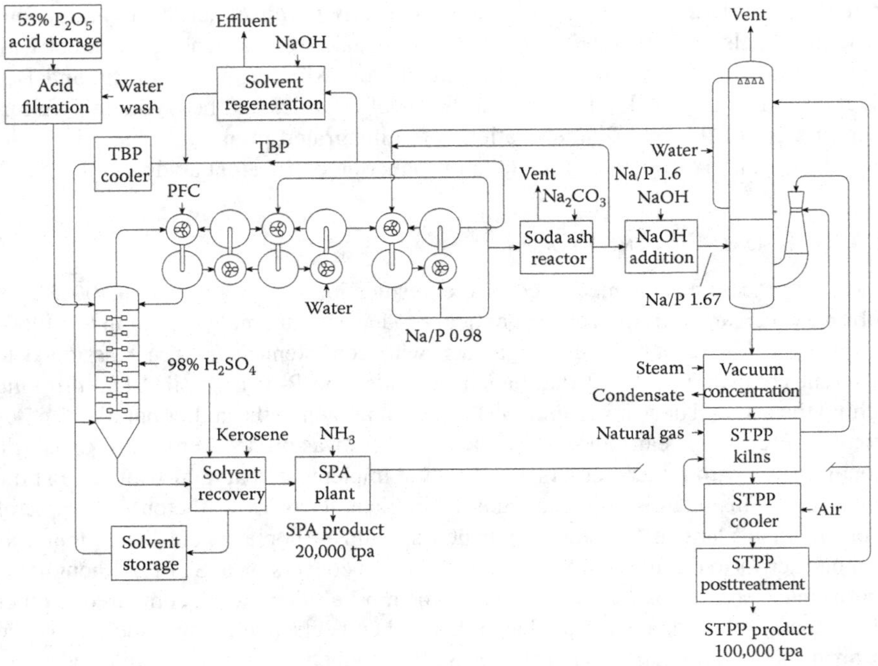

FIGURE 2.48 R–P Rouen plant block diagram.

pulverized fine carbon (PFC) was added to the first mixer–settler to aid coalescence. The aqueous stream from the first mixer–settler passes to the top end of the column at 5 m^3/h. A small amount of the monosodium phosphate (MSP) liquor circulating around the stripping mixer–settlers was added to the third scrubbing mixer–settler to aid scrubbing, and water was added to the fourth. The solvent stream was then pumped to the two stripping mixer–settlers where the phosphoric acid was extracted from the TBP with MSP liquor. The Na/P ratio of the liquor leaving this section was 0.98; some was recycled and combined with an enriched stream from the soda ash reactor. Fine Na/P control was achieved by adding NaOH to the circuit. The stripped TBP was recirculated back to the base of the extraction column; approximately 5% of this flow was continuously regenerated with NaOH as a side stream in a regeneration vessel. The NaOH both dewatered the solvent and allowed precipitation of unwanted impurities that built up in the TBP.

The MSP liquor, around 40% by weight, was transferred to the soda ash reactor where the liquor Na/P ratio was increased to 1.6 converting it into a mono-, disodium (ortho)phosphate; some of this stream was pumped to the first of the stripping mixer–settlers, and the rest was pumped forward to the gas scrubber. Just prior to entry, the liquor Na/P ratio was increased to 1.66, the level required ultimately to make STPP.

The lower section of the scrubbing column received hot, STPP dust-laden vapor from the first STPP kiln, at about 150°C. Cooling the vapor stream preconcentrated the ortho liquor. This ortho liquor was then transferred to a steam-heated concentrator and then pumped to the cocurrent fired STPP kiln. Cool dust-laden air from equipment downstream of the STPP kilns, together with vents from the SPA plant, were conveyed to the upper section of the gas scrubber and contacted with weak ortho liquor. This liquor was circulated around the upper scrubber section and to the first stripping mixer–settler: liquor concentration was controlled with water addition.

The plant was capable of producing 100,000 tpa STPP.

Built in 1991, the plant at Geismar had a nameplate capacity of 70,000 tpa P_2O_5, all of which was originally produced as 75% H_3PO_4 food grade; over time output was increased to 100,000 tpa P_2O_5. The plant was designed with 8000 h on stream time. The technology built on the experience at Rouen and early development work. Much of the minor plant and equipment is identical to that installed at Aurora and Idaho PWA plants. Similarly, the engineering, procurement, and construction were carried out by the same company, Jacobs Engineering. The initial commissioning of the plant was troublesome and lengthy with many corrosion problems; however, with time, these and others were worked through. Once fully commissioned, the plant ran well although for many years was slightly handicapped by the concentration limit on its final product.

Figure 2.49 is a block flow diagram of the plant at Geismar. The plant receives a 54% P_2O_5 wet acid via a short pipeline from its immediate neighbor, currently owned by PCS. Generally the wet acid is made from Moroccan Bu Craa rock with its typical attendant impurities; however, other rocks are feasible. The Geismar plant returns the raffinate to the wet acid plant and pays only for the P_2O_5 it exports itself as purified acid.

The wet acid is received into an agitated day tank to which NaOH is added to precipitate sodium silicofluoride that is then filtered off. The acid is then pumped to another intermediate storage, via an inline mixer. NaHS solution is also pumped to the mixer so that the sulfiding reaction takes place under pressure, precipitating arsenic and molybdenum. Air is blown through the acid in the storage tank to strip free H_2S that had not already flashed off as the acid entered the tank. The tank vapors are scrubbed with NaHS solution. The acid is then filtered using cellulose as a precoat. After passing through a guard filter, the acid is held in intermediate storage prior to decolorization. Decolorization takes place in activated carbon-filled columns (2.6 m diameter by 4.5 m high), two in series at any one time, with one regenerating with NaOH and water. The acid is preheated in a steam-heated graphite tube exchanger to 80°C prior to passing through the columns where the organic levels are reduced to less than 100 ppm.

The decolorized acid is pumped to the upper level of the 2.6 m diameter, 36-compartment Kühni extraction column at a rate of 10 m^3/h at 50°C (70 ktpa P_2O_5 rate). TBP, cooled to maintain the column temperature in the range 50°C–55°C, enters the lower part of the column through an internal distributor. TBP enters the column at 160 m^3/h and flows up through the column, countercurrent to the acid, dispersing in the aqueous stream. Acid transfers into the solvent stream that leaves the upper section of the column at 170 m^3/h. The solvent stream flows by gravity to the first of ten mixer–settlers that comprise the scrubbing section. The aqueous stream from the first mixer–settler flows by gravity to an intermediate storage and is then pumped at 10.2 m^3/h to the upper section of the column as the reflux stream. 98% H_2SO_4 enters the middle of the column at 1.15 m^3/h, and the raffinate stream leaves the base of the column at 12.5 m^3/h. The raffinate stream undergoes the same treatment process described for the Rouen plant, where kerosene and sulfuric acid strip out the TBP

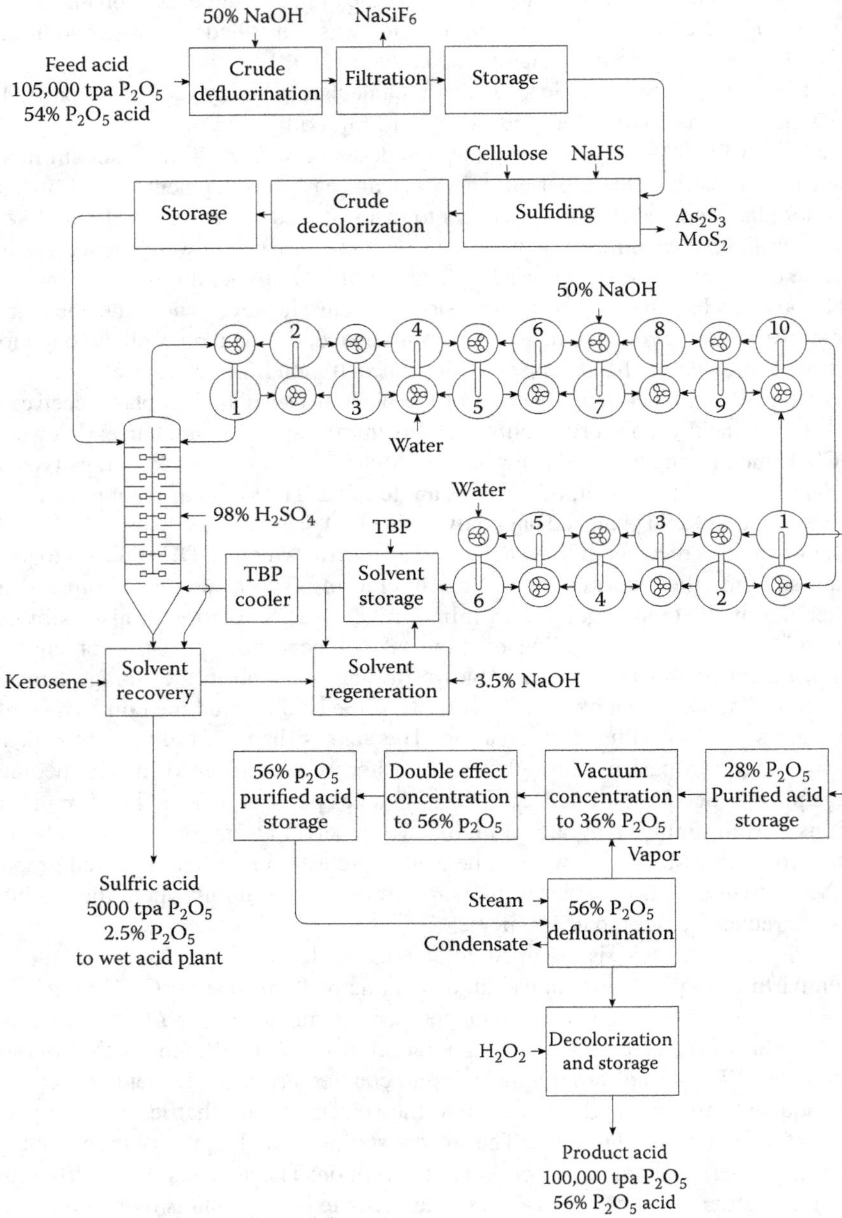

FIGURE 2.49 R–P Geismar plant block diagram.

from the acid. Once treated, the raffinate stream, comprising mostly sulfuric acid with 2.5% P_2O_5, is pumped back to the wet acid plant. The TBP flow to the column is limited by hydrodynamic considerations; the decolorized acid feed rate is controlled based on the TBP flow rate and the desired acid extraction. The sulfuric acid rate follows the acid feed rate. The lower interface level between solvent and aqueous layers is controlled by the raffinate flow out of the column. The level at the top of the column is controlled by the amount and composition of reflux returning to the column from the reflux storage. The reflux stream introduces water (along with impurities from the scrubbing section) to the column, lowering the acid concentration and therefore the amount of acid extracted into the solvent and must, therefore, be limited.

In the scrubbing section, impurities in the solvent stream are scrubbed with the countercurrent flow of what is originally pure acid introduced to the tenth mixer. There are no external pumps in the scrubbing section; flow is by gravity. The mixer–settlers are Krebs (Technip) design, with cylindrical settlers 2.5 m diameter and conceptually similar to IMI designs. The mixers are 2 m diameter and capable of pumping 340 m^3/h. The operating temperature in the scrubbing section is in the range 60°C–65°C. The aqueous/organic volume ratio is controlled in the range 1.1–1.2. Scrubbing is enhanced by the use of 50% NaOH that is added in the seventh mixer; the addition rate is cascaded from the purified acid flow rate to the 10th mixer.

Scrubbed solvent flows to the stripping section that comprises six mixer–settlers. Purified water is added to the sixth mixer at 28 m^3/h that extracts all the acid into the aqueous phase. The P_2O_5 level in the solvent is normally less than 1000 ppm. The amount of water addition governs the proportion of acid extracted from the TBP, and with sufficient water, this extraction is substantially complete; however, it decreases the acid concentration and in turn increases the amount of steam required to concentrate the acid to sales specifications. The 28% P_2O_5 purified acid flows by gravity from the aqueous line from the first stripping settler to intermediate storage. The solvent flows from the last settler to the solvent storage 160 m^3/h; 4% of this flow is side streamed to the solvent regeneration unit where the solvent is contacted with 3.5% NaOH solution in two mixer–settlers. Impurities resulting from solvent breakdown and those in the feed acid that build up in the solvent are extracted into the caustic solution.

Prior to purified acid concentration, solvent carried over with the acid is separated firstly in the intermediate storage and secondly in coalescers. The decanted acid is pumped to the first concentration stage where it is brought up to 36% P_2O_5 in a thermosyphon concentrator under vacuum. The 36% concentrator uses steam from the defluorination column; vacuum is supplied by a liquid ring pump. The 36% P_2O_5 acid is then concentrated to 56% P_2O_5 in a double-effect thermosyphon concentration unit with 7 bar saturated steam. Vacuum is provided by a liquid ring pump. Concentrated acid is cooled to 95°C in an economizer and held in intermediate storage.

The acid is then pumped via economizers to the top of the defluorination column. The column was designed and supplied by Vicarb with PTFE lining and graphite plates is 1.1 m diameter and operates at 169°C. The fluoride level in the acid is reduced from around 500 ppm to less than 5 ppm in this process step. Defluorinated acid drains from the column via an economizer to an intermediate storage where the acid is held at 130°C. Hydrogen peroxide is sparged through the acid to remove color. The acid passes through another economizer to final product storage.

The plant is located at 30°13′07.02″ N and 91°03′09.1″ W and easily viewed on a well-known search engine. The plant is compact, and the core process comprises a low structure, which houses the mixer–settlers, and a tall structure housing most of the other equipment. The single Kühni column is clearly visible, located in the left corner of the tall structure.

2.4.7 Other Processes Including Bateman (Wengfu) and Prado (AFB, Turkey)

A thorough patent search of phosphoric acid purification processes reveals that many companies have at least considered this field to the extent of developing a process in the laboratory and successfully seeking a patent. Table 2.6 gives examples of work in the 1960s and 1970s. Several have gone on to build pilot and market development–scale plants. A few have built just one industrial-scale plant.

Early examples included the Monsanto processes [56–58] based on different solvents and producing a purified MSP liquor; Pennzoil [59] used methanol as solvent and distilled this out with potassium salts to aid fluorine removal; APC (Azote et Produits Chimiques, S.A.) proposed a process [60] starting with the acidulation of phosphate rock with nitric acid with the goal of producing ammonia fertilizers and industrial phosphates; APC also developed the *Phorex* process [62], utilizing *n*-butyl or *n*-amyl alcohol that was demonstrated by Luwa, Switzerland, on asymmetric rotating disk (ARD) columns [61,62]; and a Romanian process also based on *n*-butanol was developed by the Romanian Engineering Company for Chemical Industry (IPROCHIM) and Chemical Research Institute (ICECHIM) in the 1980s; another claimed the use of a range of solvents on acid resulting from the action of nitric acid on phosphate rock.

Hoechst of Germany and Toyo Soda of Japan were both granted a number of patents in this field and built at least one full-scale industrial plant.

The Bateman Litwin group has brought together a number of engineering companies with skills in a range of industries over a number of years including solvent extraction for the mining industry. The Litwin company provided engineering services to IMI in the 1960s. Bateman has worked with Haifa Chemicals in Israel supplying two pulse columns for its phosphoric acid purification plant. Bateman was successful in winning the contract to supply a PWA plant to Guizhou Hongfu Industry and Commerce Development Corporation Ltd. The plant has a nameplate of 100,000 tpa P_2O_5 and follows a classic format of pretreatment, solvent extraction, and posttreatment. Pretreatment consists of desulfation and crude defluorination with phosphate rock and silica in the form of clay and organic reduction also with *clay*. The solvent extraction process takes place in Bateman pulse columns and mixer–settlers (IMI design). Posttreatment includes decolorization with hydrogen peroxide, concentration, and defluorination.

Both Prado [63] and MEAB [64] built plants in Turkey. The Prado plant is based on mixer–settlers, designed by Prado similar to the Davy concept used by A&W, with TBP/kerosene as solvent, similar to the FMC process. The plant has a reported capacity of 18,000 tpa 54% P_2O_5 food grade acid, most of which is directed to STPP production for the client A.B. Foods of Bandirma, Turkey.

2.5 SOLVENT EXTRACTION EQUIPMENT

There is a wide range of equipment to carry out solvent extraction. In general, all the PWA solvent extraction processes used mixer–settlers on the early plants. Once a candidate solvent is tested in a separating funnel, the next obvious step is to use laboratory mixer–settlers. In turn, their scale-up is reasonably well understood. Furthermore, at full scale, a train of mixer–settlers is easier to modify than an extraction column—it is a trivial matter to take a stream from a different outlet pipe from a settler, whereas a column requires major modification. On the other hand, the column suppliers usually offer laboratory- and pilot-scale columns, testing facilities, and scale-up parameters.

Mixer–settlers used on phosphoric acid purification processes fall into two categories: the box-type mixer–settler and the cylindrical mixer–settler. The former was used by A&W and more recently the PWA plant in Turkey, and the latter on both the IMI and R–P processes. The separate mixer–settler for both the A&W and IMI processes was subsequently developed to the *compact mixer–settler* that combines the operation in one unit.

There are several different column types used in the field of solvent extraction and most have been tested by one company or another on phosphoric acid; however, only three types have been used. The Prayon column was specified on all their plants although the most recent expansion in Brazil used a Kühni column. Bateman has sold its pulse column to the Haifa Group and installed it on the Wengfu plant in conjunction with IMI mixer–settlers. The Kühni column is in operation at both Aurora and Geismar and was used at Huelva, Idaho, Rouen, and Whitehaven and on the Phorex APC and Hoechst processes. Kühni may therefore claim the widest application.

Specifying a new plant now, and assuming no exceptional local conditions such as high import taxes, the logical choice would be to choose columns (having carried out laboratory and pilot trials). Once the plant requires more than three or four mixer–settlers, which is the case for most plants, the installed capital cost of a column is cheaper than the equivalent number of mixer–settlers. Furthermore, the electricity demand is lower as there is generally only one drive for a column, the solvent inventory is lower, and the leak paths via mechanical seals are less. If a full process license is purchased, the choice may be limited: Prayon specifies their column with their process and usually requires a stake in a joint venture; Bateman has so far specified their column for their process. With a free choice, arguably the Kühni column has the broadest résumé.

There are many books and articles describing solvent extraction equipment so only a summary of the main points of the equipment and their development for PWA processes is given in the succeeding text.

2.5.1 DAVY POWERGAS MIXER–SETTLER

Davy Powergas (which became Davy McKee in 1980) originally developed their mixer–settler [13] for the mining industry that requires volumetric throughputs two orders of magnitude greater than PWA plants. The box shape was chosen as this was seen as the easiest to construct in remote locations. When A&W started to investigate solvent extraction, they engaged Professor Carl Hanson of Bradford University as a consultant who suggested A&W should consider working with Davy as their technical center was reasonably close to Whitehaven. Davy proposed a pilot mixer–settler made from polypropylene that was used to develop the full-scale mixer–settlers used at Whitehaven.

Figure 2.26 shows a drawing of one of the mixer–settlers used at Whitehaven. The mixer section is rectangular and requires no baffles for mixing. The solvent and aqueous streams are drawn up into the eye of the mixing impellor through a draft tube. Davy tested many different impellor types but chose the backward swept, shrouded, radial flow impellor shown. The impellor is designed to mix and to pump—no external pumps are required in the train. In this context, mixing means the creation of small enough droplets, and their dispersion, to maximize mass transfer without creating droplets so small that stable emulsions are formed. The mixed fluid passes up through an *egg box* baffle that separates the agitated zone from the settling zone. The fluid then passes through a feed slot the full width of the settler and through a *picket fence* that distributes the fluid flow. In the settler, the phases start to coalesce. Uncoalesced material lies in a narrowing dispersion band that is terminated by a Hanson baffle. The lighter organic phase flows over an internal weir and through an external standpipe pot; both are adjustable; the latter sets the interface level. The heavy aqueous phase does likewise.

The mixer–settler is simple to operate and control. Weir and standpipe levels are set during commissioning. In normal operation, the organic/aqueous ratio is controlled by flow meters and adjusted depending on the impurity levels in the organic stream (Fe and SO_4 are often markers). The impellor speed is usually adjustable through a variable speed drive that is optimized during commissioning to cater for the densities of the two phases.

2.5.2 IMI Mixer–Settler

The early IMI mixer [13,65–67] differs from the Davy mixer in that the aqueous and organic components are mixed prior to pumping. The pumping is completely axial. Early IMI mixers and settlers were cylindrical primarily for ease of construction with corrosion-resistant plastics; these are the type used on the plant at Coatzacoalcos. Figure 2.50 shows this type of IMI mixer–settler. Each flow enters the mixer separately and is directed to the bottom of the mixer. Here, an agitator ensures both components are thoroughly mixed; a variable speed drive ensures the correct speed to optimize droplet formation. The mixed liquor then enters the draft tube and passes through the vanes of the inlet stator that sets up the flow direction for the axial flow impellor that is keyed on the same shaft as the agitator. The impellor provides sufficient head such that no external pumps are required. The liquor then passes through the second stator that straightens flow. The liquor continues up the widening draft tube to the deck where the flow slows and overflows into the settler. The liquor passes down a central distribution pipe and onto inclined corrugated sheets that aid separation. The lighter, usually solvent, phase passes over a fixed weir and out of the settler. The heavy phase flows out through an external interface control.

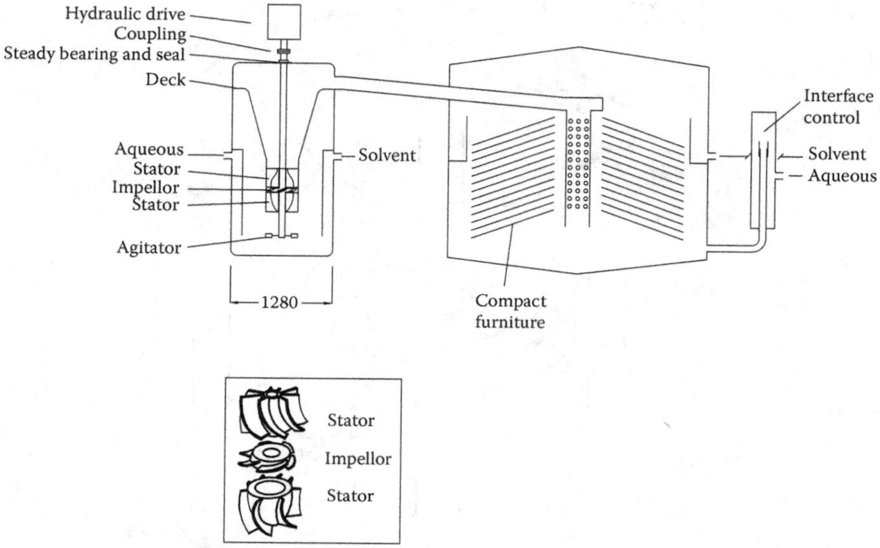

FIGURE 2.50 IMI mixer–settler.

IMI developed a turbine pump-mix mixer with some broad similarities to the Davy mixer, as well as a compact mixer–settler that combines both operations in one unit. Bateman now uses the IMI designs when supplying mixer–settlers for various solvent extraction processes.

The Krebs mixer [68] is very similar to the early IMI mixer. The patent and most documentation show a box-type settler; however, the cylindrical settlers were specified for the plants at Rouen and Geismar.

2.5.3 Kühni Column

2.5.3.1 Introduction

The Kühni column [13,42,69] has been the subject of considerable academic study and modeling [70–72], all with the objective of predicting the behavior of the full-scale unit from the results of a laboratory-scale trial. Getting the full-scale design right is critical because modification is difficult, time-consuming, and expensive. Once a column is commissioned, the main operating variables are acid and solvent feed rates and ratio, and impellor speed; however, for a given column, their range is not wide.

The Kühni column comprises a column of agitated compartments as shown in Figure 2.51. Double-entry radial flow impellors, one per compartment, are mounted on a central shaft. Stator plates, with circular holes for axial flow through the column, form the top and bottom of the compartment. Column diameters on PWA plants have ranged from 1.6 to 2.8 m with 48–72 compartments. The impellor diameter is typically 30%–70% of the column diameter and varies depending upon its location because of density changes throughout the column as acid is extracted into the solvent or stripped from it. Impellor speeds are 50–250 rpm for

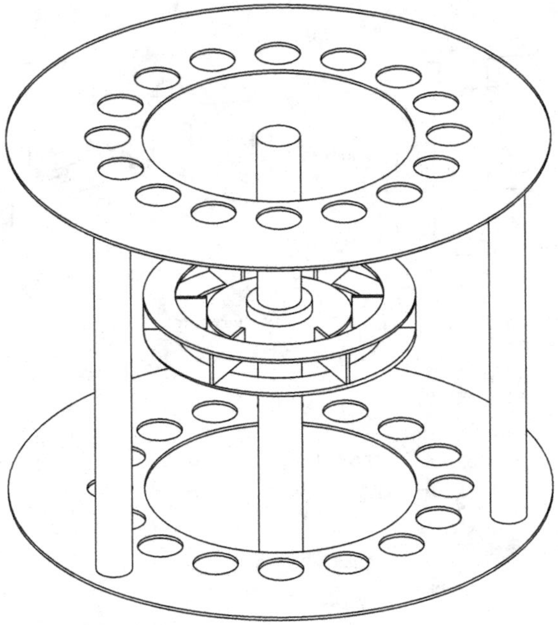

FIGURE 2.51 Kühni column internal compartment.

laboratory and pilot columns and 5–25 rpm for the full-scale unit. The free area of the stator plates is fixed by the number and diameter of holes and varies depending on the location within the column. Free area is typically 30%–45% of the column cross section. There is a disengagement space above and below the compartments of height approximately 1.5 times the column diameter; in some installations, these sections are wider than the main part of the column. A liquid distributor is located just above the top and just below the bottom compartment and comprises a hexagonal ring of drilled pipe. The main feeds come into the column through these distributors; other feeds are introduced at different locations. Normally the heavy aqueous stream flows out of the column through the bottom outlet and the light organic stream out of the top side of the column through an internal stand pipe. Occasionally, a side stream is taken from the column that requires a special separation section in place of one or two agitated compartments: A&W designed a novel device for their pilot column at Whitehaven (some referred to it as a *top hat* separator).

The column base is either dished or coned depending on the likelihood of solids.

2.5.3.2 Scale-Up

Laboratory- and pilot-scale trials usually commence with basic shake-up data as a starting point. The process is already proven chemically with separating funnels; thus, a sensible range of organic/aqueous ratios is known, the continuous phase chosen (usually the solvent), and an idea of mixer–settler stages known to achieve extraction, scrubbing, and stripping. The objectives of the pilot column program are to demonstrate the following:

a. Hydrodynamic performance, column internals configuration, and the height of mixer–settler equivalents (HTU—height transfer unit)
b. Effective separation of phases in the settling regions
c. Critical operating parameters such as temperature, concentrations, and organic impurities
d. Longer-term effects such as scaling and corrosion

Initial flow rates and column configuration are established from experience and then optimized.

Hydrodynamic similarity between the pilot and full-scale columns implies constant droplet diameter, d_m, which is linked most strongly to the following:

a. Impellor tip speed
b. Holdup, x_d; specific throughput of the continuous phase, V_c
c. Specific radial discharge flow rate Q_r/DH_c, where Q_r is volumetric radial flow rate, D is column diameter, and H_c is compartment height

Holdup is related to mass transfer, which in turn is related to droplet size and velocity through a compartment. In this case, holdup is the amount of acid in the solvent and is measurable by pressure difference as the solvent density is essentially constant throughout the column. The impellors are all mounted on the same shaft and have the same speed; therefore, droplet size is optimized by changing the impellor diameter. With droplet size optimized (1–2 mm), mass transfer and therefore holdup are optimized by fitting stator plates with different free areas.

With the pilot column optimized for droplet size and holdup, comparison of the chemical analysis of the aqueous and organic streams with either the separating funnel or laboratory mixer–settler data allows the designer to calculate the observed HTU. HTU actual is made up of HTU observed plus HDU, the height of an eddy diffusion unit that is a measure of inefficiency due to axial flows or back mixing. At pilot scale, HDU is around 30% of HTU, whereas at diameters above 2.5 m, HDU is considerably higher, which is why a three-shaft column is offered for diameters above 3.0 m. Kühni has established compartment height to column diameter ratios and column diameters for the desired volumetric throughput.

2.5.3.3 Process Control

There are differing degrees of sophistication practiced in the control of Kühni columns, and there is an irreducible minimum. Figure 2.52 shows a process flow sheet of a Kühni column with its instrumentation, carrying out the complete purification process. This is theoretically possible but practically unfeasible because the column would be very tall.

It is assumed that the instrumentation goes back to a plant computer that is not strictly essential but is always used. The hexagonal symbols indicate signals going

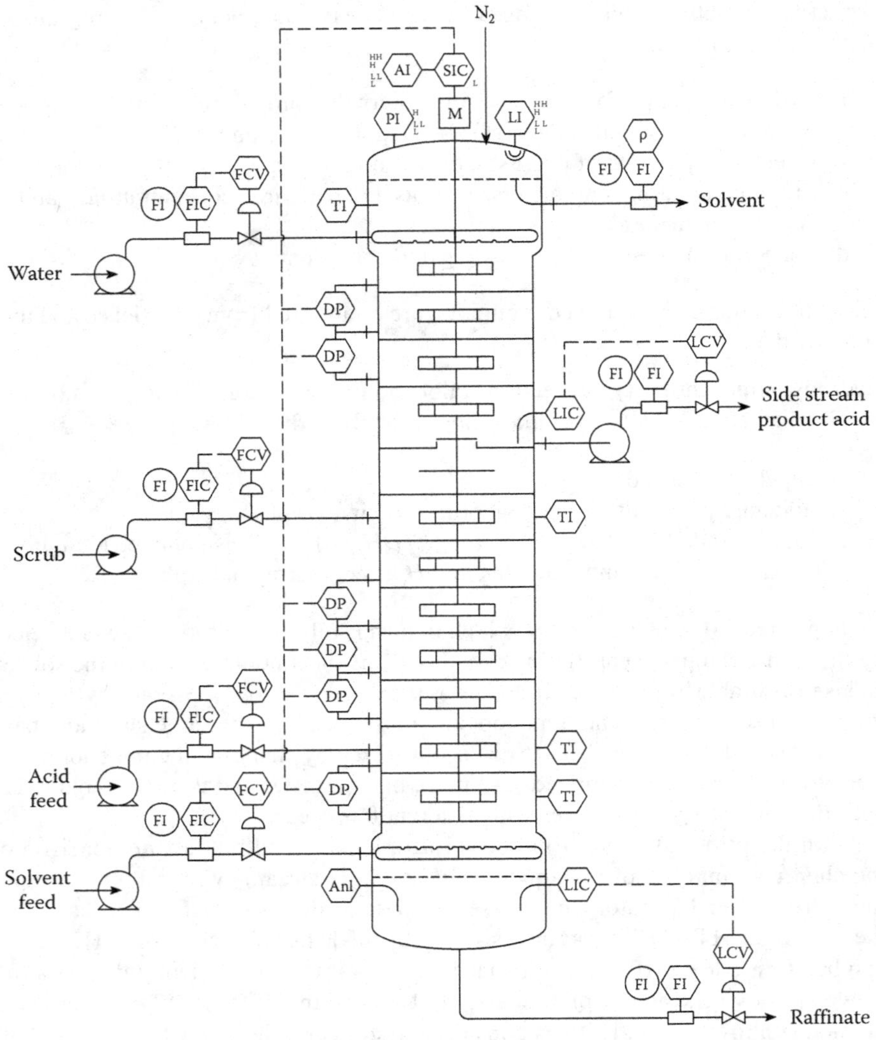

FIGURE 2.52 Kühni flow sheet diagram.

back to the computer and displayed there; circles represent local field display on the plant. All the feeds to the column are under flow control. Accurate flow measurement is essential for column control so the best available instrumentation is specified. The flow instrumentation of the water and solvent feeds is not onerous; however, acid streams are more challenging due to density variation with temperature, concentration, and composition. Mass flow measurement avoids this difficulty, and often Coriolis mass flow meters are specified that also calculate density and volumetric flow. The acid–solvent feed ratio is usually automated.

Temperature readings are taken at different points in the column. Dissolution is exothermic and solubility temperature dependent; therefore, the feed temperature of the solvent is controlled to optimize the column temperature.

The level of the settling zone in the boot (base) of the column is controlled as is the level in a side stream separating section. Flow is also measured and recorded. As shown, the flows and levels are controlled with a control valve; an alternative practice is to use speed control on the pump motor altered by the flow or level signal. At the top of the column, both pressure and level are measured and software high- and low-level alarms specified. The impellor shaft is driven by a motor with a variable speed drive. The speed is set according to calculated holdup that is derived by taking differential pressure readings at several points up the column. The motor current is also monitored. In normal operation, holdup is around 30% of the flooding condition; however, blockages are not unknown leading to local flooding, and the differential pressure sensors are helpful in locating the problem.

2.5.4 Bateman Pulsed Column [13]

The pulsed column was first invented in 1935 [74] and has been the subject of considerable development and academic study [75]. Columns diameters are in the range 0.5–3.0 m and up to 35 m tall. Figure 2.53 shows the compartments of a pulsed column. The operation is straightforward in concept; the feed acid is introduced into the upper decanter of the column and the solvent into the lower

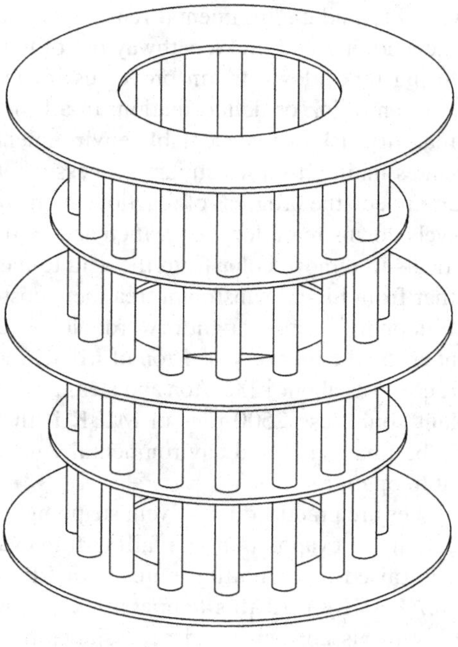

FIGURE 2.53 Pulse column internals.

decanter. A suitable gas or mechanical means create a pulse in the pulse leg that is located adjacent to the column. The pulse leg is connected to the lower decanter and the pulse is translated to the liquid column. As the liquid column moves up and down, the two phases are intimately contacted. The solvent phase rises and extracts acid. Holdup is measured by pressure difference and feeds back to the pulse control analogous to the speed control on the Kühni shaft. The interface between aqueous and solvent is controlled in the lower decanter by pressure difference.

The claimed benefits of the pulsed column include low axial mixing, uniform distribution of the dispersed phase over the cross-sectional area, and economical process. Scale-up is considered predictable.

2.6 POSTTREATMENT

2.6.1 Solvent Stripping

The first operation following solvent extraction is always solvent stripping if the solvent solubility is any more than a few hundred ppm in the acid. For processes that use either MIBK or IPE, the product and raffinate streams from the solvent extraction section contain 1%–2.5% solvent. TBP solubility is two orders of magnitude lower at around 0.03% and is treated differently. Both points are clear by inspection of the tertiary diagrams of Figures 2.19 through 2.21. The stripping operation objective is to reduce the solvent content to less than 30 ppm in the acid stream.

The solvent is stripped for safety, environmental, productivity, product quality, and cost reasons. The safety and environmental reasons are similar: If the solvent is not stripped from the acid, it could find a pathway out of the plant while in intermediate storage or during other downstream processes involving raising the acid temperature such as concentration or defluorination; in all these cases, the solvent could present a flammability risk or unacceptable environmental impact. The productivity challenge relates mainly to downstream processes for the raffinate where high organic levels can impede the progress of reactions in the manufacture of fertilizers. High organic levels downstream for the purified acid will necessitate high levels of treatment chemicals to remove color from the acid to meet specification. The cost reasons come either from high downstream treatment cost or increased solvent cost due to losses. The economics are straightforward: Not accounting for increased downstream treatment costs, the recovery of 1 ton of MIBK requires 1.5 tons of LP steam. In 2012, MIBK cost was about \$2300/ton and steam cost \$20/ton. Over a year, a 100,000 te P_2O_5 plant could lose 2500 tons of MIBK if there were no stripping operations; this would be financially and environmentally unacceptable—the plant would not get a permit to operate.

Broadly two approaches are practiced for solvent stripping: either steam stripping in a plate or packed column or evaporation in a falling film evaporator. In all cases, the acid temperature is raised to its boiling point with LP steam (approximately 107°C for acid with 0.7% MIBK). At this temperature, and with potentially 0.1% fluoride content, the regime is corrosive and necessitates the use of graphite and PTFE-lined equipment.

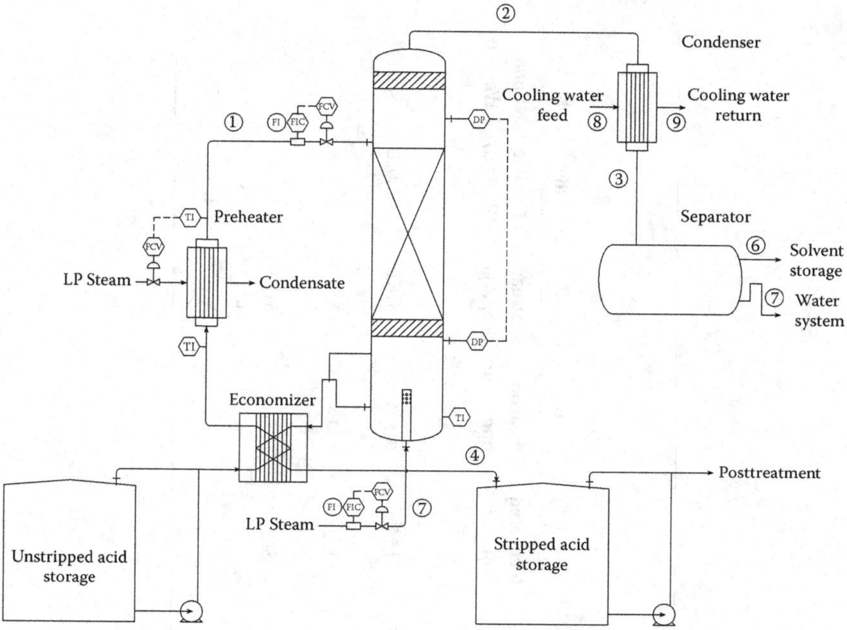

FIGURE 2.54 Acid stripping flow sheet.

Figure 2.54 shows a flow sheet and Table 2.15 a simple mass balance of a solvent stripping unit based on MIBK. Acid is pumped from the stripping section of solvent extraction to an intermediate tank. Of course, the concentration of the acid depends on the solvent used although the purified acid temperature is usually near to 50°C. The acid is pumped toward the stripping column under flow control and passes through the economizer where the feed acid is heated by the hot stripped acid emerging from the column base. As drawn, the economizer is a plate heat exchanger, specified for reasons of utility and economy. The preheater is usually made with graphite tubes because of the higher temperatures making the acid more corrosive to metals.

The heated acid partially flashes on entry to the column. Liquor phase passes down the column over graphite Raschig rings. At the rates shown, the column internal diameter is about 700 mm and the packing height 5 m. Heat is maintained in the column by a steam sparge. An alternative operational mode is to use the sparge to provide all the heat input to the column (as chosen in Table 2.14). The stripped acid flows to an intermediate storage from where it is pumped to further posttreatment. The overhead vapor is condensed against cooling water in a graphite tube exchanger. A small amount of acid is carried over together with trace fluoride. The condensate forms two phases in the separator. The upper solvent phase is returned to the solvent system and the lower aqueous phase to the solvent extraction section.

The solvent stripping unit is repeated for the raffinate and all product streams. The operation of the raffinate stripper is more maintenance intensive as a consequence of solids formation due to the high levels of impurity in the acid. Consequently, A and

TABLE 2.14
Solvent Stripper Mass Balance

		1	2	3	4	5	6	7	8	9
		Feed Acid	Column Vapor	Condensate	Stripped Acid	Recovered Solvent	Aqueous Condensate	Stream Sparge	Cooling Water (f)	Cooling Water (r)
P_2O_5	kg/h	6,000	1.5	1.5	5,998.5	0	1.5			
	%	41.0	0.02	0.02	41.6					
H_2O	kg/h	8,298	392.8	392.8	8,422.7	15.7	377.1	518.0	45,000	45,000
	%	56.7	53.8	53.8	58.4			100.0		
MIBK	kg/h	337	336.4	336.4	0.2	332.0	4.4			
	%	2.3	46	46	14 ppm					
Total		14,634	731	731	14,421	348	383	518.00		
Temperature, °C		108	102	40	113			134	25	31.2
Phase		L	V	L	L	L	L	V	L	L

B strippers are installed to allow maintenance off line. Cleaning is usually carried out with purified acid and steam.

TBP processes do not undergo solvent stripping per se largely because the TBP solubility in the acid is so low at 0.02%–0.04%. The R–P process removes TBP from the raffinate by dissolving it in kerosene and then separating the organic and aqueous streams.

2.6.2 Dearsenication

Dearsenication as a posttreatment process is little different to crude dearsenication as discussed in Section 2.3.4. The only major plant posttreating was the A&W process at Whitehaven. The advantage of posttreating is that a smaller quantity of acid requires treatment. If the As_2S_3 and accompanying filter aid and, for the A&W process, activated carbon powder require special disposal measures, then the smaller the quantity the better. On the other hand, if a sodium sulfide is the sulfide provider, then sodium levels in the purified acid may be raised outside specification.

2.6.3 Decolorization

Food grade acid is described as clear and water white and there are several patents that teach processes decolorization to achieve this specification including [76–79]. Analytically, total organic compounds (TOCs) are less than 10 ppm. Color, measured on the APHA/Hazen/Pt–Co color scale, is 5–10 and often less than 5. For an acid that has undergone satisfactory purification, the only cause of color is the presence of residual organic compounds; however, even at low levels, an apparently clear acid emerging from solvent extraction may discolor following heating during final concentration or defluorination. The sources of these compounds are both those found in the crude feed acid to solvent extraction and compounds picked up, including the solvent itself, through the extraction process. Ideally, acid entering the solvent extraction process contains less than 100 ppm organics; to that end, many processes carbon-treat the feed acid [80] (and discussed previously). For MIBK and IPE processes, organic content after solvent stripping is around 20 ppm. Purified acids from TBP processes do not undergo solvent stripping although the solvent content is minimized, in the R–P case, by a coalescer. Nevertheless, acids from TBP processes also undergo decolorization, and in general the goal is to reduce organic concentrations from up to 400 ppm to less than 10 ppm and end up with a water white acid.

To that end, some processes use an oxidizing agent such as hydrogen peroxide, sodium chlorate, or potassium permanganate [81] alone (the latter only practiced on WPA but evaluated for PWA), and others use activated carbon together with an oxidizing agent.

In terms of equipment, carbon treatment is relatively straightforward. The most convenient operation is continuously running granular active carbon-filled columns. Most usually three columns are installed with two running and one on standby or undergoing washing and regeneration with warm water and sodium hydroxide solution. Figure 2.55 shows a flow sheet of a typical three-column active carbon treatment unit. The active carbon is granular and usually chemically activated (using phosphoric acid in the manufacturing process); suppliers include Calgon and Norit.

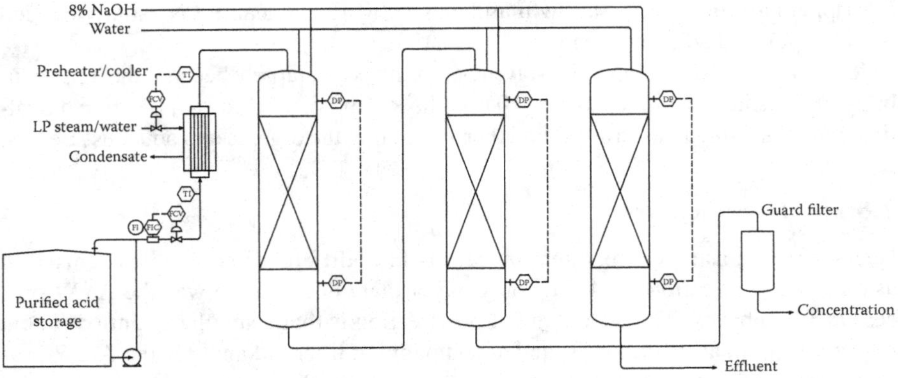

FIGURE 2.55 Carbon treatment flow sheet.

The columns are sized for a residence time of 2–3 h. Acid is pumped to the first column via a heat exchanger that either heats or cools the acid to achieve a feed temperature of 70°C. The acid flows down through the carbon bed. Organic species are adsorbed in the pores of the carbon granules. The acid passes out of the first column through an internal filter that minimizes carbon loss from the column. The acid enters the top of the second column and passes down the column and out via the internal filter to an external guard filter. The guard filter picks up carbon carryover and other particulates above 5 μm. Over a period of about one week, the active carbon becomes saturated with organics (if samples are taken down the bed an advancing front of inactivity is observed). Before *breakthrough*, the column is taken offline and emptied of acid. It is then washed with warm water, then with weak, 8% or so, sodium hydroxide solution, then with another water wash. The column is then refilled with treated acid before coming back online.

Treated acid is sampled and is color tested in the plant laboratory at least once per shift.

Most processes include a final decolorization step after defluorination prior to final product storage. Commonly 35% hydrogen peroxide solution is mixed in a 0.3%–0.5%/ton P_2O_5 proportion in an agitated tank at 130°C with a residence time of 1 h. Sodium chlorate is also used as final oxidizer on some plants, although this does introduce both sodium and chloride to the acid. The chloride is managed by operating this process step at over 165°C whereby some HCl vapor is liberated.

2.6.4 Concentration

The concentration of purified acid is a little more straightforward than crude acid concentration due to the low levels of impurities and consequent absence of precipitation and resultant blockages. Nevertheless, the presence of fluoride and the high temperatures required make this a corrosive process necessitating the use of graphite for the heat exchangers and rubber, PTFE, or polypropylene linings for vessels and piping. As with crude acid, purified acid concentration is carried out under vacuum to lower the acid boiling point and steam requirement. Vacuum is usually supplied by steam ejectors. The process objective is simply to bring the acid up to sales concentration of 61.6% P_2O_5.

Two approaches are practiced: the A&W process uses a single forced circulation concentrator, selecting the circulation pump and exchanger and specifying the rest of the equipment for fabrication; some other processes use multiple-effect evaporation and specify a process package from a specialist supplier.

The flow sheet of the single forced circulation concentrator is essentially a simplification of those shown in Figures 1.21 and 2.22. The impurity level is low; therefore, precipitation is negligible; therefore, there is no requirement for the settling or filtration equipment shown in Figure 1.21. Similarly, the fluoride level is too low to make recovery worthwhile, so that facility is absent (which obviates the central section of Figure 2.22). Figure 2.56 is a photograph of the A&W concentrator on the *MO* plant at Whitehaven. The vapor body is front and center and connected to the primary condenser at the right upper part of the photograph.

Multiple-effect evaporation trades an increase in number of plant items, and therefore capital cost, against a reduced energy demand, and therefore variable operating cost. Essentially, a proportion of the energy is the vapor coming off the

FIGURE 2.56 Final product concentrator. (With permission from Solvay.)

first-effect evaporator is then used to heat the acid flowing into the second-effect evaporator. The boiling point of the acid in the second evaporator is lowered below that of the first evaporator by vacuum. One plant design pumped stripped, decolorized, and preheated acid into the first-effect heat exchanger where it was heated to 116°C with low-pressure steam and achieved 42.7% P_2O_5 concentration. Vapor from the vapor head passed to the second-effect heater shell side, heating the liquid from the first-effect concentrator to 74°C. The second-effect evaporator operated under a vacuum system shared with the third, single-effect evaporator. Acid emerging from the second-effect evaporator was at 54.3% P_2O_5 concentration. The third evaporator operated with pump recirculation and raised the acid temperature to 94°C (with low-pressure steam) that with an operating pressure of 109 m bar corresponds to an acid concentration of 61.6% P_2O_5. Vapor from this system ends up either in the dedicated cooling water provision or part of a larger plant hot water system.

Because the vapor from each evaporation step is not simply water vapor but also acidic components, the shell side design of the heat exchangers and the attendant piping and other ancillary equipment must be designed to withstand the corrosion challenge. This further adds to the capital cost relative to the single-effect solution and potentially increases the maintenance cost in operation.

2.6.5 Defluorination

Depending on the WPA impurity profile, purified acid emerging from solvent extraction might contain 1000 ppm fluoride. Concentrating the acid to 58%–61.6% P_2O_5 would typically reduce the fluoride level to 200 ppm. This level is acceptable for most technical applications but is an order of magnitude too high for food grade use.

In order to achieve food grade standards, the fluoride levels must be reduced to less than 10 ppm. This is done in a defluorination step. On a PWA plant, there are two slightly differing approaches:

1. Steam stripping where live steam is injected into the base of a packed stripping column and contacts fluoride-bearing acid moving down the column—the operating temperature range for this approach is 160°C–180°C.
2. The A&W approach whereby instead of live steam a steam/hot air mix is injected into the column. In this approach, the operating temperature is slightly lower; however, in both cases, the acid must be maintained at or above its boiling point, 155°C, for the desired product concentration, 61.6% P_2O_5, at or near atmospheric pressure.

At these temperatures, both liquid and vapor are highly corrosive to metals. Therefore, the defluorination column, heat exchangers, and pipework tend to utilize PTFE liners and graphite as principal materials of construction. PTFE has excellent corrosion resistance but goes soft above 250°C and is not normally capable of withstanding high vacuum. Graphite also has excellent corrosion resistance in this field of application but lacks mechanical strength, compared to metals. Consequently, the operating regime for either approach is somewhat limited and dictates a column operating at or just below vacuum and less than 200°C.

Column sizing, for a desired throughput, requires knowledge of the vapor liquid equilibrium of HF, P_2O_5, and water. Published data of pure systems are a good first approximation [82] although cannot fully account for some minor interactions that will vary with the acid impurity profile—boron, for example, is present in some acids as a very minor impurity that does interact slightly. Armed with an understanding of the VLE, the column designer must specify the column internals. The R–P process designers chose graphite trays [83]; A&W chose graphite pall rings in four beds 3 m high each. Column diameter dictates volumetric throughput, and for one installation A&W designed a 1 m diameter column to handle an 8000 kg/h feed acid rate. The corresponding gas stripping rate was 3600 kg/h with a steam/air ratio of 2.5:1.

All designs utilize some or most of the recoverable heat. As a minimum, the hot, defluorinated acid is partially cooled in an economizer that heats the feed acid to the column.

2.7 CRYSTALLIZATION

2.7.1 Introduction

Early papers describing the crystallization of phosphoric acid were published in the 1870s by Thomsen, Joly, and Berthelot and continue to be published today [84,85]. Many of the academic papers and patents of the last 40 years cite Ross [86–88]. The use of crystallization for phosphoric acid has not taken off as a mainstream process because either thermal acid or PWA has provided sufficiently pure acid for most applications as they arose. Compared to solvent extraction, achieving the necessary purity via crystallization from a typical WPA [89,90] is nontrivial and still requires many of the supporting processes that PWA needs. Nevertheless, crystallization of either thermal acid that has undergone arsenic removal or a food grade PWA is practiced on an industrial scale to provide tens of thousands of tons of acid for very high-purity applications in the pharmaceutical and especially the electronic markets. Much recent study, and patents, has focused on the recycling of etchants used for the LCD market [91,92]. The main components of these etchants are high-purity phosphoric, nitric, and acetic acids that become contaminated with a small number of elements such as aluminum, molybdenum, and potassium. The nature and extent of the contamination point toward crystallization as the obvious recycling process. The general field of crystallization is large, and the reader is referred elsewhere for a more comprehensive treatment of the topic [93,94].

2.7.2 Freezing Point Curve of Phosphoric Acid

The freezing point curve of phosphoric acid and water is shown in Figure 2.57 adapted from Ross [88]. The advantage of transporting 75% H_3PO_4 during winter is immediately apparent; at this concentration, the freezing point is −17.5°C. The freezing point of 85% H_3PO_4 acid is 21°C, which forms the hemihydrate, $H_3PO_4 \cdot ½H_2O$. As the acid becomes more concentrated, the freezing point rises to a maximum of 29.3°C. Pure crystals of the hemihydrate are 91.6% H_3PO_4. With increasing

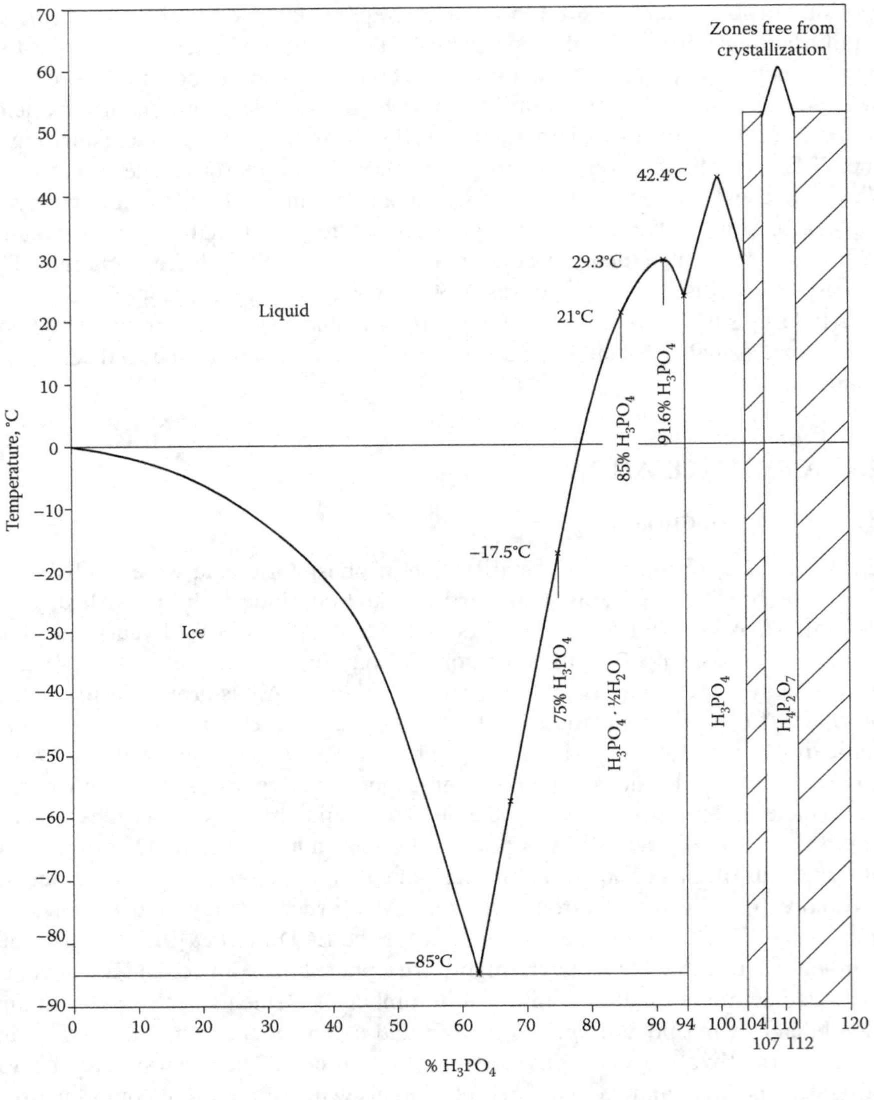

FIGURE 2.57 H_3PO_4–H_2O crystallization phase diagram. (Adapted with permission from Ross, W.H. and Jones, R.M., The solubility and freezing point curves of hydrated and anhydrous orthophosphoric acid, *J. Am. Chem. Soc.*, 47(8), 2165–2170. Copyright 1925 American Chemical Society.)

concentration, the freezing point then falls to a eutectic at 94.8% H_3PO_4 where both the hemihydrate and the anhydrous crystal form. Above this concentration and up to 104% H_3PO_4, only the anhydrous form is present, and strictly speaking other forms of phosphoric acid are present. Between 104% and 107% H_3PO_4, there are no crystals, but between 107% and 112% H_3PO_4, crystals of pyrophosphoric acid form at a relatively high temperature that is relevant for the manufacture and handling of polyphosphoric acids.

2.7.3 Crystallization Design Considerations

While it is possible to carry out processes at very low temperatures in the laboratory with liquid nitrogen, it is less practical to do so at an industrial scale. Both capital and operating cost increase with decreasing temperature. It is obvious therefore that any industrial crystallization process for phosphoric acid is likely to operate with feed acid concentration greater than 75% H_3PO_4 and ideally around 85% H_3PO_4 that permits the use of chillers operating above 0°C rather than refrigeration at subzero ranges. Logically, operating at concentrations above 94% H_3PO_4 would seem better still, and several processes have been patented using superphosphoric acid as a feed acid [95,96]; however, as the acid becomes more concentrated, it also becomes more viscous, hindering the crystallization process. At least one patent exists for the purification of pyrophosphoric acid [97] that is sufficiently viscous to require a kneader for mixing.

Given the general aim is to improve an already pure, commercial acid, the usual feed concentration is 85% H_3PO_4. Figure 2.58 shows the relevant portion of the phase diagram. In practice, the crystallization curve varies with impurity profile and is depressed by 2°C–3°C in the case of typical etchant blends. As feed acid is introduced to the crystallization system, it is cooled, graphically passing down the 85% line. Either side of the crystallization curve is a metastable zone; above the curve, crystallization may occur in this zone; below the acid, it may subcool and remain a liquor. The width of the metastable zone is dependent on the impurity profile [84]. Below the metastable zone, the acid will crystallize spontaneously. Uncontrolled crystallization leads to a vessel full of an unfilterable magma of tiny crystals incorporating impurities within their crystal lattice and is undesirable. Therefore, in practice, the rate of cooling and the degree of supersaturation (the difference between the crystallization curve and the operating temperature) are carefully controlled. For continuous crystallizers, once stable, feed acid is well mixed with the magma so that existing crystals provide the crystallization sites. For batch operation, the acid is cooled to just below the crystallization curve, then seed crystals are introduced and the batch is then steadily reduced in temperature. When the acid reaches the desired operating temperature, it is possible to calculate the theoretical crystal yield from the phase diagram and the concentration of the mother liquor. In Figure 2.58, the crystal yield at 11°C is given by the following equation:

$$\text{Crystal yield} = \frac{a}{a+b} = 33.7\% \quad (2.35)$$

Clearly, a higher feed concentration or lower operating temperature will achieve a higher crystal yield. Given the crystals and mother liquor must flow to a separation device and are usually pumped, the yield is kept in the 35%–40% range.

Crystal separation is either by centrifuge or wash column. Either way, good crystal size and morphology is essential for effective separation. Crystal size,

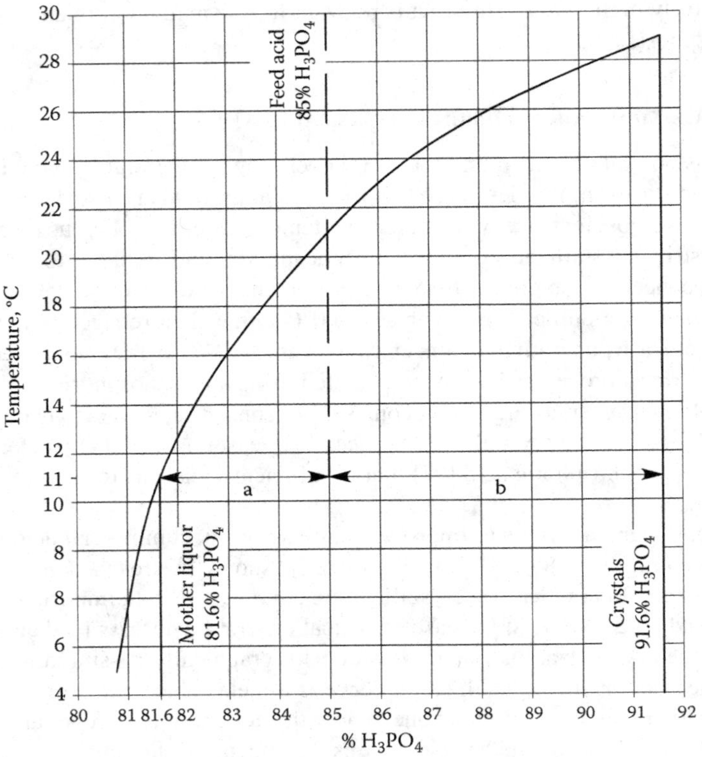

FIGURE 2.58 H_3PO_4–H_2O freezing curve for crystallization design.

morphology, and purity all depend on the right conditions being created in the crystallizer. By limiting supersaturation, optimizing mixing, and allowing plenty of residence time, it is possible to grow easily separable crystals of high purity. To achieve these conditions, continuous crystallization is advantageous; as fresh acid is added, the process of net crystal growth includes a degree of crystal melting, allowing the construction of a more pure crystal. In batch crystallization, there is inevitably, and even under the best controlled systems, a brief period of rapid crystal growth that can cause occlusion of impurities. There are several different crystallizer types proposed for phosphoric acid including falling film, cooling disk, crystallizing tube [98,99], and crystallizing tower [100]; however, the agitated tank with cooling coils and jackets has proved itself on an industrial scale, and more detailed comments are based on this type.

Crystal yield and growth rate and therefore desired residence time allow a preliminary sizing of a crystallizer. Agitator design must account for crystal properties, ensuring good mixing and uniform suspension of crystals and heat transfer—that is, crystal slurry rheology. Heat transfer design is critical as the removal of heat is the basis of the whole operation. Clearly, sufficient heat transfer area is required both to cool the acid to achieve the desired crystal yield and remove heat of crystallization. Heat transfer area is minimized by high differential temperature; however, this does cause excessive heat transfer surface fouling due to local crystallization; thus, a minimal temperature differential is balanced with vessel size and coil space requirements. The lowest temperature differential disclosed in the patent literature is 4°C.

Having made highly pure, easily separable crystals, they must be separated. The earlier patents [41,89] and common industrial practice make use of centrifuges. These maximize removal of mother liquor. Multiple-stage pusher centrifuges allow an internal crystal wash to be applied to reduce mother liquor in the inter-crystal void and attached to the crystal surface. Figure 2.59 shows a crystal washing effectiveness curve that applies both to centrifuging and filtration in general. The assumptions made are that the crystals are incompressible and that washing (more accurately displacing) mother liquor does not result in net reduction in crystal size through dissolution. The graph plots *Wash Ratio* against *Percent Remaining.* The wash ratio is defined as the ratio of the volume of the wash liquid to the volume retained in the voidage in the crystal mass. Percent remaining means percentage of mother liquor remaining with crystals after washing. Theoretically, one wash volume equal to the mother liquor in the crystal mass removes 100%. This asymptote is not achieved in practice so tests are carried out to optimize the wash ratio. The wash comes from the purified acid stream and is usually diluted to 75%–80% H_3PO_4 with deionized water. The higher the wash ratio, the greater the product purity; however, this generates a bigger recycle and lowers the overall yield. Well-grown crystals typically retain 2%–3% mother liquor by weight after centrifuging. Chemical analysis of the mother liquor, the melted product (including retained mother liquor), and some crystallized product washed in ethanol (to give a more accurate analysis of impurity within the crystal) gives the weight percentage of mother liquor with

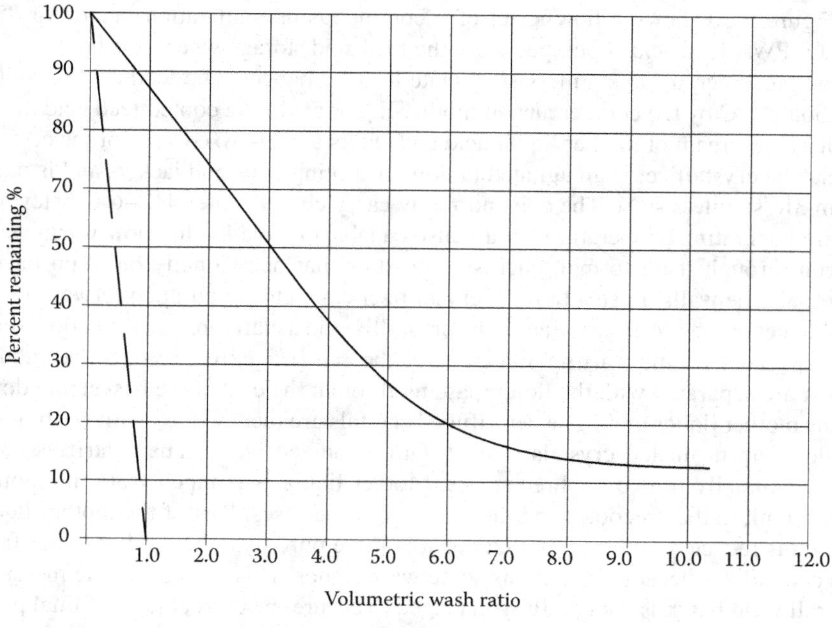

FIGURE 2.59 H_3PO_4–H_2O crystallization wash effectiveness curve.

the crystals after centrifuging. The following equation shows this calculation using sulfate as the marker:

$$\text{Retained mother liquor \%} = \frac{\text{Product ppm} - \text{Crystal ppm}}{\text{Mother liquor ppm} - \text{Crystal ppm}}$$

$$= \frac{410 - 100}{10{,}500 - 100} = 3.0\% \tag{2.36}$$

By repeating this calculation for different wash ratios and for one or two more markers (such as iron or sodium), the wash effectiveness curve is drawn and an operating point chosen. This allows completion of the mass balance of the system.

An alternative to centrifuging is an elutriation or a wash column [101]. Variants include a crystallization tower [102] that is agitated and may have cooling and heating zones. In the elutriation column, the crystallizer slurry enters the top of the column and the mother liquor overflows to downstream storage. The crystals fall under gravity but against a gentle upflow of fine crystals and, as the crystals proceed down the column, increasingly pure wash acid that is pumped into the base of the column. At the bottom of the column, the washed crystals flow out to the melt tank. A proportion of the acid in the melt tank is mixed with deionized water and pumped back as the wash. The crystallization tower combines the crystallizer and wash column duty in one item.

The wash column is mechanically simpler than the centrifuge and cheaper to install, and of itself requires less electrical energy. On the other hand, the wash flow is higher; therefore, the overall plant yield is lower, so to achieve the same yield, the wash column may consume more energy through making more crystals.

Figure 2.60 shows a flow sheet of a continuous crystallization plant. An 85% H_3PO_4 PWA feed acid is pumped from the feed acid storage where it is held at 45°C to the precooler, a 316 stainless steel plate heat exchanger. The feed acid is cooled to about 30°C by the countercurrent mother liquor acid. The cooled feed acid mixes with a side stream of mother liquor acid and enters a well-mixed zone of the crystallizer. The crystallizer is an agitated tank with cooling coils and jackets and is made from 316 stainless steel. The coils normally carry chilled water 4°C–6°C below the desired operating temperature but are also capable of isolation to allow warm water to run through them to melt encrusted crystals that occasionally build up on the external tube walls. In steady state, crystallizer contents are pumped forward to the pusher centrifuge. A return line to the crystallizer is usually installed to allow recycle especially during start-up. On entering the centrifuge, the crystals and mother liquor are separated with the liquor passing through the centrifuge basket and down to the mother liquor tank. The centrifuged crystals are then washed with wash liquor made up from melted crystals diluted with deionized water. The centrifuge does this continually in two or three stages. Mother liquor is pumped from the mother liquor tank to the precooler and then away for other uses. Part of the mother liquor stream is recycled to the crystallizer as it is more pure than the mother liquor from the crystallizer because it contains some wash liquor. This partial recycle increases overall yield but must be carefully managed to ensure the correct level of final product purity. Washed crystals pass by gravity to the crystal melt tank and are mixed

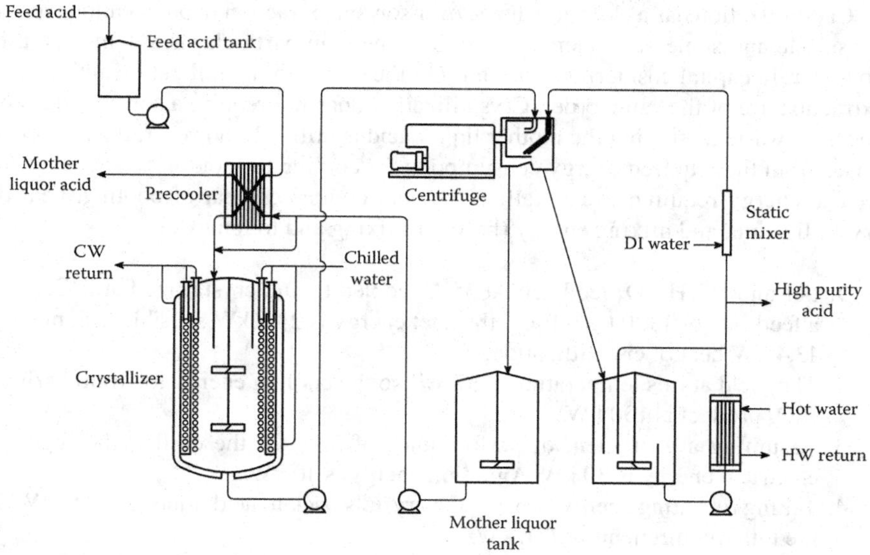

FIGURE 2.60 Crystallization flow sheet.

with circulating warm acid by agitation, thus melting to form acid at about 90% H_3PO_4. A weir is installed in the tank to reduce the amount of crystals passing into the pump and heat exchanger. The heat exchanger uses hot water to heat the high-purity acid. From the exchanger, some acid is recycled to the tank to maintain the melting temperature of 35°C, some passes through a static mixer and is blended with deionized water to make up the crystal wash, and the rest is exported.

While the temperatures in the system are below 40°C, the amount of metal pickup and therefore contamination from the equipment and piping built in 316 stainless steel is very small. For product impurity levels just below 1 ppm, this is satisfactory. For the purest grades, below 50 ppb, equipment and piping after the centrifuge should, as far as practicable, be made from or lined with plastics.

The achievable stage purification differs depending on the individual element, the overall mix of impurities, and whether the process is batch or continuous. Broadly speaking, heavy metals have low purification factors, usually less than 10 and some less than 5, where purification factor is defined in the following equation:

$$\text{Purification factor} = \frac{\text{Crystal melt }\%P_2O_5}{\text{Feed acid }\%P_2O_5} \times \frac{\text{Feed impurity ppm}}{\text{Crystal melt impurity ppm}} \quad (2.37)$$

Sulfate, sodium, and others of the lighter Group 1 and 2 elements have high purification factors, the latter around 80, and the former above 100. Other elements fall in the range 20–50. Purification factors calculated from some patents are much lower than these figures reflecting small batch experiments with short residence times and low wash ratios. One patent [41] achieved a very respectable purification factor of 100 for sulfate at laboratory scale with a centrifuge.

Crystallization has a clear advantage over solvent extraction in purification factors for sulfate and some other elements, but for others, in particular heavy metals, it is no better. In capital cost terms, per ton P_2O_5, the costs for crystallization and solvent extraction are of the same order. Crystallization does not require a solvent and produces no waste (assuming the mother liquor acid is utilized); however, the challenge arises from the required energy consumption for cooling the acid.

The energy required to crystallize the acid comprises heat of cooling, heat of crystallization, and mixing energy (both for mixing and to remove it):

1. For an 85% H_3PO_4 feed acid at 35°C, cooled to and crystallized at 8°C, at a feed rate of 1000 kg/h P_2O_5, the heat energy is 24.2 kW sensible heat plus 13.4 kW heat of crystallization.
2. The yield at this temperature is 37.5%, so the cooling energy for 1000 kg/h P_2O_5 product is 100 kW.
3. Assuming a coefficient of performance of 3.33 for the chiller, the input electrical energy is 30 kW. Agitation energy is 10 kW.
4. Taking reheating energy to bring the crystals back to acid liquor as 37.6 kW, the total requirement is 77.6 kW.
5. At $0.10/kWh, the cost is $7.76/ton P_2O_5.
6. By comparison, the best practice solvent extraction requires 2 kg of solvent per ton P_2O_5 product. Taking TBP at $1500/ton, the solvent cost is $3/ton P_2O_5.

Theoretically, solvent extraction given enough stages can achieve the levels of purity needed for the purest grades of acid; however, industry has found that the practical compromise is for solvent extraction to do the *heavy lifting* and crystallization the fine refining.

2.8 MEMBRANE SEPARATION

Membrane separation and ion exchange have been evaluated in the laboratory by several corporations. Simplot patented [103] a process to purify WPA to food grade acid using membranes.

In the Simplot process, WPA is filtered at 0.5 μm, diluted to 20% P_2O_5 concentration, and then pumped through a polyamide nanofiltration unit at 25°C. Simplot teaches a flux rate of 0.12 m^3/m^2/day through the membranes and a desirable crossflow feed rate of 11.3 m^3/h at 60 bar pressure. The membranes are effective in removing many impurities, and after two stages, the permeate is good enough for many technical derivative products. With a different feed acid and pretreatment and with posttreatment for fluoride and organic removal, the membranes are capable of producing a near food grade acid with limited application.

Simplot offers this process commercially and claims a 70% split and the requirement for 270 kWh and 3.9 tons of LP steam per ton of purified P_2O_5. This process, like crystallization, could only be compared to the solvent extraction stage of a PWA plant, as posttreatment for defluorination, concentration, and organics removal is still necessary. Furthermore, in normal operation, both solvent extraction and crystallization plant are not high maintenance, whereas the longevity of membranes operating at high pressures is questionable and has spawned many recent patents [104,105].

Nevertheless, membranes have the advantage of simplicity and respectable purification factors for some elements and will pass acids. Consequently, they are beginning to find a place in the etchant area, whether in LCD manufacture, printing, or metal treatment. The metal pickup is quite specific, and if the manufacturer can extend the life of the etchant by just removing the built up elements, then that is much cheaper than replacing the etchant inventory. The etchant user is not a chemical manufacturer, and while there are several chemical processes involved in the output of an LCD screen, printing plates, or metal panels, any process introduced should be simple.

2.9 PURIFICATION TECHNOLOGY COMPARISON

The purification of wet process phosphoric acid began as a way of providing a cheaper alternative to thermal acid for the manufacture of bulk technical industrial phosphates, in particular STPP. To this end, chemical purification by precipitation was not a bad process, and if the precipitates are, in the main, recycled back to the WPA reaction and used for fertilizer manufacture, the P_2O_5 efficiency is reasonable. The chemical purification process can never achieve the purity of thermal acid. Solvent extraction processes were developed for several reasons and have their respective advantages and disadvantages; however, they have demonstrated that they are capable of purifying acid to the same purity as thermal acid. As phosphoric acid purification has developed, so has the technology in acid markets. This has led to the need for higher levels of purity and the development of heavy metal removal and crystallization and, more recently, membrane filtration. Making detailed comparisons between these various technologies is not easy; the cost and extent of pre- and posttreatment are highly complicating factors, and the quality of the feed acid and the desired impurity profile of the product are another.

Table 2.15 shows a comparison of purification factor for the different technologies; it should be considered carefully as indicative rather than definitive because the feed qualities are different, the product qualities are different, and the scale of operation is different and ranges from laboratory to industrial.

The first column shows the purification factors from data in the Simplot patent. The second column is calculated from crystallization data in a Kemira patent. The third and fourth columns are crystallization data from the author's work. The fifth and sixth columns are purification factors from industrial-scale plants operating with MIBK and TBP as solvent. The first, second, fifth, and sixth columns are purifying different wet process acids with different impurity profiles. The third and fourth columns are operating in pure and highly pure regimes. Arsenic (and lead) is not included in the solvent extraction data as the plants have dearsenication processes. The cadmium figures for solvent extraction are affected to some degree by dearsenication and may be high.

TABLE 2.15
Purification Factor Comparison

	Purification Factor Comparison					
	Simplot	Kemira				
	Membrane	Crystallization			Solvent Extraction	
	US5945000	US6814949			MIBK	TBP
SO_4	1.2	16.4	120		70	648
Al	61.7	12.6	20		239	1715
As	0.9	3.2	1	24		
Ca	19.0	11.7	20	2.8		24
Cd	15.9	3.2	5	12.7	102	691
Co	1.9		2			
Cr	47.4	6.3	20		809	281
Cu	9.3	1.8	2		41	194
Fe	119.9	7.9	20	2.1		1306
K	2.1		12	33		360
Mg	49.8	16.4	28	16.1	11,159	561
Mn	9.3	12.0	6	18.1	21	50
Mo	2.1			12	5	
Na	1.8		110	40	18	13
Pb	10.2		5			
V	18.5	12.6	18		570	1458
Zn	18.5		75	14.4	861	864

With the caveats mentioned earlier in mind, broad observations are possible:

a. Solvent extraction is effective in removing many impurities.
b. There are indicative differences in the effectiveness of MIBK and TBP.
c. Crystallization is the most effective in removing sodium.
d. Membranes hardly touch sulfate (which is indicative of their potential for cleaning up etchants *in situ*).

REFERENCES

1. ASTM Standard D1209, 2011, Standard Test Method for Color of Clear Liquids (Platinum-Cobalt Scale), ASTM International, West Conshohocken, PA, 2011, DOI: 10.1520/D1209-05R11, www.astm.org.
2. K. V. Darragh and M. R. Irani, Preparation of detergent grade sodium tripolyphosphate from wet process phosphoric acid and soda ash, US Patent 4209497, June 1980.
3. J. Marty, Manufacture of alkali orthophosphates, US Patent 3081151, March 1963.
4. G. E. Taylor, Production of sodium phosphates, US Patent 2390400, December 1945.
5. A. Wehrstein, Method of producing commercial alkali phosphates, US Patent 2162657, June 1939.
6. F. Pofliez, Anhydrous colorless phosphates, US Patent 2977191, March 1961.
7. A. E. Lewis, Review of metal sulphide precipitation, *Hydrometallurgy*, 104(2), 222–234, September 2010.

8. J. M. Coulson, J. F. Richardson, J. R. Backhurst, and J. H. Harker, *Chemical Engineering*, Butterworth-Heinemann, Oxford, U.K., 2002.
9. C. J. Geankoplis, *Transport Processes and Separation Process Principles: Includes Unit Operations*, Prentice Hall Professional Technical Reference, Upper Saddle River, NJ, 2003.
10. G. Towler and R. K. Sinnott, *Chemical Engineering Design: Principles, Practice and Economics of Plant and Process Design*, Elsevier, Boston, MA, 2012.
11. J. Rydberg, M. Cox, C. Musikas, and G. R. Choppin, (Eds.) *Solvent Extraction Principles and Practice*, Marcel Dekker, New York, 2004.
12. V. S. Kislik, *Solvent Extraction*, Elsevier, Burlington, MA, 2011.
13. T. C. Lo, M. H. I. Baird, and C. Hanson, *Handbook of Solvent Extraction*, Wiley & Sons, New York, 1983.
14. R. E. Treybal, *Liquid Extraction*, McGraw-Hill, New York, 1963.
15. L. L. Frederick and J. F. McCullough, Purification of wet-process phosphoric acid with methanol and ammonia, US Patent 3975178, August 1976.
16. P. Ananthanarayanan and P. B. Rao, Ternary liquid equilibria of the water-phosphoric acid-isoamyl alcohol, cyclohexanol, or methyl isobutyl ketone systems at 35°C, *J. Chem. Eng. Data*, 13(2), 194–196, 1968.
17. M. Feki, M. Fourati, M. M. Chaabouni, and H. F. Ayedi, Purification of wet process phosphoric acid by solvent extraction liquid-liquid equilibrium at 25°C and 40°C of the system water-phosphoric acid-methylisobutylketone, *Can. J. Chem. Eng.*, 72(5), 939–944, 1994.
18. D. B. Hand, Dineric distribution, *J. Phys. Chem.*, 34(9), 1961–2000, January 1929.
19. J. S. C. Chiang and W. S. Moore, Method of regenerating phosphoric acid extraction solvent, US Patent 4311681, January 1982.
20. S. Ubal, C. H. Harrison, P. Grassia, and W. J. Korchinsky, Numerical simulation of mass transfer in circulating drops, *Chem. Eng. Sci.*, 65(10), 2934–2956, May 2010.
21. W. R. Parish, Fluorine recovery, US Patent 3091513, May 1963.
22. A. W. Petersen and J. M. Stewart, Co-current absorber for recovering inorganic compounds from plant effluents, US Patent 4152405, May 1979.
23. W. R. Parish, Removal of fluorine compounds from phosphoric acid, US Patent 3273713, September 1966.
24. J. Martinez, Method of recovering fluorine compounds from vapors exiting wet process phosphoric acid evaporators, WO9801214, January 1998.
25. A. V. Slack, (Ed.) *Phosphoric Acid, Part 2*, vol. 1, Marcel Dekker, New York, 1968.
26. R. J. Hurka, Purification of phosphoric acid, US Patent 2287683, June 1942.
27. D. W. Mueller and D. J. M. Michel, Herstellung chemisch reiner phosphorsaeure, German Patent 503202, July 1930.
28. J. M. Bisang, F. Bogado, M. O. Rivera, and O. L. Dorbessan, Electrochemical removal of arsenic from technical grade phosphoric acid, *J. Appl. Electrochem.*, 34(4), 375–381, April 2004.
29. Z. Twardowski, Removal of arsenic from acids, US Patent 4692228, September 1987.
30. K. Hotta and F. Kubota, Method for purification of phosphoric acid high purity polyphosphoric acid, US Patent 6861039, March 2005.
31. F. Bassan, G. Calicchio, and N. M. Ito, Process for the removal of cadmium from solutions of phosphoric acid, US Patent 5246681, September 1993.
32. R. Gradl and G. Schimmel, Process for decontaminating phosphoric acid, US Patent 4629614, December 1986.
33. R. Gradl, G. Heymer, and G. Schimmel, Process for removing heavy metal ions and arsenic from wet-processed phosphoric acid, US Patent 4466948, August 1984.
34. L. E. Segrist, Method for extracting arsenic and heavy metals from phosphoric acid, US Patent 4824650, April 1989.

35. D. Perron and L. Winand, Purification of wet-process phosphoric acid, US Patent 4769226, September 1988.
36. K. Schrodter, Process for freeing mineral acids from heavy metals, US Patent 4777028, October 1988.
37. R. E. Hall, F. B. Jueneman, and P. H. Zeh, Process for purifying phosphoric acid for electrical semiconductor use, US Patent 4804526, February 1989.
38. A. Baniel and R. Blumberg, Purification of phosphoric acid, Swiss Patent 1199041, July 1970.
39. A. H. Rooij and J. Elmendorp, Process for recovering phosphoric acid from aqueous solutions containing nitric acid and phosphoric acid, US Patent 3363978, January 1968.
40. T. Sakomura, M. Kikuchi, Tsuno-gun, Yamaguchi-ken, and H. Shimizu, Process for refining phosphoric acid preparations, US Patent 3297401, January 1967.
41. T. A. Williams and F. M. Cussons, Purification of phosphoric acid, US Patent 3912803, October 1975.
42. A. Mögli, Kolonneneinrichtung zur Gegenstromextraktion von Flüssigkeiten, Swiss Patent 470903, April 1969.
43. K. Beltz, K. Frankenfeld, and K. Gotzmann, Process for the extractive purification of phosphoric acid containing cationic impurities, US Patent 4018869, April 1977.
44. K. Beltz and K. Frankenfeld, Process for separating a phosphoric acid-solvent solution, US Patent 4117092, September 1978.
45. D. S. Bunin, F. J. Kelso, and R. A. Olson, Phosphoric acid purification, US Patent 3410656, November 1968.
46. D. P. Brochu, J. C. Dore, and R. E. Hall, Process for purifying phosphoric acid, US Patent 4780295, October 1988.
47. N. K. Khanna, Manufacture of sodium tripolyphosphate from wet acid, US Patent 4676963, June 1987.
48. J. Mizrahi, *Developing an Industrial Chemical Process: An Integrated Approach*, 1st edn. CRC Press, Boca Raton, FL, 2002.
49. J. J. Aman, Improvements in or relating to the thermal decomposition of certain chlorides and sulphates, Swiss Patent 793700, April 1958.
50. A. Baniel and R. Blumberg, Process for the preparation of phosphoric acid, US Patent 2880063, March 1959.
51. M. C. Mew, *World Survey of Phosphate Deposits*, 4th edn. British Sulphur Corporation, London, U.K., 1980.
52. E. W. Pavonet, Method for purifying phosphoric acid, US Patent 3970741, July 1976.
53. R. Champ, M. Martin, and L. Winand, Purification of phosphoric acid derived from phosphate rock, US Patent 3397955, August 1968.
54. J. R. Goret and A. L. M. Winand, Continuous process for solvent purification of phosphoric acid, US Patent 3607029, September 1971.
55. L. Winand and J. R. Goret, Extraction of phosphoric acid at saline solutions state, US Patent 3767769, October 1973.
56. E. L. Koerner and E. Saunders, Purifying phosphoric acid using an amine extractant, US Patent 3367749, February 1968.
57. R. A. Ruehrmein, Purification of phosphoric acid, US Patent 2955918, October 1960.
58. E. L. Koerner, Purifying phosphoric acid, US Patent 3479139, November 1969.
59. E. K. Drechsel and J. B. Sardisco, Method of producing fluoride-free phosphoric acid, US Patent 4055626, October 1977.
60. R. Amanrich, G. Cousserans, and A. Mahe, The "Apex" process for the extraction of phosphoric acid for the production of phosphatic fertilizers and industrial phosphates, *ISME Technical Conference*, Sandefjord, Norway, 1970, p. 39.
61. R. Amanrich, Purification d'acide phosphorique de voie humide, French Patent 2093372, January 1972.

62. J. Bergdorf and R. Fischer, Extractive phosphoric acid purification, *Chem. Eng. Prog.*, 74(11), 41–45, 1978.
63. F. L. Prado, Food grade phosphoric acid in Turkey, Presented at the *American Institute of Chemical Engineers, Central Florida Section, Annual Meeting*, Clearwater Beach, FL, 2007, p. 6.
64. The SAEC Process [Online] Available: http://www.meab-mx.com/pdf/The_SAEC_process_pdf. [Accessed: April 24, 2013].
65. R. Blumental, Mixer with six bladed impeller, Israeli Patent 34096, January 1973.
66. J. Mizrahi and E. Barnea, A gravitational settler, Israeli Patent 43693, July 1976.
67. E. Barnea and J. Mizrahi, Liquid-liquid mixer, US Patent 3973759, August 1976.
68. I. Szanto, Mixer-settler apparatus having a submerged chute, US Patent 4844801, July 1989.
69. A. Mögli, Kolonneneinrichtung zur Gegenstromextraktion von Flüssigkeiten, Swiss Patent 470902, April 1969.
70. S. Mohanty, Modeling of liquid-liquid extraction column: A review, *Rev. Chem. Eng.*, 16(3), 199–248, January 2000.
71. D. M. Attarakih and P. H.-J. Bart, LLECMOD: An effective simulation tool for liquid extraction columns using bivariate population balance–TVT University of Kaiserslautern. [Online]. Available: http://www.uni-kl.de/en/tvt/tvtforschung/extraction/llecmod-an-effective-simulation-tool-for-liquid-extraction-columns-using-bivariate-population-balance/. [Accessed: April 24, 2013].
72. M. Attarakih, S. Al-Zyod, M. Abu-Khader, and H. Bart, PPBLAB: A new multivariate population balance environment for particulate system modelling and simulation, *Procedia Eng.*, 42, 1574–1591, 2012.
73. W. J. D. van Dijck, Process and apparatus for intimately contacting fluids, US Patent 2011186, August 1935.
74. B. Grinbaum, Review article: The existing models for simulation of pulsed and reciprocating columns—How well do they work in the real world? *Solvent Extr. Ion Exch.*, 24(6), 795–822, 2006.
75. L. Winand, Decolorization and purification of crude wet-process phosphoric acid, US Patent 4330516, May 1982.
76. R. E. Barker, E. E. Borchert, and R. J. Urban, Method of decolorizing wet process phosphoric acid, US Patent 4643883, February 1987.
77. M. Nakatani, Method for removing organic materials dissolved in purified phosphoric acid, US Patent 4820501, April 1989.
78. G. Schimmel, G. Bettermann, G. Heymer, and F. Kolkmann, Continuous process for complete removal or organic impurities from and complete decolorization of prepurified wet-process phosphoric acid, US Patent 4906445, March 1990.
79. H. L. Allen and W. W. Berry, Purification of phosphoric acid, US Patent 4341638, July 1982.
80. G. A. Gorman, K. E. Kranz, D. D. Leavitt, and J. N. L. Stewart, Oxidation-reduction process for enhancing the color of and stabilizing wet process phosphoric acid, US Patent 4808391, February 1989.
81. T. Yamada, M. Haruta, and K. Itou, The liquid-vapor equilibrium of the system $HF–P_2O_5–H_2O$, *Bull. Nagoya Inst. Technol.*, 23, 481–487, 1971.
82. J. R. Costes and R. Botton, Nouveau plateau pour colonne de contact entre des liquides et des gaz ou vapeurs, French Patent 2267812, November 1975.
83. L. Dang, H. Wei, Z. Zhu, and J. Wang, The influence of impurities on phosphoric acid hemihydrate crystallization, *J. Cryst. Growth*, 307(1), 104–111, September 2007.
84. B. Wang, J. Li, Y. Qi, X. Jia, and J. Luo, Phosphoric acid purification by suspension melt crystallization: Parametric study of the crystallization and sweating steps, *Cryst. Res. Technol.*, 47(10), 1113–1120, 2012.

85. W. H. Ross, R. M. Jones, and C. B. Durgin, The purification of phosphoric acid by crystallization, *Ind. Eng. Chem.*, 17(10), 1081–1083, October 1925.
86. W. H. Ross, C. B. Durgin, and R. M. Jones, Process for the purification of commercial phosphoric acid by crystallization, US Patent 1451786, April 1923.
87. W. H. Ross and R. M. Jones, The solubility and freezing point curves of hydrated and anhydrous orthophosphoric acid, *J. Am. Chem. Soc.*, 47(8), 2165–2170, August 1925.
88. J. Aaltonen, S. Riihimäki, P. Ylinen, and A. Weckman, Process for production of phosphoric acid by crystallization of phosphoric acid hemihydrate, US Patent 6814949, November 2004.
89. V. C. Astley, P. D. Mollere, and J. J. Taravella, Process for seed crystal preparation for phosphoric acid crystallization, US Patent 4657559, April 1987.
90. K. J. Kim, S. Y. Kim, and J. K. Kim, Recovery of phosphoric acid from semiconductor waste etchant, Korean Patent 20050106825, November 2005.
91. R. S. Ruemekorf and R. U. Scholz, Purification of phosphoric acid rich streams, US Patent 8034312, October 2011.
92. J. Mullin, *Crystallization*, Butterworth-Heinemann, Oxford, U.K., 2001.
93. A. Myerson, *Handbook of Industrial Crystallization*, Butterworth-Heinemann, Boston, MA, 2002.
94. J. D. Jernigan and B. M. Whitehurst, Process for obtaining pure orthophosphoric acid from superphosphoric acid, US Patent 4083934, April 1978.
95. R. Pahud, Process for manufacturing crystallized pure and anhydrous phosphoric acid, and installation for carrying out said process, US Patent 2847285, August 1958.
96. G. Heymer and H.-D. Wasel-Nielen, Process for the manufacture of pyrophosphoric acid by crystallization, US Patent 3853486, December 1974.
97. Y. Yamazaki, S. Tabei, and K. Negishi, High purity phosphoric acid, Japanese Patent 2000026111, January 2000.
98. Z. Jian, Method for preparing electronic grade phosphoric acid by fusion crystallization method, Chinese Patent 1843900, October 2006.
99. J. Wang, X. Jiang, B. Hou, M. Zhang, Q. Yin, Y. Bao, Y. Wang, and J. Gong, Static multistage melting crystallization method for preparing electronic grade phosphoric acid, Chinese Patent 102198937, September 2011.
100. J. P. Roodenrijs, Installation for separating and purifying solids, US Patent 6241101, June 2001.
101. J. Li, B. Wang, Y. Jin, J. Luo, K. Zhou, C. Ma, Y. Qi, and T. Wang, Method for preparing electronic-grade phosphoric acid via melting-crystallization, Chinese Patent 102583283, July 2012.
102. K. J. Hutter and H. J. Skidmore, Methods of purifying phosphoric acid, US Patent 5945000, August 1999.
103. C. J. Kurth, S. D. Kloos, J. A. Peschl, and L. T. Hodgins, Acid stable membranes for nanofiltration, US Patent 7138058, November 2006.
104. T. J. Fendya, S. A. Geibel, M. F. Hurwitz, X. Shen, and T. ul Haq, Fluid treatment arrangements and methods, US Patent 8048315, November 2011.
105. K.-P. Ehlers and G. Heymer, Production of pure alkali metal phosphate solutions from wet-processed phosphoric acids, US Patent 4134962, January 1979.
106. S. A. Ficner, A. J. Klanica, and T. F. Korenowski, Process for making sodium tripolyphosphate from wet process phosphoric acid, US Patent 4251491, February 1981.
107. K.-P. Ehlers, R. Mulder, and K. Schrodter, Production of alkali metal phosphate solutions free from zinc, US Patent 4299803, November 1981.

3 Polyphosphoric Acid

3.1 INTRODUCTION

Polyphosphoric acid links inorganic and organic chemistry. While phosphoric acid is used widely as a pH buffer, polyphosphoric acid catalyzes ring modifications in a wide range of organic reactions, many of which are carried out in industrial processes.

As orthophosphoric acid is condensed and becomes more concentrated as water is removed, it reaches a maximum concentration of 93% H_3PO_4. As more water is removed, higher acids start to form commencing with pyrophosphoric acid, $H_4P_2O_7$. As condensation continues, more pyrophosphoric acid is formed and triphosphoric acid, $H_5P_3O_{10}$, appears, then tetraphosphoric acid, $H_6P_4O_{13}$, and so on according to the formula $H_{n+2}(P_nO_{3n+1})$. These acids are also described with the shorthand, P_n, so H_3PO_4 is P_1, $H_4P_2O_7$ is P_2, and so on. At equilibrium the constituent components of a particular acid depend on its total P_2O_5 content. This mélange of acids are collectively known as polyphosphoric acid. Typically, producers offer polyphosphoric acid at about five concentrations, although special arrangements are always possible. Perhaps confusingly, these products are known in terms of % H_3PO_4 calculated from their P_2O_5 content; for example, the highest concentration standard grade is *118*, although it actually contains only 2% H_3PO_4. Phosphoric acids are sometimes classified by their molar H_2O/P_2O_5 ratio; for orthophosphoric acids, the H_2O/P_2O_5 ratio is greater than 3:1; polyphosphoric acids are in the range 3:1–1:1; and ultrapolyphosphoric acids less than 1:1.

Polyphosphoric acid is made either by dissolving solid phosphorus pentoxide in orthophosphoric acid or condensing orthophosphoric acid through direct heating. There are several ways of accomplishing both routes, and the industry continues to refine different production technologies.

Polyphosphoric acid has many uses, and its extended application remains an active area of industrial research and development. The greatest volume of P_2O_5 as polyphosphoric acid is directed to high-quality liquid fertilizer. While technically a polyphosphoric acid, it is more commonly known as *superphosphoric* acid and is typically available at 70% P_2O_5 concentration, which equates to 96.6% H_3PO_4. At this concentration, it contains less than 4% $H_4P_2O_7$ and no higher acids. Superphosphoric acid is made by concentrating wet process phosphoric acid (WPA) and has the logistical advantages of a high P_2O_5 content and reduced propensity to precipitate solids in storage compared to WPA. This chapter addresses the more concentrated, higher purity polyphosphoric acids, which are manufactured from either phosphorus pentoxide or PWA in smaller volumes than superphosphoric acid. These *polyacids*, as they are known in the industry vernacular, are directed toward catalytic applications, as an acidity modifier, in the manufacture of pigments, pharmaceutical intermediates, fine chemicals, synthetic fibers, ceramics, and in water treatment.

3.2 CHEMISTRY

The study of the chemistry of condensed phosphoric acids goes back 200 years but was limited to pyrophosphoric acid until the latter half of the nineteenth century because of a lack of understanding of polymerization [1–4]. As orthophosphoric, or more precisely monophosphoric, acid is condensed, water molecules are stripped from between the PO_4 tetrahedra, which in turn is joined by a shared oxygen atom, as shown in Figure 3.1a. As condensation continues, more tetrahedra are joined in a linear chain (see Figure 3.1b and c) and so on to acids of $(n + 2)$ P atoms (see Figure 3.1d). Linear polyphosphoric acid chains may extend to great length before branching occurs. Analysis of a polyphosphoric acid may be accomplished with gradient elution ion chromatography, x-ray diffraction (XRD), pH titration, infrared spectroscopy, P tracer studies, and nuclear magnetic resonance (NMR) [5]. Acids of up to 14 phosphorus atoms have been separated; Figure 3.2 shows a plot of the data from de Klerk et al. [5], which is similar to Jameson [1] and Huhti and Gartaganis [6]. Only linear forms of polyphosphoric acid are present in the range 68%–81% P_2O_5, above this level cyclic metaphosphoric acids start to form. Their structure is shown in Figure 3.3. At higher levels of condensation, branching occurs and cross-linked acids are formed; Figure 3.4a through c shows those of isopolyphosphoric, isometaphosphoric, and ultraphosphoric acid. The classification of condensed phosphoric acids is completed with anhydrous phosphorus pentoxide (Figure 3.4d).

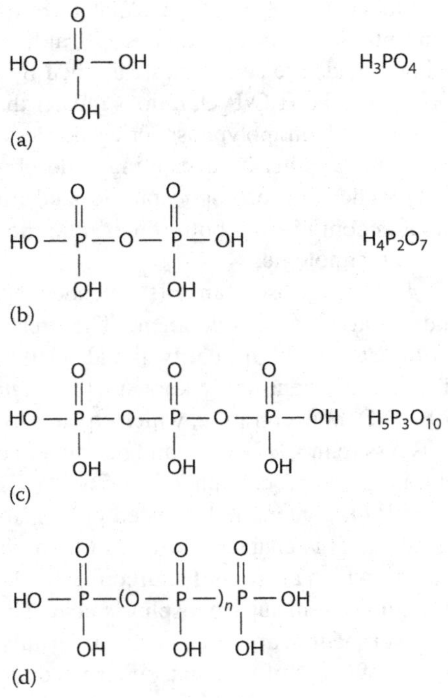

FIGURE 3.1 Semistructural model of linear polyphosphoric acids: (a) mono- (ortho-)phosphoric acid, (b) di- (pyro-)phosphoric acid, (c) triphosphoric acid, and (d) linear polyphosphoric acid.

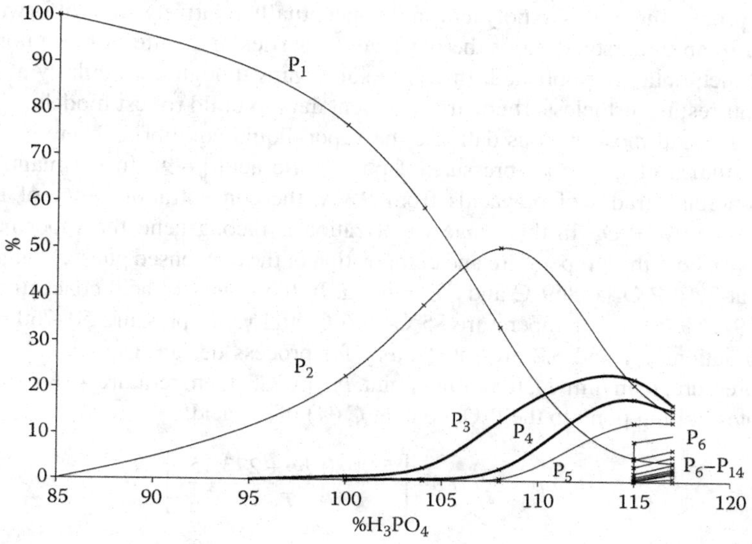

FIGURE 3.2 Composition of polyphosphoric acid.

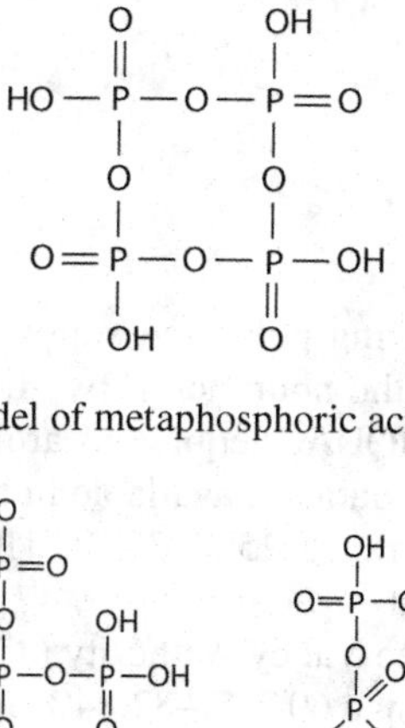

FIGURE 3.3 Semistructural model of metaphosphoric acid.

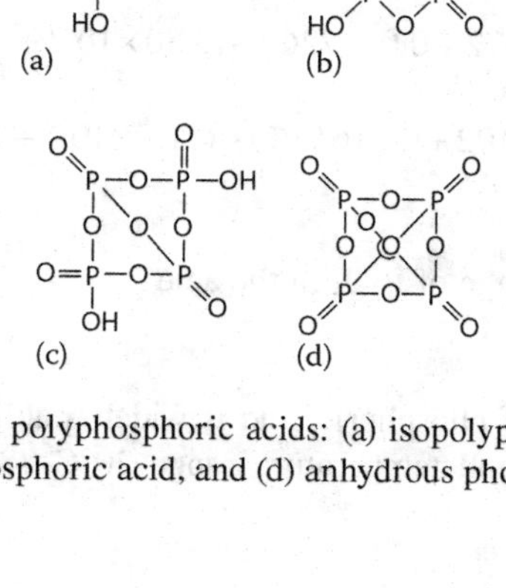

FIGURE 3.4 Cross-linked polyphosphoric acids: (a) isopolyphosphoric acid, (b) isometaphosphoric acid, (c) ultraphosphoric acid, and (d) anhydrous phosphorus pentoxide.

The processing routes to polyacids are conceptually relatively straightforward but do require an understanding of the physical properties of a suite of these non-ideal acids. Much data are published; there are some gaps though, particularly at higher temperatures; nevertheless, there are sufficient data to build robust models.

The first and most obvious data are the vapor–liquid equilibria. There have been several studies of the vapor pressure of phosphoric acid [7–9]. In the manufacture of commercial grades of polyacids from PWA, the concentration range of interest is 61.0%–86.9% P_2O_5. In this range vaporization is incongruent; the vapor pressure depends on both the temperature and composition of the condensed phase. Congruency occurs at 92% P_2O_5 at 869°C and 753 mmHg. In the ranges of acid concentration of 61.6%–92.7% P_2O_5, of temperature 85°C–866°C, and vapor pressure 50–760 mmHg, both Equations 3.1 and 3.2 are satisfactory for process design. Equation 3.1 relates vapor pressure, *p*, in mmHg, to boiling point, *bp*, in °C, at temperature *T* (K). Equation 3.2 relates boiling point to the P_2O_5 content *C* (%) in the acid:

$$\log p = 8.7788 - \left[\frac{5.9080(bp + 273.15)}{T}\right] \tag{3.1}$$

$$bp = aC^4 + bC^3 + cC^2 + dC - 31821.4 \tag{3.2}$$

where

$a = -0.0011267784$
$b = 0.3389473525$
$c = -37.255696$
$d = 1791.2$

Figure 3.5 shows both the boiling point and vapor curves of polyphosphoric acid. For modeling purposes, both the nonrandom-two-liquid (NRTL) equation and the universal quasi chemical (UNIQUAC) equations are satisfactory.

Heat capacity and enthalpy data are available in the concentration range 61.8%–83.0% P_2O_5 and temperature range 25°C–367°C [10]. An earlier paper [11] gives equations for enthalpy and heat capacity.

Density data from Tennessee Valley Authority (TVA) [12] give two equations for density in the concentration range 60.75%–82.84% and 85.61%–122.52% H_3PO_4 and temperature range 25°C–160°C, shown here as follows:

$$\rho_{60.8-82.8} = (0.7972 + 0.01472C) - (3.30\times10^{-4} + 8.00\times10^{-6}C)T \tag{3.3}$$

$$\rho_{85.6-122.5} = (0.7102 + 0.01617C) - (11.7\times10^{-4} - 6.00\times10^{-6}C)T \tag{3.4}$$

where

C is the weight percentage of P_2O_5 in the acid
T is the temperature in °C

The kinematic viscosity of phosphoric acid is widely published in the concentration range 3.6%–85.5% P_2O_5 and temperature range 20°C–180°C and is based on the

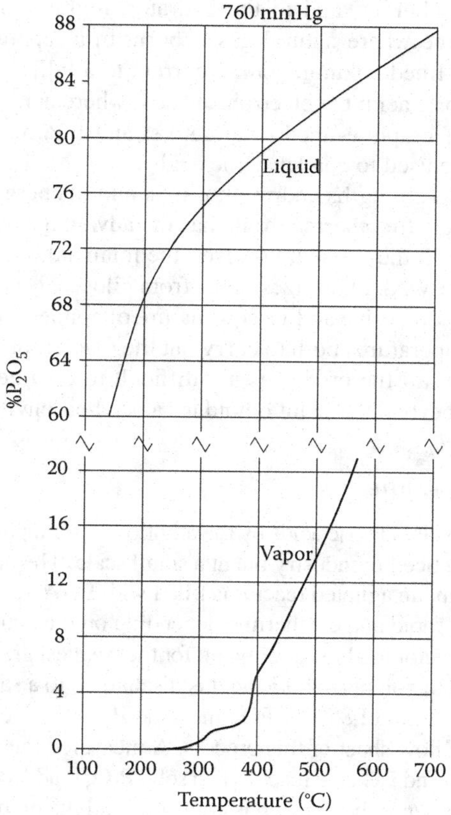

FIGURE 3.5 Vapor–liquid graphs for P_2O_5 and water.

Monsanto Technical Bulletin No. P-26 [12] and is in a TVA patent for superphosphoric acid [13]. The absolute viscosity of ultraphosphoric acids is published in the concentration range 88.7%–92.9% P_2O_5 and temperature range 198°C–505°C [14]. The application of the de Guzman, or more commonly known as the Andrade, equation to model viscosity outside the published ranges is a good approximation [15]. The viscosity of polyphosphoric acid is notable, in the laboratory it will sit in a flask as a highly viscous almost unpourable syrup. It must be heated to over 100°C to start to be useable. By comparison, oleum has a lower viscosity and many similar applications. It may be the case that this simple relative inconvenience has led to less research work and consequently greater opportunity for new discoveries.

3.3 PRODUCTION PROCESSES

There are broadly five production processes for the manufacture of polyphosphoric acid: the solid phosphorus pentoxide route, whereby solid P_2O_5 is added to orthophosphoric acid; the thermal routes where by restricting the water content of the air to the phosphorus burner, and to the hydrator, highly concentrated acids are made; the hot gas route where natural gas is burnt in a separate burner and flows to a packed, graphite-lined column, countercurrent to a falling flow of increasingly concentrated phosphoric acid; the electroheat route where acid undergoes resistance heating in a series of graphite-lined evaporators; and finally the microwave route where microwaves are used to condense the acid.

Each process has its strengths and weaknesses, and in energy terms there is little to choose between them (the thermal route has the advantage of not requiring a heat input as this comes with the P_4 feed; however, the financial benefit is negated by the relative cost of P_4 and PWA). Plant sizes range from pilot scale up to 20,000 tons/year. Because of the highly corrosive and viscous nature of condensed acids, and the need to operate at high temperatures both to carry out the condensation and just pump the acid around the plant, all the processes are difficult to operate. Nevertheless, their safety impact is low beyond the plant boundary as is their environmental footprint.

3.3.1 Solid P_2O_5 Route

The production of polyphosphoric acids by the addition of solid phosphorus pentoxide to phosphoric acid is practiced in industry but at a small scale. The operation is straightforward; in a batch system, an agitated reactor is filled with PWA and then the solid P_2O_5 is added. The reaction is rapid and exothermic depending on the relative feed quantities. A jacket is fitted to the reactor to allow cooling or, for the weakest grades, heating. Once the reaction is complete, the polyphosphoric acid is discharged to a sales package. Table 3.1 shows the relative quantities of 85% H_3PO_4 and solid P_2O_5 for five standard grades.

Figure 3.6 shows a flow sheet of this process. A measured quantity of 85% H_3PO_4 is pumped to the agitated and jacketed reactor then solid P_2O_5 is added from a charge hopper. Agitation is maintained until the reaction is complete, and the product acid is sufficiently cool to be discharged to either an intermediate bulk container (IBC) or 25 liter packages.

Depending on the product quality requirements, a small quantity of either hydrogen peroxide or nitric acid is added to reduce color. If solids content is important, the acid is pumped through a filter.

Superficially, this operation seems simple; however, the product acid is highly corrosive to stainless steel at temperatures above 80°C, which are easily achieved

TABLE 3.1
Feed Proportions

Polyphosphoric Acid Concentration		Feed Proportions kg/kg Product	
%H_3PO_4	**%P_2O_5**	**85% H_3PO_4**	**P_2O_5**
100	72.4	0.717	0.283
105	76.1	0.623	0.377
110	79.7	0.529	0.471
115	83.3	0.434	0.566
118	85.5	0.378	0.622

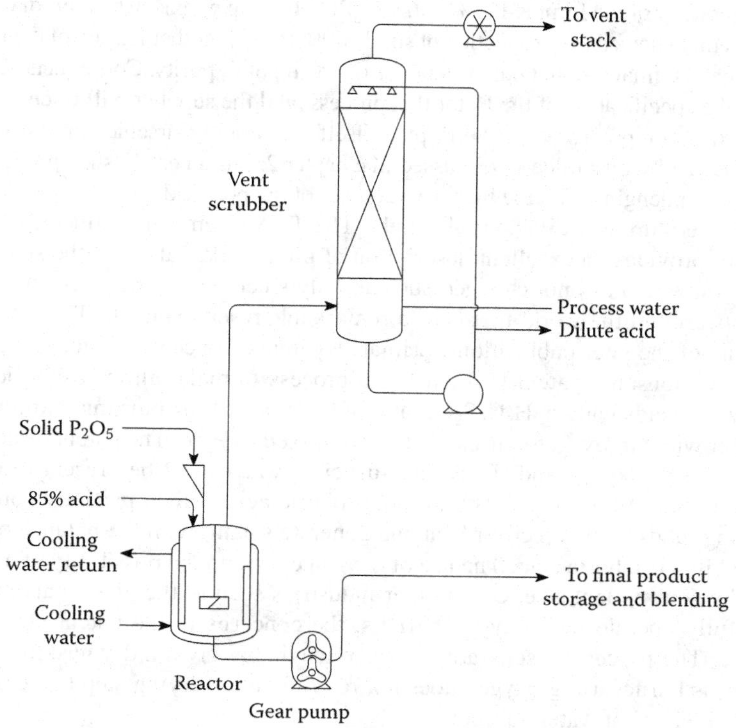

FIGURE 3.6 Solid phosphorus pentoxide flow sheet.

during reaction. A graph of the corrosion of extra low carbon 316 stainless steel is available [13]; it shows a strong link between temperature and corrosion and the highest corrosion rates for acid concentrations of 105%–107% H_3PO_4. Materials of construction options include polytetrafluoroethylene (PTFE) (although the reactor must be maintained below 250°C), graphite, or tantalum lining. The other challenge with this operation is that solid P_2O_5 is hygroscopic and can quickly become a sticky mess as particle surfaces are hydrated by moisture in air to metaphosphoric acid and then stick to each other and equipment, thus causing blockages. A small-scale operation manually charging solid P_2O_5 must be carefully designed to minimize blockages. A larger, potentially continuous operation, linked directly to the P_2O_5 production unit can be designed to avoid manual handling and water vapor ingress.

3.3.2 Thermal Route

The thermal route to orthophosphoric acid was outlined in Section 1.2.10 and a generic flow sheet shown in Figure 1.22; in essence, P_4 is burnt in air to P_2O_5 vapor, which is then condensed in water to form the acid; the amount of added water regulates the concentration of the product acid. Most plants are set up to produce acid in the 75%–85% H_3PO_4 range and can easily extend this to 90% H_3PO_4. By careful control of the water addition to the system, higher acids are produced, although without going to air drying the practical concentration limit is 115%–116% H_3PO_4. Examples exist where air drying was employed that allowed the production of solid 100% P_2O_5. Another feature of the thermal route is that the final product purity depends on the input P_4 purity. Consequently, care is taken in the specification of the P_4 for this process, and the supplier will often carry out some purification processes on the P_4 plant itself. Commonly, arsenic is removed from the product acid with sulfide as discussed in Chapter 2; it is an established process but a little more challenging with the higher viscosities of the polyacids.

Much literature is available in this field. The TVA Chemical Engineering Report No. 2 [16] provides an excellent description of pilot work, data, and the state of the art in 1948, which has not changed substantially since then—there are still operating plants around the world that bear a remarkable resemblance to TVA acid plant 2M; many of the later publications, primarily patents, concentrate on heat recovery [17–19]. A Monsanto patent [14] teaches a process to make ultraphosphoric acids, that is, those acids with an H_2O/P_2O_5 ratio of less than 1:1, by burning phosphorus in a chamber with air, oxygen-enriched air, or admixed oxygen. The patent discourages the use of pure oxygen and states that sufficient water must be present to achieve the required H_2O/P_2O_5 ratio. The ultraphosphoric acid is then partially condensed on cooling tubes (which recover heat and generate steam) with the balance of vapor scrubbed in phosphoric acid. The use of oxygen opens up the possibility of a significant reduction in plant size. Given other industries such as the glass manufacturers successfully operate with oxygen burners, the concerns in the patent are perhaps overdone. The process of separating oxygen inevitably gives a dry gas; therefore, a phosphorus burner using oxygen does not require an air drying step and can easily regulate the level of water addition.

A further feature of the thermal route is its facility in handling recycled acid. Polyphosphoric acid is used in several processes as a catalyst. In many of these processes, it is separated from the desired product at lower concentration and contaminated with organic solvents. If the recycle acid is returned to the producer and co-fed to the P_4 burner, where flame temperatures are in the range 1650°C–2760°C, the organic species are converted to carbon dioxide and water vapor. Depending on the residence time and the configuration of the thermal acid plant, the acid is concentrated and partially vaporized and becomes indistinguishable from the acid derived from the phosphorus. In cases where the recycle acid carries high levels of metal impurity, for example, resulting from etching processes in the electronics industry, the impurities precipitate out on the burner walls as metal metaphosphates. The thermal acid plant at Nashville (see Section 3.3.3) recycled return acid from polyphosphoric acid customers as a co-feed with P_4. One patent describes this application to etchant, which contains nitric and acetic acids [20] and is practiced in Taiwan. These acids are separated by distillation at 80°C and 100°C then the phosphoric acid impurities are removed in the P_4 burner.

3.3.3 Hot Gas Route

The obvious financial difficulty with the thermal route is the cost per ton of the P_2O_5 as P_4 compared to P_2O_5 as H_3PO_4 with routes that condense PWA. Set against the cost advantage of starting with the cheaper acid is the energy cost of removing water; this advantage is further reduced if energy recovery is installed on the thermal plant.

As part of the post-integration rationalization program resulting from the integration of Albright & Wilson (A&W) into Rhodia in 2001, Rhodia closed their old, small polyphosphoric plant operating at Morrisville, PA, and built a new plant at the Nashville site. The old plant comprised a 600 mm diameter graphite column, packed with graphite pall rings and with a demister at the top of the column. Feed acid in the concentration range 75%–85% H_3PO_4 entered the top of the column and passed down it contacting a mixture of hot air and acid vapor passing up the column. Hot air from a small natural gas burner entered the lower part of the column. Acid up to 118.5% H_3PO_4 passed out of the base of the column and was cooled, filtered, and packed. Any acidic mist passing the demister was scrubbed. The old plant lacked any engineering refinement but was functional and produced up to 5000 tpa 118.5% H_3PO_4.

The new plant is described in a patent granted to Rhodia in 2003 [21] and was sized to replace the capacity at Morrisville, replace the sales capacity of the old thermal plant at Nashville, and allow for market growth. The plant has a capacity of 21,000 tpa P_2O_5. Technically, the new plant had to be capable of processing return acid from a range of customers.

A flow sheet of the plant is shown in Figure 3.7. Natural gas is burnt in air in the air heater. The hot air leaves the heater at 900°C–1000°C and enters a duct into the base of the column. If the plant is processing recycle acid containing phosphate esters and

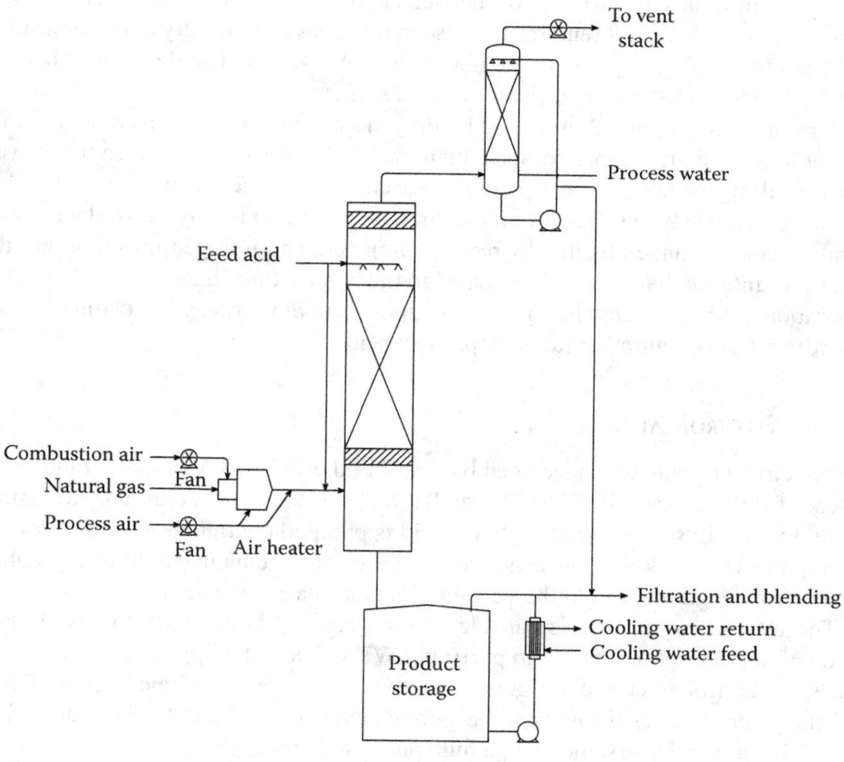

FIGURE 3.7 Polyphosphoric acid hot gas process flow sheet.

organic solvents, this acid is sprayed into the hot air stream. The duct is designed to ensure there is sufficient residence time, around 2 s, to break down the organic species. Cooling air is added in the duct after this stage to lower the column entry temperature to 600°C–700°C. The entry temperature is varied depending on the concentration of the combined feed acids and the designed product acid concentration, Equation 3.2 indicates that 118% H_3PO_4 is produced at 614°C and 120% at 667°C.

Hot gas enters the graphite-lined column and passes upward through the graphite packing support plate and on up through the Raschig ring packing. The packing height is 6 m. Liquid acid passes down the column becoming more concentrated as it does so. The main acid feed enters the top of the column and is distributed uniformly over the packing. The main feed to recycle acid feed is in the range 9–1:1.

Acidic vapor passing through the demister flows to a mist eliminator, then a stack demister, and finally up a vent stack. Weak acid is used around this vent management system and is used in the product dilution system.

The product acid passes out of the base of the column into a receiving tank. A pump recirculates a proportion of the acid through a cooler and back to the receiving tank. The rest of the acid goes forward through a filter to the final product storages. Different sales concentrations are blended here.

The patent teaches a hot air requirement of 200–1200 ft^3/gal of (75%) feed acid. Taking the upper figure, and natural gas at \$4/kft^3, the gas cost is about \$16 per ton P_2O_5. The difference in variable cost between P_4 and PWA, measured in tons P_2O_5, is approximately \$300, so it remains the case that the gas cost penalty of concentrating PWA is very small compared to the higher cost of P_4-derived acid and goes back to the difference in cost between electricity and sulfur.

A patent application [22] was filed by Prayon for a different hot gas route in 2010. The process is slightly more complex than that of the Nashville plant. In the Prayon process, all the orthophosphoric acid is sprayed into a vertical combustion chamber into a flame made by burning air and natural gas. The acid is polymerized with temperature control made possible by recirculating acid and nitrogen injection into the flame. An intermediate vessel is attached to the combustion chamber to aid acid/gas separation, and the gases then pass through a scrubbing tower. No claim is made regarding the concentration range of product acid.

3.3.4 Electroheat Route

The electroheat route was developed by A&W in the 1970s. A&W were granted two patents for the process in 1980 [23] and 1981 [24], the latter being a continuation of the former. In summary, phosphoric acid is pumped continuously into a cascade of graphite vessels, known as evaporators; electricity is conducted down a graphite electrode through the acid to the vessel walls thus heating the acid.

The electroheat process is capable of both concentrating PWA to any desired grade of polyphosphoric acid and purifying WPA. The latter approach initially produces an acid of intermediate purity between WPA and a technical grade PWA, and the patent teaches if this polyphosphoric acid is diluted and further purified by crystallization and dearsenication, a high purity acid results.

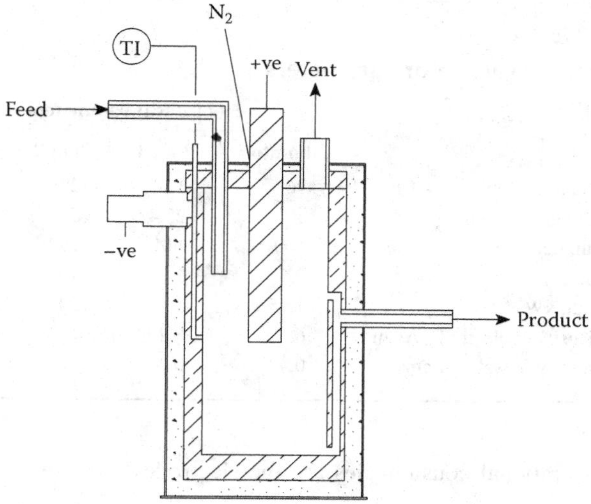

FIGURE 3.8 Electroheat acid process evaporator cross section.

Figure 3.8 shows a section through an evaporator. The larger production scale units comprise a graphite cylinder with a 100 mm wall, a 100 mm graphite base, and roof. The graphite is surrounded by a 100 mm *pug*, an anthracite-based flexible putty. The outer casing is made from stainless steel for acid containment purposes. The pug is designed to handle thermal expansion; its purpose is to provide an acid-impermeable layer around the graphite. The graphite is porous because the operating temperatures do not permit the use of resin fillers. Fresh acid is introduced into the evaporator through a graphite tube. The feed tube terminates just above the liquid surface to minimize splashing. The vessel wall is drilled axially to make the overflow outlet, which is the same diameter as the inlet. Other axial drill holes are made for a thermowell and pressure measurement (for level). A vapor outlet hole is made in the roof the same diameter as the electrode. A&W experimented with different numbers of electrodes and found a single central electrode satisfactory.

The evaporator went through a steady increase in size as it developed. The first laboratory scale evaporator had an internal diameter of 76 mm and utilized a single 19 mm diameter electrode consuming 1.75 kW electrical load. Several pilot scale evaporators were operated; their parameters are shown in Table 3.2.

The largest industrial scale evaporators had an internal diameter of 700 mm, electrode diameter of 150 mm, and a power input of 250 kW. An evaporator of this scale is capable of processing a feed rate of 1200 kg/h of 85% H_3PO_4. The electrical power required depends on the desired temperature and concentration and ranges from 0.325 kWh/kg acid to achieve 73.4% P_2O_5 to 0.953 kWh/kg acid to achieve 82.3% P_2O_5 acid. Taking industrial electricity cost at $0.07/kWh, then the electricity cost for an electroheat plant making 118 grade acid from an 85% H_3PO_4 feed is around $80 per ton P_2O_5.

TABLE 3.2
Electroheat Evaporator Parameters

	Electroheat Parameters			
	Laboratory	Pilot 1	Pilot 2	Pilot 3
Internal diameter, mm	76	127	228	305
Electrode diameter, mm	19	25.4	51	63.5
Acid depth, mm	100	114	190	254
Acid volume, L	0.5	1.7	7.8	18.6
Power input, kW	1.75	3	24	21
Current density of electrode, A/cm²	6.2	6.2		5.9
Current density of walls, A/cm²	0.42	0.25		0.22

The main operational consumables are the electrodes that are eroded at around 2 mm/h.

Carbon from the electrodes as well as any organic components in the feed acid, and any precipitated metal metaphosphates, is filtered after the acid is cooled. Filtration is carried out at 130°C–140°C so that the acid viscosity does not impair filtration. Once filtered the acid is held in final product storage and then packed.

Other aspects of purification taught in the A&W patents are the substantial removal of sulfate above 460°C and the nil effect of concentration on arsenic impurity levels. The example given in the patent starts with a WPA with sulfate content of 9930 ppm. At 380°C, the sulfate level is reduced to 2786 ppm; at 420°C, 904 ppm; and at above 500°C, it is less than 100 ppm.

The removal of arsenic from polyphosphoric acid using sulfide and filtration is hampered because of its high viscosity at moderate temperatures. The A&W patent teaches that 2000 ppm NaCl added at 120°C reduces arsenic from 10 ppm to less than 1 ppm but leaves 86 ppm chloride, at 150°C the chloride is reduced to less than 5 ppm. A recent Japanese patent to Toyo [25] teaches that blowing gaseous hydrogen chloride through polyphosphoric acid above 130°C reduces the arsenic content from 58 ppm to less than 1 ppm. A Hoechst patent [26] teaches the addition of sodium hypophosphate, $NaH_2PO_2 \cdot H_2O$.

Figure 3.9 shows a simplified flow sheet of the electroheat route. Feed acid, typically 85% H_3PO_4 technical or food grade, is received into the plant day tank. The feed acid is pumped continuously under flow control to the first evaporator. The acid enters the first evaporator through a graphite dip pipe and flows to the liquor surface where it boils. As acid flows into the evaporator, it displaces concentrated acid through an internal overflow. As the acid boils, fumes of phosphoric, sulfuric, and hydrofluoric acid are given off that pass out of the evaporator via the vent branch to the vent header. As the central electrode erodes it is lowered automatically to maintain the desired gap with the evaporator wall. A small gap around the electrode as it penetrates the evaporator roof allows free movement. To avoid air ingress and consequent oxidation, or uncontrolled egress of acidic fumes, a low pressure nitrogen purge is applied to the electrode nozzle. The concentration of the acid is varied in each evaporator by controlling the

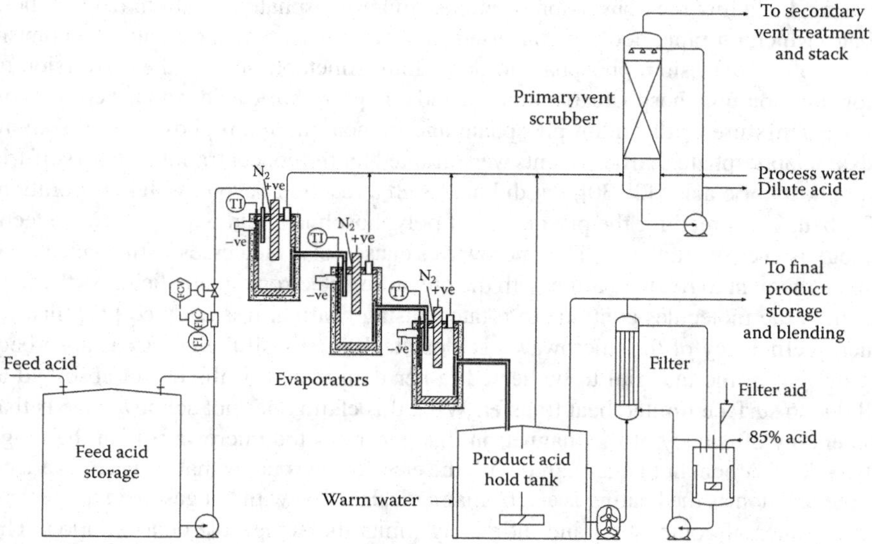

FIGURE 3.9 Electroheat process flow sheet.

acid temperature, which is done by regulating the voltage and electrode/wall gap. The acid overflows to the second evaporator, which in turn overflows to the third.

Hot vent gases pass into a vertical scrubbing column. These gases are scrubbed with a dilute acid spray. Scrub liquors are cooled and pumped away and used for phosphate processes.

The concentrated acid overflows from the third evaporator into the product acid hold tank, an agitated, jacketed vessel. Hot water is pumped around the jacket; this and some recycled product acid cools the concentrated feed from the evaporator.

The acid is pumped, with a gear pump, through a filter. The operating temperature rules out most filter cloths and a metal candle filter is used. The filter is pre-coated with diatomaceous earth mixed with 85% H_3PO_4. Following filtration, some acid is recycled back to pre-filtration to aid cooling and the rest is transferred forward to final product storage. From final product storage, the acid is either packed or blended. Depending on the concentration of acid produced, and the desired final product concentrations, blending is done with either 85% H_3PO_4 or solid P_2O_5. Thus, if the plant normally operates at 116% H_3PO_4 concentration and is required to produce a 118 grade, then solid P_2O_5 is added; for a 105 or 110 grade, 85% H_3PO_4 is blended.

3.3.5 Microwave Route

Polyphosphoric acid has been made at pilot scale with a microwave unit. A&W developed this route and piloted it in Canada in the late 1990s and was granted a patent in 1999 [27]. More recently, this route was assessed in an academic setting for energy efficiency and processing rate compared to conventional heating [28]. The patent also gives examples of phosphate products made through the application of microwave heating, including the conversion of monosodium phosphate to sodium trimetaphosphate, which on rapid cooling converted to sodium hexametaphosphate; the conversion of monopotassium phosphate to potassium trimetaphosphate; the conversion of monoammonium phosphate to ammonia and polyphosphoric acid; and the conversion of a 1:1 mixture of disodium phosphate and monoammonium phosphate to sodium hexametaphosphate. Earlier patents were granted for the concentration of both sulfuric and phosphoric acids [29,30], but did not teach a reaction beyond water evaporation. The benefit claimed for the production of polyphosphoric acid with microwave technology is energy efficiency. The microwaves cause polar molecules with an electrical dipole moment to rotate to align with the prevailing electromagnetic field. As the field changes the molecules continue to rotate causing heating. It is claimed [27] that the energy efficiency of the microwaves is close to 100% as all the energy is absorbed, by design, in the material to be heated rather than reactor walls or not absorbed at all due to surface limited heat transfer. What this claim does not acknowledge is that the energy efficiency of the magnetron that generates the microwave is in the range 60%–70%. A benefit that is shared with the electroheat route is that no water is added to the reaction as the heating is electromagnetic. Heating with hot gas inevitably introduces water of combustion and ultimately limits the achievable concentration. The microwave patent gives examples of concentrations up to 125% H_3PO_4.

A fascinating patent of academic, if not as yet industrial, interest teaches the reduction of phosphoric acid and carbon by microwave heating to elemental phosphorus [31] according to the reaction in Equation 3.5. This reaction is well known (see M.M. Coignet in Chapter 1) and is described in a patent of 1876 [32]; the difference is that rather than *white heat* (1370°C–1480°C), with microwave heating only 540°C is needed:

$$4H_3PO_4 + 16C \rightarrow 6H_2 + 16CO + P_4 \qquad (3.5)$$

The microwave patent claims that phosphorus can be made with much less waste and lower energy than the traditional route. It opens up the possibility of users of relatively small amounts of phosphorus making rather than buying their feed material.

3.4 USES

Space does not permit an exhaustive coverage of all the uses of polyphosphoric acid, a list of which anyway continues to grow. Nevertheless, a summary of much of the organic chemistry functionality and a few well-known and industrially significant applications are set out in the following.

3.4.1 Polyphosphoric Acid as a Reagent in Organic Chemistry

Although polyphosphoric acid was first studied in the nineteenth century, and as an example the industrially significant application of polyphosphoric acid in solid phosphoric acid (SPA) catalysts for polymerization of olefins to gasoline was developed in the 1930s [33], it was in the 1950s that many new applications were discovered. Two reviews were published in 1958 and 1960, which still cover much of the ground [34,35]; the following summary follows the structure of these papers.

3.4.1.1 Cyclization Reactions

Cyclization reactions are grouped into three groups: the preparation of heterocyclic compounds, intramolecular acylation, and the cyclodehydration reactions of aldehydes, ketones, and alcohols.

Under the preparation of heterocyclic compounds, examples include the following: Fischer indole synthesis, Pomeranz–Fritsch reaction, synthesis of nitrogen heterocycles, and synthesis of sulfur and oxygen heterocycles.

Indole is an aromatic heterocyclic organic compound consisting of a benzene ring and a five-membered pyrrole ring. Indoles are precursors to fragrances and antimigraine drugs. The oldest indole synthesis is that discovered by Fischer for which different acid catalysts are used, including polyphosphoric acid.

Isoquinoline [36] is an aromatic heterocyclic organic compound consisting of a benzene ring and a pyridine ring. Isoquinolines are precursors to anesthetics, antihypertension drugs, antifungal agents, disinfectants, and vasodilators. Isoquinoline is prepared by the Pomeranz–Fritsch reaction with polyphosphoric acid catalysis.

Hexahydrophenanthrenones [37] are intermediates for the synthesis of morphines and D-homosteroids. These molecules are prepared via the polyphosphoric acid–catalyzed acylation–cycloalkylation of aromatics with cyclohexene-1-acetic acid, as shown in Figure 3.10.

The Bradsher reaction is an example of polyphosphoric acid–catalyzed cyclodehydration where a diarylmethane containing a carbonyl group is cyclized to anthracene derivatives, as shown in Figure 3.11.

3.4.1.2 Rearrangements

There are many arrangements catalyzed by polyphosphoric acid; the following examples are given: Beckmann, Lossen, and Wagner–Meerwein.

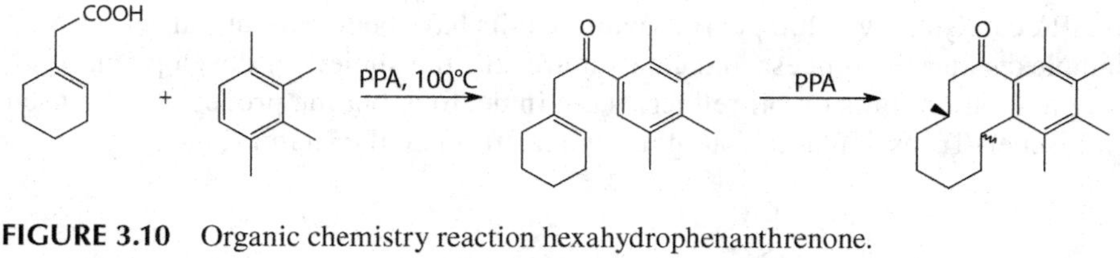

FIGURE 3.10 Organic chemistry reaction hexahydrophenanthrenone.

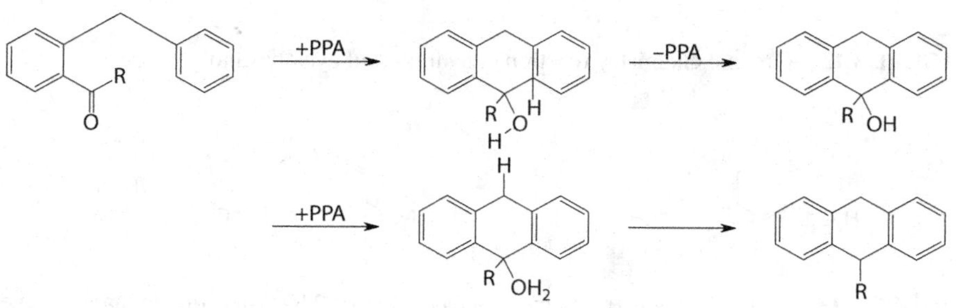

FIGURE 3.11 Organic chemistry reaction Bradsher reaction.

The Beckmann rearrangement is used to make caprolactam from cyclohexanone, which in turn is made into nylon 6 or polycaprolactam. In general, the Beckmann rearrangement is the acid-catalyzed rearrangement of an oxime to an amide. At industrial scale, sulfuric acid is used for the caprolactam process.

Hydroxamic acids undergo Lossen rearrangement to an isocyanate when treated with polyphosphoric acid.

An example of a Wagner–Meerwein rearrangement is the conversion of spiroindolenine into a carbazole derivative by polyphosphoric acid.

3.4.1.3 Dehydration

The dehydration of cyclohexanol to cyclohexene with polyphosphoric acid is a classic example of dehydration of alcohols with an acid catalyst; see Figure 3.12. Polyphosphoric acid can be used to catalyze the production of motor fuel in the gasoline range from alcohols such as ethanol, propanol, normal and iso-butanol, normal and iso-pentanol, and various hexanols [38].

3.4.1.4 Hydrolysis

The hydrolysis of nitriles to carboxylic acids with orthophosphoric acid is classic as is reductive decyanation with polyphosphoric acid, for example, Figure 3.13, which shows the decyanation of 2,3,6,7-tetramethylnaphthalene-1,4-dinitrile.

3.4.1.5 Polymerization

Oil refineries produce high volumes of alkenes (olefins), which are used as a feedstock for petrochemical processes. For example, ethylene is polymerized to polyethylene. Not all oil refineries have petrochemical units; therefore, an alternative is to polymerize the olefins to larger olefins in the gasoline range. This process was invented by Vladimir Ipatieff [33] and assigned to UOP (formerly Universal Oil Products), an oil processing technology company. The process used an SPA catalyst. Over 200 polymerization units have been built as part of refinery complexes and the process and catalyst are still the subject of development [39]. Polymerization units on oil refineries are in decline, but the process is also used on Fischer–Tropsch plants that make syngas from coal or natural gas.

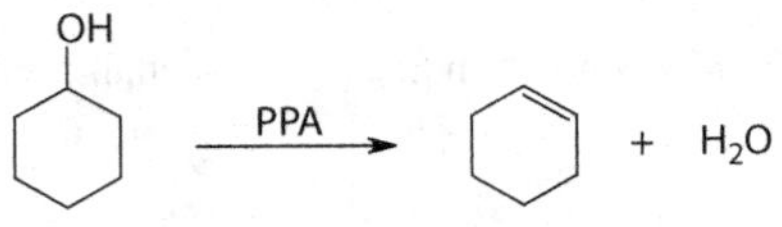

FIGURE 3.12 Organic chemistry reaction dehydration of cyclohexanol.

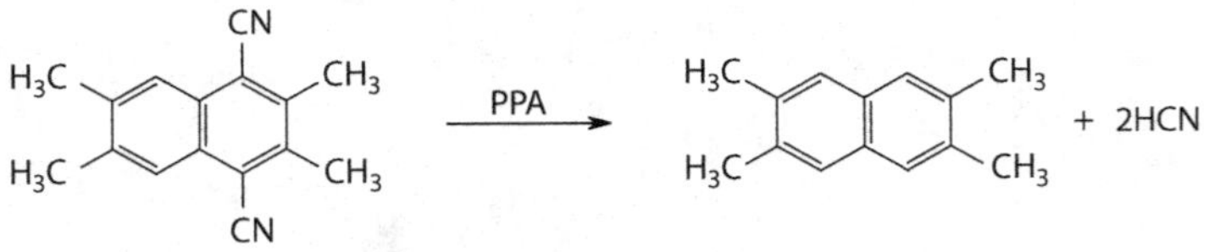

FIGURE 3.13 Organic chemistry reaction decyanation of 2,3,6,7-tetramethylnaphthalene-1,4-dinitrile.

3.4.2 SPA: Solid Phosphoric Acid Catalyst

SPA catalysts are made by mixing a siliceous or aluminous substrate and higher phosphoric acids, extruding the mixture, cutting it into pieces, and drying the pieces at about 300°C [40,41]. Later patents teach drying, calcining, and steaming at 120°C, 400°C, and 200°C, respectively [42]. As SPA manufacture has continued to develop, the main teaching of patents granted in the last 20 years is about porosity control [38]. The main competition for SPA is zeolite-based catalysts; certainly porosity can be positively designed into a zeolite structure whereas an SPA based on diatomaceous earth is inevitably irregular.

In operation, older SPA formulations have tended to break up; thus, one aspect of their design is robustness. Chemically SPA catalysts are damaged by sulfur and poor hydration control [43]. This is because the active phase of the catalyst is the viscous acidic layer on the substrate surface (rather than the silicon phosphates that are formed during catalyst manufacture), and the activity is reduced as the acid concentration is lowered by increased hydration. The effect of hydration and temperature on product selectivity for butane oligomerization was investigated by de Klerk et al. [5].

SPA continues in use because changing catalyst on an oil refinery is a nontrivial decision because of the scale and interdependence of so many operations. SPA development continues because it is a competitive catalyst in terms of performance and is also environmentally friendly. Spent SPA is neutralized with ammonia and then available as a fertilizer.

3.4.3 Polyamide Yarns

Kevlar is a famous example of an aromatic polyamide yarn. Exceptionally high tensile properties and strength-to-weight ratio make it useful for bulletproof vests and containment shielding on the fan section of commercial jet engines. Once the polymerization is carried out, the polymer is dissolved in a dope (solvent) to allow it to be spun into a continuous yarn. Kevlar uses sulfuric acid but two other polyamides of similar properties, polyareneazole [44] and polybenzazole (PBZ) [45] polymers, use polyphosphoric acid as the solvent.

3.4.4 Quinacridone Pigments

Quinacridone pigments are known for their red to violet shades and high performance. They are used in automotive, industrial, and powder coatings, for coloring various plastics and in inks. The basic quinacridone molecule was first discovered in 1935 but only came to industrial importance in the late 1950s. There are at least 13 polymorphs of quinacridone (α, β, six γ-forms, δ, Δ, ε, and ζ) [46]. The final step of the industrial synthesis is ring closure with polyphosphoric acid.

3.4.5 Modified Bitumens

Bitumen is used in over 200 applications including road surfacing and roofing. In many countries, surface breakdown due to thermal expansion and contraction is a real problem. Bitumen is a residue of the distillation of crude oil, the composition of which varies with its source. Different bitumens behave somewhat differently; nevertheless, in order to improve their surfacing performance, they are frequently modified with an additive. Polyphosphoric acid, either alone or with another polymer, has proved to be highly effective at improving the surface performance of bitumen. Many patents have been granted in the last 10 years, and this field remains an active area of industrial development.

REFERENCES

1. R. F. Jameson, The composition of the "strong" phosphoric acids, *J. Chem. Soc. Resumed*, 1959(0), 1752–759, January 1959.
2. R. F. Jameson, M. M. Striplin, W. C. Scott, R. A. Shetler, and J. M. Williams, Polyphosphoric acid, in *Phosphoric Acid*, vol. 1 Part II, 2 vols., A. V. Slack, Ed., Marcel Dekker, New York, 1968, pp. 983–1101.
3. E. Thilo, Condensed phosphates and arsenates, in *Advances in Inorganic Chemistry and Radiochemistry*, vol. 4, H. J. Emeléus and A. G. Sharpe, Eds., Academic Press, New York, 1962, pp. 1–75.
4. F. Rashchi and J. A. Finch, Polyphosphates: A review their chemistry and application with particular reference to mineral processing, *Miner. Eng.*, 13(10), 1019–1035, 2000.
5. A. de Klerk, D. O. Leckel, and N. M. Prinsloo, Butene oligomerization by phosphoric acid catalysis: Separating the effects of temperature and catalyst hydration on product selectivity, *Ind. Eng. Chem. Res.*, 45(18), 6127–6136, August 2006.
6. A.-L. Huhti and P. A. Gartaganis, The composition of the strong phosphoric acids, *Can. J. Chem.*, 34(6), 785–797, June 1956.
7. B. J. Fontana, The vapor pressure of water over phosphoric acids, *J. Am. Chem. Soc.*, 73(7), 3348–3350, July 1951.
8. E. H. Brown and C. D. Whitt, Vapor pressure of phosphoric acids, *Ind. Eng. Chem.*, 44(3), 615–618, March 1952.
9. A. K. Chowdhury, M. B. Liu, and A. Gulbenkian, Mass spectrometric studies of vaporization of phosphoric acids, *Ind. Eng. Chem. Res.*, 32(5), 989–994, May 1993.
10. B. B. Luff, Heat capacity and enthalpy of phosphoric acid, *J. Chem. Eng. Data*, 26(1), 70–74, January 1981.
11. Z. T. Wakefield, B. B. Luff, and R. B. Reed, Heat capacity and enthalpy of phosphoric acid, *J. Chem. Eng. Data*, 17(4), 420–423, October 1972.
12. T. D. Farr, *Phosphorus—Properties of the Element and Some of Its Compounds*, Chemical Engineering Report No. 8, Tennessee Valley Authority, Muscle Shoals, AL, 1950, reprinted 1966.
13. H. Y. Allgood, Process for the production of highly concentrated phosphoric acid, US Patent 3442611, May 1969.
14. R. B. Hudson, Method of preparing ultraphosphoric acid, US Patent 4309394, January 1982.
15. B. E. Poling, J. M. Prausnitz, and J. P. O'Connell, *The Properties of Gases and Liquids*, McGraw Hill Professional, New York, 2001.
16. M. Striplin, *Development of Processes and Equipment for Production of Phosphoric Acid*, Tennessee Valley Authority, Wilson Dam, AL, 1948.
17. W. Klemm, B. Kuxdorf, P. Luhr, U. Thummler, and H. Werner, Process for making phosphorus pentoxide and optionally phosphoric acid with utilization of the reaction heat, US Patent 4603039, July 1986.
18. H. T. Spruill and T. A. Webster, Heat recovery through oxidation of elemental phosphorus in a fluidized bed, US Patent 4618483, October 1986.
19. H. Rosenhouse and J. F. Shute, Heat recovery in the manufacture of phosphorus acids, US Patent 4713228, December 1987.

20. Y. H. Hsu and S. Hsu, Processes for purifying phosphoric acid and for preparing phosphoric acid, US Patent 7166264, January 2007.
21. D. L. Myers, R. W. Hudson, N. P. Mills, D. M. Razmus, and V. Payen, Method for making polyphosphoric acid, US Patent 6616906, September 2003.
22. A. Germeau and B. Heptia, Method and device for producing polyphosphoric acid, US Patent 0058038, March 2012.
23. E. J. Lowe, M. W. Minshall, and A. Wilson, Purification of wet process phosphoric acid, US Patent 4215098, July 1980.
24. E. J. Lowe, M. W. Minshall, and A. Wilson, Purification of wet process phosphoric acid, US Patent 4296082, October 1981.
25. K. Hotta and F. Kubota, Method for purification of phosphoric acid high purity polyphosphoric acid, US Patent 6861039, March 2005.
26. G. Heymer, W. Scheibitz, and H. Spott, Process for the dearsenication of polyphosphoric acid, US Patent 3991164, November 1976.
27. J. P. Godber and H. Jenkins, Condensation polymerization of phosphorus containing compounds, US Patent 5951831, September 1999.
28. N. A. Pinchukova, V. A. Chebanov, N. Y. Gorobets, L. V. Gudzenko, K. S. Ostras, O. V. Shishkin, L. A. Hulshof, and A. Y. Voloshko, Beneficial energy-efficiencies in the microwave-assisted vacuum preparation of polyphosphoric acid, *Chem. Eng. Process. Process Intensif.*, 50(11–12), 1193–1197, 2011.
29. G. M. J. Masse, Purification and reconcentration of waste sulphuric acid, US Patent 4671951, June 1987.
30. C. Y. Cha, Process for microwave catalysis of chemical reactions using waveguide liquid films, US Patent 5451302, September 1995.
31. R. R. Severns, Method of preparing phosphorus, US Patent 6207024, March 2001.
32. A. G. Hunter, Improvement in retorts for distilling phosphorus, US Patent 171813, January 1876.
33. V. Ipatieff, Treatment of hydrocarbons, US Patent 1993513, March 1935.
34. F. D. Popp and W. E. McEwen, Polyphosphoric acids as a reagent in organic chemistry, *Chem. Rev.*, 58(2), 321–401, April 1958.
35. F. D. Popp and W. E. McEwen, Polyphosphoric acid as a reagent in organic chemistry, II, *Trans. Kans. Acad. Sci. (1903–)*, 63(3), 169–193, Autumn 1960.
36. G. Grethe, *The Chemistry of Heterocyclic Compounds, Isoquinolines*. John Wiley & Sons, New York, 2009.
37. G. Sartori and R. Maggi, *Advances in Friedel-Crafts Acylation Reactions: Catalytic and Green Processes*, CRC Press, Boca Raton, FL, 2010.
38. J. W. Jean, Process for synthesizing hydrocarbons from alcohols with tetraphosphoric acid catalyst, US Patent 2373475, April 1945.
39. J. L. Braden, G. A. Korynta, W. Turbeville, and L. Xu, Solid phosphoric acid with controlled porosity, US Patent 8063260, November 2011.
40. V. Ipatieff, Treatment of hydrocarbons, US Patent 2020649, November 1935.
41. J. C. Morrell, Kaolin-phosphoric acid polymerization catalyst, US Patent 2586852, February 1952.
42. J. L. Braden, G. A. Korynta, W. Turbeville, and L. Xu, Solid phosphoric acid with controlled porosity, US Patent 7557060, July 2009.
43. R. R. D. Graff, Process for catalyst hydration control, US Patent 3520945, July 1970.
44. C. W. Newton, Processes for hydrolyzing polyphosphoric acid in shaped articles, US Patent 7977453, July 2011.
45. K.-S. Lee, Polybenzazole fibers and processes for their preparation, US Patent 7189346, March 2007.
46. E. F. Paulus, F. J. J. Leusen, and M. U. Schmidt, Crystal structures of quinacridones, *Cryst. Eng. Comm.*, 9(2), 131, 2007.

4 Sodium Phosphates

4.1 INTRODUCTION

Sodium phosphates have been and remain the most prolific phosphate derivatives of orthophosphoric acid. Of these products, sodium tripolyphosphate (STPP), although in steep decline, remains the largest single product. The washing effectiveness of STPP-based products was so much better than anything else on the market that its sales growth in the United States in the late 1940s and early 1950s was nearly vertical. By the early 1950s, it had gone from nowhere to three times the combined volume of all the other sodium phosphate products in the United States. Sales continued to grow and the drive to reduce its manufacturing cost in turn drove the development of PWA technology. Nevertheless, even as new plants were built around the world in the 1960s and in all the decades up to the end of the twentieth century, voices questioning its environmental impact began to be heard. The industry worked with independent bodies to study the full environmental life cycle impact of STPP compared to its principal competitor zeolite (discussed further in Chapter 7). Although the conclusions were that STPP was overall slightly the better product, sales volumes crashed as politically lead bans took place. Greatly reduced sales volumes inevitably lead to plant closures and were one cause of industry rationalization.

Sodium acid pyrophosphates (SAPP) are used as raising agents in bakery products. In recent years, health concerns about the level of sodium in the diet have lead to a move to reduce sodium and so switch away from sodium phosphates. In the main, this has lead to a move toward both calcium and potassium phosphates as replacements. Despite this apparent decline, the sodium phosphates field, particularly in applications rather than manufacturing, remains an active area of industrial research and development as evidenced by the number of patents granted in the last 10 years.

4.2 CHEMISTRY

4.2.1 Sodium Orthophosphates

The chemistry of the pure sodium orthophosphates is best described with reference to the Na_2O–P_2O_5–H_2O phase diagram. Studies of the phase equilibria of sodium orthophosphates in aqueous solutions have shown the existence of at least 15 sodium salt solid phases in the temperature range 25°C–100°C [1]. These crystalline sodium phosphates are the following:

NaH_2PO_4	Na_2HPO_4	Na_3PO_4
$NaH_2PO_4 \cdot H_2O$	$Na_2HPO_4 \cdot 2H_2O$	$Na_3PO_4 \cdot ½H_2O$
$NaH_2PO_4 \cdot 2H_2O$	$Na_2HPO_4 \cdot 7H_2O$	$Na_3PO_4 \cdot 6H_2O$
$2NaH_2PO_4 \cdot Na_2HPO_4 \cdot 2H_2O$	$Na_2HPO_4 \cdot 8H_2O$	$Na_3PO_4 \cdot 8H_2O$
$NaH_2PO_4 \cdot Na_2HPO_4$	$Na_2HPO_4 \cdot 12H_2O$	$Na_3PO_4 \cdot ca12H_2O$
$NaH_2PO_4 \cdot H_3PO_4$ (also expressed $NaH_5(PO_4)_2$)		

Sodium triphosphate dodecahydrate does not exist as a pure crystal and always contains an excess of sodium hydroxide. More accurately, the dodecahydrate comprises two isomorphous salts of the formula $Na_3PO_4 \cdot ¼NaOH \cdot 12H_2O$ and $Na_3PO_4 \cdot \frac{1}{7}NaOH \cdot 12H_2O$ [1], although there is also evidence for $Na_3PO_4 \cdot \frac{1}{5}NaOH \cdot 12H_2O$ [2] and others claim there is a range of composition between 0 and ¼ [3]. Figure 4.1 shows the solubility curves at 25°C, 40°C, 60°C, and 100°C in the Na_2O–P_2O_5–H_2O orthophosphate system and is redrawn from Wendrow and Kobe [1]. The salts shown are labeled as follows: TSP·$n$$H_2O$—trisodium phosphates; DSP·$n$$H_2O$—disodium phosphates; MSP·$n$$H_2O$—monosodium phosphates; HSP—sodium pentahydrogen

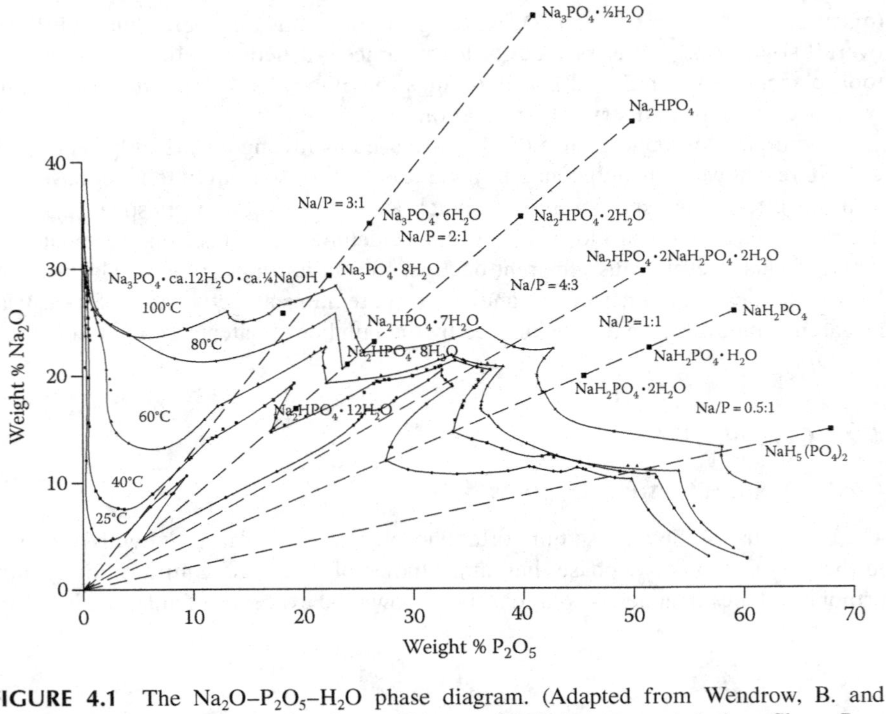

FIGURE 4.1 The Na_2O–P_2O_5–H_2O phase diagram. (Adapted from Wendrow, B. and Kobe, K.A., The alkali orthophosphates. Phase equilibria in aqueous solution, *Chem. Rev.*, 54(6), 891–924. Copyright 1954 American Chemical Society.)

TABLE 4.1
Summary of Sodium Phosphate Compositions

Name	Formula	MW	Na/P	Na_2O/P_2O_5	$\%Na_2O$	$\%P_2O_5$	$\%H_2O$
HSP	$NaH_5(PO_4)_2$	218	0.50	0.22	14.2	65.1	20.7
MSP	NaH_2PO_4	120	1.00	0.44	25.8	59.2	15.0
MSP1	$NaH_2PO_4 \cdot H_2O$	138	1.00	0.44	22.5	51.4	26.1
MSP2	$NaH_2PO_4 \cdot 2H_2O$	156	1.00	0.44	19.9	45.5	34.6
DSP	Na_2HPO_4	142	2.00	0.87	43.7	50.0	6.3
DSP2	$Na_2HPO_4 \cdot 2H_2O$	178	2.00	0.87	34.8	39.9	25.3
DSP7	$Na_2HPO_4 \cdot 7H_2O$	268	2.00	0.87	23.1	26.5	50.4
DSP8	$Na_2HPO_4 \cdot 8H_2O$	295	2.00	0.87	21.0	24.1	54.9
DSP12	$Na_2HPO_4 \cdot 12H_2O$	367	2.00	0.87	16.9	19.3	63.8
TSP	Na_3PO_4	164	3.00	1.31	56.7	43.3	0.0
TSP0.5	$Na_3PO_4 \cdot 0.5H_2O$	173	3.00	1.31	53.8	41.0	5.2
TSP6	$Na_3PO_4 \cdot 6H_2O$	272	3.00	1.31	34.2	26.1	39.7
TSP8	$Na_3PO_4 \cdot 8H_2O$	317	3.00	1.31	29.3	22.4	48.3
A	$Na_3PO_4 \cdot$ ca.$12H_2O \cdot$ ca.¼NaOH	390	3.25	1.42	25.8	18.2	56.0
B	$Na_2HPO_4 \cdot NaH_2PO_4$	262	1.50	0.65	35.5	54.2	10.3
C	$Na_2HPO_4 \cdot 2NaH_2PO_4 \cdot 2H_2O$	418	1.33	0.58	29.7	50.9	19.4

bis phosphate, $NaH_5(PO_4)_2$; A = $Na_3PO_4 \cdot$ ¼$NaOH \cdot 12H_2O$; B = $NaH_2PO_4 \cdot Na_2HPO_4$; and C = $2NaH_2PO_4 \cdot Na_2HPO_4 \cdot 2H_2O$. At any point in the diagram, the composition is the sum of the weight percentage of sodium and phosphorus expressed as Na_2O and P_2O_5 with the balance to 100% expressed as H_2O. The water constituent might be intermolecular, water of crystallization or water of dilution, and combinations of each. The dotted lines represent the molar ratios of sodium to phosphorus—in both research and development and production circles, the Na/P ratio is a common term. All compounds of the same Na/P ratio lie on the same line regardless of the level of hydration. The location of each pure salt is marked by a square, and in the area between these points and the solubility curve, this is the salt that precipitates. Table 4.1 sets out the molecular weight, the Na/P molar ratio, the Na_2O/P_2O_5 weight ratio, and the weight percentage Na_2O, P_2O_5, and H_2O.

Figure 4.2 shows the Na_2O–P_2O_5–H_2O system at 40°C. The 40°C curve is chosen because industrial processes that isolate the desired sodium salt via crystallization do so in the 40°C–60°C range, and also all the compounds in Table 4.1 are still present at 40°C. As a modest simplification, it is noted that below 1% P_2O_5 the phase diagram becomes fairly complex but since industrial processes do not operate in this area, it requires no further discussion. The zones with tie lines are two phase, that is, the pure crystal at the head of the zone and its mother liquor. Numbered zones comprise three constituents, a mother liquor and two crystalline phases of the composition of the compounds bounding the zone.

Not all these stable sodium salts are sold as commercial products. The principal pure sodium orthophosphates of commerce are MSP (food grade and technical

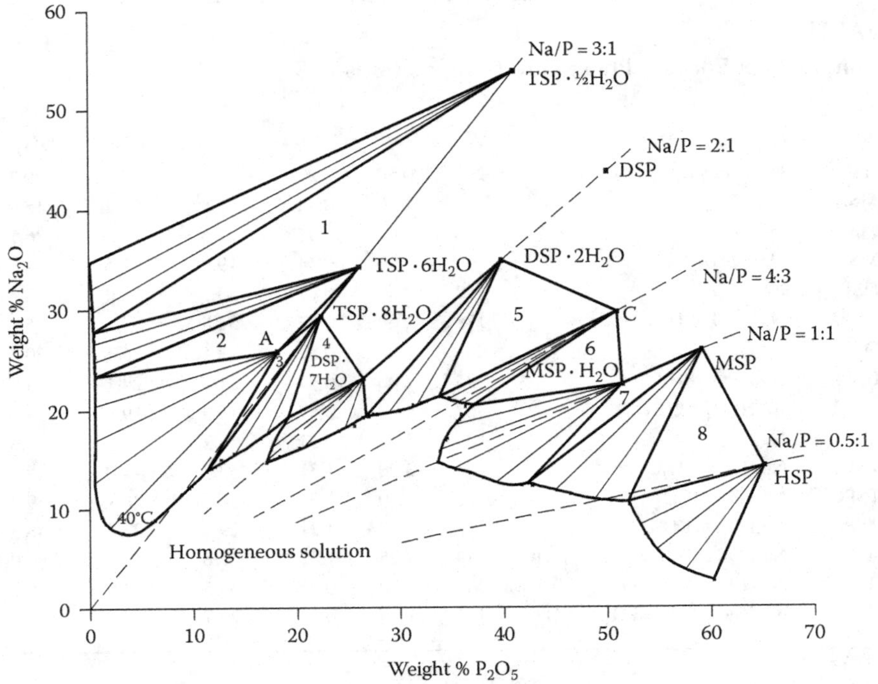

FIGURE 4.2 The Na_2O–P_2O_5–H_2O system at 40°C.

grade), DSP (food grade and technical grade) and DSP2 (food grade), TSP (food grade and technical grade), and TSP12 (food grade). There are broadly two possible approaches to manufacturing these products: crystallization followed by drying and dehydration, or rehydration, for the anhydrous or hydrated products respectively, or just drying (and dehydration as necessary). The crystallization route affords the opportunity of some product purification through crystallization and less energy for drying. This is the route that was used traditionally and while most food grade sodium orthophosphates were made from thermal acid, the crystallization route was also used to make these salts from wet process acid (discussed in Chapter 2 under chemical purification). The current approach is to make these products in one step in a spray drier, with TSP requiring additional calcining. Both processes start by making a neutral liquor of the required Na/P ratio for the desired product. This neutralization is exothermic and when the following step is drying it makes sense for the reaction to be controlled in the range 90°C–110°C, which is why Figure 4.3 is of interest that shows the solubility line at 100°C. This system is clearly a little more straightforward than those at lower temperatures. Both soda ash (Na_2CO_3) and 50% sodium hydroxide are used as the sodium source. In their raw state, they sit on the Na_2O axis either side of the 40% Na_2O point. By using the lever arm rule, relative proportions of the sodium source and phosphoric acid can be estimated graphically. For example, the production of

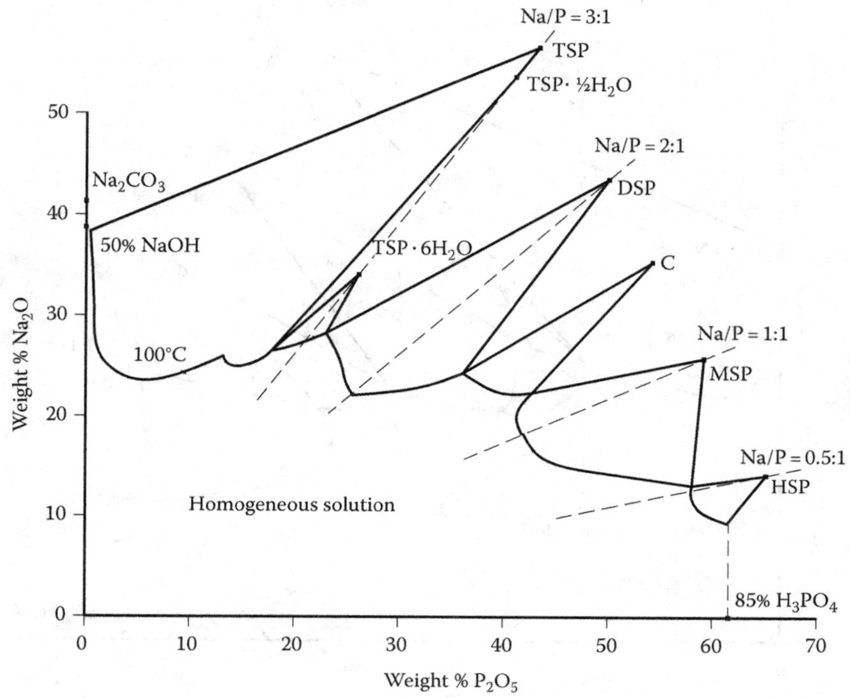

FIGURE 4.3 The Na_2O–P_2O_5–H_2O system at 100°C.

DSP is calculated both analytically and graphically. The reaction of 50% NaOH and 85% H_3PO_4 is given in the following equation:

$$(H_3PO_4 + 0.96H_2O) + 2(NaOH + 2.22H_2O) \rightarrow Na_2HPO_4 + 7.4H_2O \quad (4.1)$$

Using molecular weights, the weight of the acidic solution and the caustic soda solution are 115.3 and 160, respectively. Therefore, the weight proportion of the dilute acid component is 115.3/275.3 = 41.9%.

Turning to Figure 4.4, a straight line is drawn from the 50% NaOH point to the 85% H_3PO_4 point. Solving graphically the proportion of dilute acid to make DSP is the distance from the 50% NaOH point to the intersection with the Na/P = 2:1 line divided by the distance between the 50% NaOH point and the 85% H_3PO_4 point. In this example, the dimensions are 60.95 and 145.52, so dividing we have 41.9%. Similarly, once the reaction is complete, the weight proportion of water that must be evaporated for a dry product is 7.4 × 18/275.3 = 48.4%. Graphically (following the line from the origin to DSP), the proportion of dry DSP is given by 68.48/132.76 = 51.6%; therefore, the water for evaporation is 48.4%.

Two other sodium orthophosphates of commercial interest are sodium aluminum phosphate (SALP) and chlorinated TSP (TSP-chlor). There are many forms of SALP, $x Na_2O \cdot y Al_2O \cdot 8P_2O_5 \cdot z H_2O$, where x is in the range 0.5–15, y 1.5–4.5, and z 0–50 [4,5], and many patents claiming different functionalities and reactivities depending on composition.

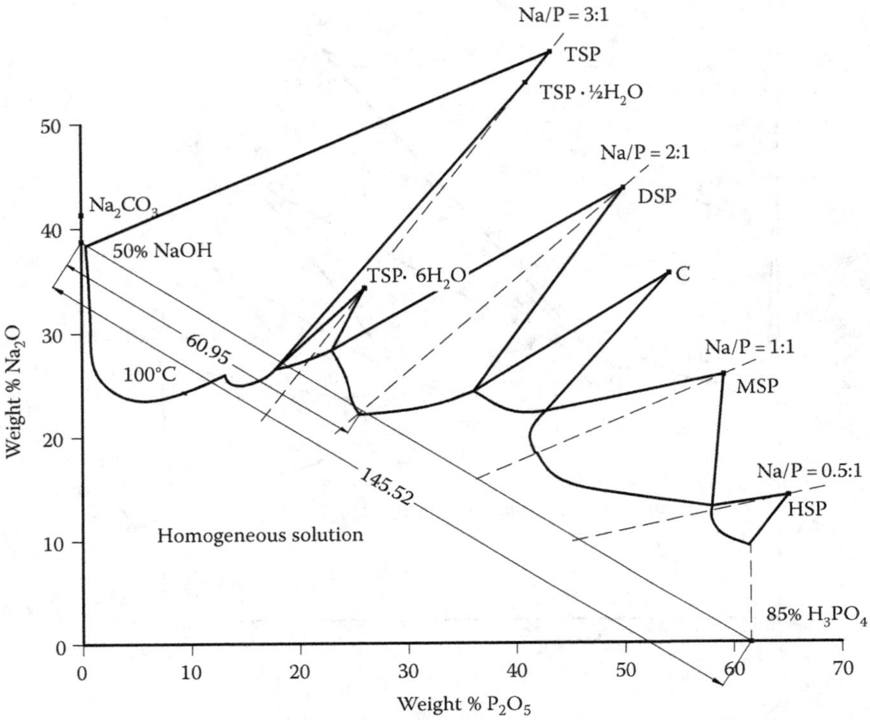

FIGURE 4.4 Graphical solution to the production of DSP.

SALP has three areas of application: as a bakery agent, as a cheese emulsifier, and as a meat-binding agent. SALP falls into two categories, acidic (used in bakery applications) and alkaline (used in cheese emulsifying). The acidic group are lower in sodium than the alkaline group, a classic example of the former is a baking agent, $NaAl_3H_{14}(PO_4)_8 \cdot 4H_2O$, the first SALP to be patented [6], and sometimes referred to as a 1:3:8 SALP.

Sodium compounds such as sodium nitrite, sodium permanganate, sodium hypochlorite, sodium chloride, and sodium nitrate form complexes with trisodium hydrates [2]. Chlorinated TSP is a useful cleaning agent and has the generally accepted and simplified formula $Na_3PO_4 \cdot \frac{1}{4}NaOCl \cdot 11H_2O$. The name chlorinated TSP, like SALP, is inaccurate and covers a crystalline material consisting of Na_3PO_4, Na_2HPO_4, NaOCl, NaCl, and H_2O with an overall Na/P ratio of 2.60:2.85, Na_2O present in the range 26.0%–28.0% and available chlorine in the range 3.4%–4.2%. Several patents have been granted in this field, the earliest in 1925 [7], the most recent in 1983 [8].

4.2.2 Sodium Pyrophosphates

The extensive study of the equilibria between crystalline solids and their respective melts in the Na_2O–P_2O_5 system has shown that there are only three compositions corresponding to crystalline compounds: pyrophosphate ($2Na_2O \cdot P_2O_5$), tripolyphosphate ($5Na_2O \cdot 3P_2O_5$), and metaphosphate ($Na_2O \cdot P_2O_5$). The following sodium pyrophosphates are known [9]:

$NaH_3P_2O_7$		
$Na_2H_2P_2O_7$	$Na_2H_2P_2O_7 \cdot 6H_2O$	
$Na_3HP_2O_7$	$Na_3HP_2O_7 \cdot H_2O$	$Na_3HP_2O_7 \cdot 9H_2O$
$Na_4P_2O_7$	$Na_4P_2O_7 \cdot 10H_2O$	

Of these compounds, only $Na_2H_2P_2O_7$ and $Na_4P_2O_7$ are of commercial importance. The former, anhydrous disodium pyrophosphate, is commonly known, especially in the United States, as SAPP. The latter, tetrasodium pyrophosphate, TSPP is named consistently.

At the industrial scale, SAPP is made by reacting phosphoric acid and a sodium source to form a dry MSP and then heating the MSP to a temperature in the range 225°C–240°C [10]. Care is required not to exceed 240°C as at around 250°C SAPP starts to convert to different forms of sodium metaphosphates:

$$nNa_2H_2P_2O_7 \xrightarrow{>250°C} (NaPO_3)_n + nH_2O \qquad (4.2)$$

Different grades of SAPP are sold with different performance characteristics based on their rate of reaction with neutralizing agents. This property is of particular relevance to bakery products and is governed by minor impurities added to the acid in the initial reaction to make MSP (K^+, Ca^{++}, and Al^{+++}) or blended as powder with the dry product and co-milled [11].

At the industrial scale, TSPP is made by calcining DSP at temperatures greater than 300°C:

$$2Na_2HPO_4 \xrightarrow{>300°C} Na_4P_2O_7 + H_2O \qquad (4.3)$$

Thermal-gravimetric studies in the laboratory showed a condensation step between 318°C and 355°C [9]. There are two to five crystalline forms of TSPP [12,13] according to the following equation:

$$Na_4P_2O_7\ \text{V} \xrightleftharpoons{400°C} \text{IV} \xrightleftharpoons{510°C} \text{III} \xrightleftharpoons{520°C} \text{II} \xrightleftharpoons{545°C} \text{I} \xrightleftharpoons{985°C} \text{melt} \qquad (4.4)$$

Form V is the one of interest as the others are reportedly unstable at room temperature.

4.2.3 Sodium Polyphosphates

STTP (or formally sodium triphosphate, normally referred to as sodium tripolyphosphate) is formed by calcining an intimate mixture of MSP and DSP such that the Na/P ratio is 5:3:

$$NaH_2PO_4 + 2Na_2HPO_4 \xrightarrow{>250°C} Na_5P_3O_{10} + 2H_2O \qquad (4.5)$$

Anhydrous STPP occurs in two crystalline forms, form (or phase) I and form (or phase) II. Form II starts to form first at about 250°C [14,15]. The phase transition to form I occurs at 417°C. When predominantly form I product is desired, heating continues to attain a temperature in the range 470°C–550°C. The operating temperatures in the industrial setting are linked to residence time in the kiln that is usually at least 60 min. A wide range of STPP grades have been developed by the industry for different applications, and one key parameter is the proportion of form I and form II in the product. A hydrate of STPP, $Na_5P_3O_{10} \cdot 6H_2O$, is also of interest commercially and is made by spraying water onto form I or II material in a rotating drum.

The solubility of STPP in water is a function of temperature; forms I and II are more soluble in water than the hydrate. In water, form I starts to dissolve but then hydrates rapidly to the hydrate that in turn dissolves to 13 g STPP/100 g solution. Form II dissolves to 32 g STPP/100 g solution and hydrates much more slowly. The consequence for these different solubilities is that in detergent manufacture, where STPP constitutes 30%–50% of a phosphate-based formulation, during mixing/formulation and prior to spray drying, solids of $Na_5P_3O_{10} \cdot 6H_2O$ can form and cause blockages. The proportion of forms I and II is determined by several techniques including x-ray diffraction and IR spectra; however, in the industrial plant laboratory, a more simple method is the temperature rise test. In summary, form II hydration is inhibited by the presence of glycerol; therefore, when a mixture of forms I and II is hydrated in a water–glycerol mixture, the percentage form I is given by the empirical formula

$$\% \text{ Form I} = 4\,(TR - 6) \quad \text{where } TR = \text{temperature rise in °C}$$

In one variant of the test method, 50 g of STPP is added and stirred with 50 g of glycerol for 3 min in a 180 mL beaker at 240 rpm. The agitator speed is then reduced to 90 rpm and run for 65 s during which the highest temperature over 30 s is recorded. Twenty-five milliliters of water is then added and the mix agitated at 240 rpm for 60 s. After 60 s, the agitator is stopped and the temperature measured. The difference between the first and second temperatures is the temperature rise, *TR*, which allows the % form I to be calculated.

A related property of STPP is the *rate of hydration*. The temperature change generated as a result of adding STPP to a sodium sulfate–water solution can be continuously plotted against time. The slope of the resulting curve correlates with the effect of STPP on the processing of synthetic detergents. One test method is to add 50 g anhydrous sodium sulfate to 200 mL of boiling deionized water in an agitated 500 mL flask. As the sulfate is added, the temperature is recorded. When the temperature reaches 80°C, 150 g of STPP is added quickly. The temperature is recorded and plotted. The temperatures recorded after 1 and 5 min are often used to quantify the test.

STPP, in common with other phosphate salts, has other product specifications that are more generic and include density, dryness, particle size distribution, and impurity levels. In common with other salts or acid for food grade applications, the levels of heavy metals are restricted. Also common to white powder products are restrictions on impurities that impart unwanted color to the product; in general, these are addressed by ensuring the feed acid (and the sodium source) is sufficiently pure (see Table 2.5).

4.2.4 Vitreous Sodium Phosphates

As orthophosphates, including those of sodium, are heated, they condense to form polyphosphates in a manner analogous to the heating of orthophosphoric acid to the higher acids. The linear polyphosphate anions are represented by the following formula:

$$P_nO_{3n+1}^{(n+2)-} \tag{4.6}$$

For pyrophosphates, $n = 2$; for tripolyphosphate, $n = 3$. Crystalline polyphosphate salts up to $n = 6$ have been characterized. When n is large, the polyphosphate approaches the cyclic metaphosphate composition $(P_nO_{3n})^{n-}$. Three high molecular weight crystalline compounds are known as Maddrell's salt ($NaPO_3$—II), low-temperature Maddrell's salt ($NaPO_3$—III), and Kurrol's salt ($NaPO_3$—IV). These are all linear polyphosphates. The first true metaphosphate is Knorre's salt, sodium trimetaphosphate $(NaPO_3)_3$, also known as $NaPO_3$—I.

Thomas Graham first designated the name *metaphosphate of soda* in 1833 [16] to describe the fused salt resulting from heating *biphosphate of soda* (MSP) in a platinum crucible to low redness. Fleitmann tentatively assigned the vitreous sodium metaphosphate $(NaPO_3)_x$ a degree of polymerization of six because of the composition of some *double salts* that were undoubtedly amorphous [12]. The name hexametaphosphate, although now known to be inaccurate, has stuck with a particular class of commercial phosphates made by many companies on an industrial scale. Thus, one may purchase Graham's salt, sodium hexametaphosphate, Calgon®, tetrasodium tetrametaphosphate, pentadecasodium tridecametaphosphate oxide, SHMP, etc., and be buying, broadly speaking, the same product. So, apart from sodium trimetaphosphate, which is a true crystalline metaphosphate and product in its own right, the industry offers a small range of amorphous, vitreous (or glassy) sodium phosphates commonly known as SHMP or *hexameta*. The product sold under the Calgon brand name was for many years a powdered SHMP but the formulation and product range has developed and changed. The Calgon name is derived from its prime function, the sequestration of calcium ions, or *Calcium gone*. SHMP is sold in the form of an unmilled glass, a granular form, or powder and in three chain lengths, where chain length is synonymous with the number of $NaPO_3$ units. The products tend to be known as short (5 or 6), regular (10–14), and long (14–30) chain and different suppliers have slightly differing recipes. Chain lengths correspond to the modified molar Na/P ratio after the formula of the following equation:

$$\frac{(Na_2O + H_2O)}{P_2O_5} = \frac{(n+2)}{n} \tag{4.7}$$

Water content is always low but influential. For a pure, dry phosphate and exact Na/P ratio of 1, the chain length tends to infinity corresponding to a 69.6% P_2O_5 composition. Figure 4.5 shows the relationship between Na/P ratio and chain length given a 0.1% H_2O content. In practice the curve is influenced by the operating temperature, particularly in the region where the Na/P ratio is in the range 1.00–1.10 and the desired product is a long chain *hexameta*.

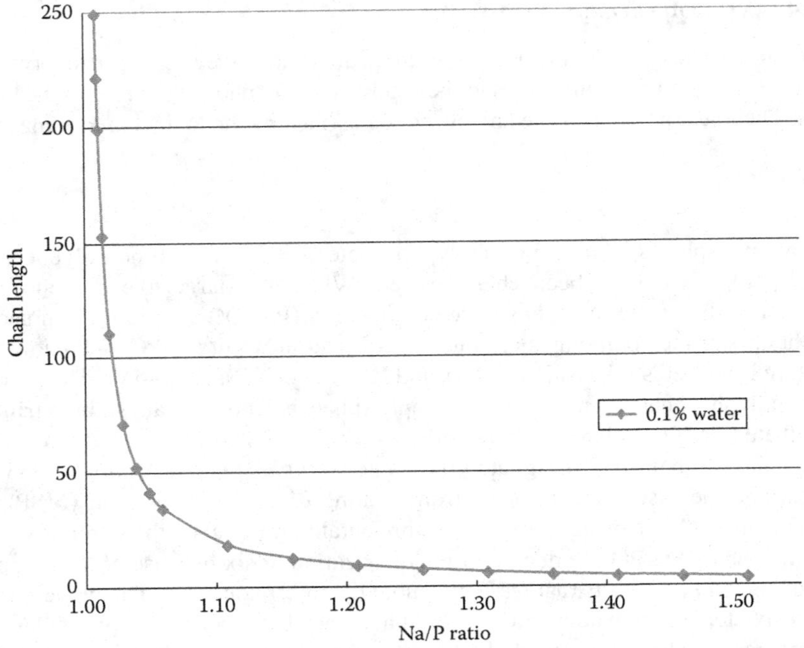

FIGURE 4.5 The relationship between Na/P ratio and chain length of sodium hexametaphosphate.

As an example calculation, the following weight percentage composition is assumed: 62.9% P_2O_5, 0.1% H_2O, and 36.98% Na_2O.

The total molar mass of this composition is 141.95/62.9% = 225.62 g mol

Therefore,

The molar mass of H_2O is 225.62 × 0.1% = 0.226 g mol

The molar mass of Na_2O is 225.62 × 36.98% = 83.446 g mol

Therefore,

Moles of H_2O = 0.226/18.02 = 0.0125

Moles of Na_2O = 83.446/61.979 = 1.346

Therefore,

$(Na_2O + H_2O)/P_2O_5 = 1.358$

Therefore,

Chain length $n = 2/(1.358 - 1) = 5.6$

By comparison, for 0% H_2O, $n = 5.7$; for 0.5% H_2O, $n = 5.1$; and for 1.0% H_2O, $n = 4.6$.

For a product with a Na/P ratio of 1.111, the chain length is 18. The reaction to make this compound from sodium hydroxide commences with that set out in the following equation:

$$18H_3PO_4 + 20NaOH \rightarrow 16NaH_2PO_4 + 2Na_2HPO_4 + 17H_2O \quad (4.8)$$

Depending on the detail process, the MSP/DSP solution is either dried, pre-concentrated, or fed as dilute liquor to the SHMP furnace melt operating in the temperature range 650°C–750°C. Free water is quickly boiled off and the MSP/DSP undergoes polymerization reactions. In the polymerization reaction, MSP extends the chain and DSP terminates it. The following equation shows the simplified reaction to SHMP:

$$16NaH_2PO_4 + 2Na_2HPO_4 \xrightarrow{650°C-750°C} (NaPO_3)_{18}Na_2O + 17H_2O \quad (4.9)$$

In addition to Na/P ratio and trace water, three other principal factors are known that influence chain length, namely, melt operating temperature, residence time, and the feed material.

In the normal Na/P operating range of 1.04–1.40, corresponding to chain lengths in the range 5–55, the variation in melt temperature from 650°C to 750°C lowers chain length by around 10% from the lower to higher temperature. As the Na/P ratio is lowered further toward 1.00, the effect is more profound.

Residence times in the range 15–60 min are ideal; above this range, chain length decreases.

An SHMP plant is often used on a phosphates site to recycle out of specification sodium phosphates. The high operating temperature tends to destroy organic impurities and vaporizes others. For many technical applications, there is a higher toleration of both cation and anion impurities than in food grade products. Thus, the feed may comprise a mixture of different sodium phosphates, usually dissolved in phosphoric acid, and with the requisite Na/P ratio. At low Na/P ratios, the feed material does influence chain length; however, in the normal operating range, the effect is negligible.

In common with all other sodium phosphates used for food grade applications, the industry has sought to retain the utility of SHMP but reduce its sodium impact by, for example, developing a mixed sodium–potassium polyphosphate glass [17].

4.3 USES

4.3.1 INTRODUCTION

Despite the current drive to reduce sodium going into food grade products, the range of sodium phosphates remains extensive and is continuously developing. The industry tends to group the products by market, and this approach is followed in the succeeding text. Of course, individual compounds often have many applications across different markets. Some uses are covered not only by many academic papers and patents but also by whole books; consequently, only a brief summary of the application and its scientific principle is given together with a small number of references that should facilitate further investigation if the reader so wishes.

4.3.2 Industrial Uses

4.3.2.1 Cements, Ceramics, Clay, and Drilling Fluids

A factor common to cements, ceramics, clay, and drilling muds is that at several stages in their manufacture and use they are pumped as a suspension in water. Often these are fine suspensions and their flow characteristics fall well within the study of colloids. A colloid is simply a mixture in which small particles of one material, usually less than 1 µm, are evenly dispersed in another material that is often a fluid. A hydrocolloid is a mixture where the dispersant is water. Thomas Graham, who proposed the term metaphosphate, is also credited as the father of colloid chemistry. In discussing the pumping or agitation of colloids, we immediately enter the world of rheology or the study of how things flow because when pumping a suspension we are often not dealing with what is known as a *normal* fluid such as water.

In rheology, fluids are classified as normal or non-Newtonian. In normal fluids, shear rate is proportional to shear stress, with apparent viscosity the proportionality constant. In non-Newtonian fluids, viscosity varies with shear rate in different ways depending on the fluid. A Bingham plastic (or simply plastic) fluid will not flow until a threshold shear stress is achieved; a shear thinning (or pseudoplastic) fluid has a nonlinear shear rate to shear stress relationship where shear stress falls with increasing shear rate or, in other words, the fluid appears to thin with increased stirring; a shear thickening (or dilatant) fluid exhibits the opposite behavior to a shear thinning fluid. Shear thickening behavior is most common with highly concentrated suspensions including starch, paint, and ink. In these suspensions, the interparticle space is increased to allow particles to move under shear forces; thus, the suspension expands or dilates, hence dilatants. The shear thinning or thickening suspensions are also known as power law fluids as the curves approximate to a power law.

Portland cement is an invention of Joseph Aspdin of Leeds, England, for which he was granted a patent, GB5022, in 1824. In 2008, the United States consumed 93.6 m tons of cement. Cement is made by either the wet or dry process, the latter tending to be the process of choice for new plants. In both processes, the feed materials, of which silica and lime make up 85%, are crushed and milled prior to calcining in kilns at about 1500°C. The main difference between wet and dry processes is whether the milling and subsequent transport are carried out wet or dry. Wet milling generally demands less power than dry milling and creates less dust. (Phosphate rock is often wet milled for this reason.) The next stage for milled phosphate rock is the reaction step so the water content is not critical and must simply be accounted for. Any water content cement production consumes energy through evaporation. Therefore, on wet process cement plants, the goal is to minimize water content in the raw-cement feed slurry while still permitting it to remain pumpable. The slurry behaves as a Bingham plastic or shear thinning suspension depending on water content. The addition of as little as 0.05% STPP is transformational. The use of STPP to reduce the water content of raw-cement slurry was patented in 1955 [18] although several patents since have taught different deflocculants.

Ceramic comes from the Greek κεραμικός meaning *of pottery*; however, it covers a wide range of technology from simple earthenware to high-tech applications such as jet engine turbine blade coatings, biomedical coatings, and ballistic protection. Both phosphoric and polyphosphoric acids as well as several phosphates are used in different ceramic applications as processing aids as binders, viscosity modifiers, dispersants, and pH control. SHMP is used as a refractory binding agent and is a component in diverse products such as armor plating [19] and chromia–magnesia ramming cement [20] and also as a dispersant in the preparation of slurries for whiteware production and electrical circuit substrates [21]. STPP is taught in the composition of a furnace cement for patching the hot lining in steelmaking furnaces [22] and molding refractory shapes [23]. TSPP and SAPP are also used in this area.

Upstream of ceramic production clay is mined and refined to remove coarse impurities such as quartz, mica, titania, and iron oxides. Some of the refining processes are chemical such as oxidation but many are physical such as centrifuging, hydrocycloning, and settling—the physical processes depend on the fluidity of the clay suspension, in particular its viscosity. Geologically, kaolin or china clay is kaolinite, a mineral with composition $Al_2Si_2O_5(OH)_4$. The mineral crystallizes in a hexagonal form of distinct platelets 0.1–0.5 μm thick with hexagonal sides and 1–2 μm long. Purified suspensions of kaolin exhibit a negative zeta potential signifying a negative charge on the platelet faces. Likewise, kaolin has exchangeable cations at the plate edges. Clays adsorb phosphate anions, with relative adsorption increasing from orthophosphate to condensed phosphate, such that a regular SHMP is adsorbed 10 times more than an orthophosphate. When a small amount of SHMP is added to a kaolin suspension, the phosphate anion attaches to the cations on the platelet edges, usually aluminum, rendering an increase in negative charge and corresponding increase in zeta potential. The net effect is that all the particles repel each other and therefore the whole suspension viscosity is reduced and flows more easily. The effect in terms of pumping design and capital and operating cost is significant when one considers that without viscosity reduction kaolin suspensions are pumpable at about 40% concentration, whereas with 0.3% SHMP the pumpable concentration is about 70%. The earliest patent in this field was in 1922 [24] to "William Feldenheimer, a subject of the King of England," and has been the subject of continued research and invention [25]. In practice, kaolin miners and processors test clay samples in the laboratory and produce deflocculant demand curves and viscosity concentration curves to control their operations because just like phosphate rock kaolin composition varies slightly even within a mine.

Drilling fluids are added to the wellbore to facilitate the drilling process. Drilling fluids are also known as *mud* after drillers in Spindletop, Texas, used a clay/water suspension created when cattle stirred up the ground by a new wellbore to flush out drill cuttings and found that it stabilized the wellbore walls. The function of drilling fluids is to suspend cuttings, control pressure, stabilize exposed rock, and provide cooling and lubrication to the drill head. The nature of the drilling fluid should be shear thinning so that when the drill bit is withdrawn to extend the drill length the hole does not collapse, nor the fluid particles settle, but when the drill is turning it is not wasting energy fighting a thick fluid. Consequently, the control of viscosity is critical; thus, different viscosity modifiers are added to the drilling fluid formulation that is predominantly bentonite, including SAPP, STPP, and SHMP.

4.3.2.2 Metal Finishing

Sodium phosphates are used in alkaline metal cleaners, the electroplating of tin and iron phosphatizing.

During manufacture, metal components are often contaminated with oils and grease. Similarly, equipment requiring maintenance often must be cleaned first. Prior to the ban on many halogenated hydrocarbons by the Montreal Protocol in 1987, degreasing was usually carried out with chlorofluorocarbons (CFCs). The use of CFCs as a degreasing agent has an environmental impact if not used correctly; there were also health hazards for those working with them. An alternative approach to grease removal is to use aqueous systems with high pH. Unfortunately, aqueous systems have the potential to cause corrosion, rusting, and so on. Therefore, formulations were developed that incorporated phosphates as corrosion inhibitors as part of alkaline metal cleaners. These phosphates include STPP and SHMP but also zinc phosphates [26,27].

Sodium phosphates are added to the plating bath formulation when tin and other metals are electroplated onto various substrates [28,29].

MSP is a mild phosphatizing agent on steel surfaces and provides an undercoating for paints. When the steel plate or component is placed in an aqueous solution containing dissolved MSP, surface iron forms an adherent and protective coating of iron phosphate that can be painted. Some phosphatizing solutions use accelerators although some patents teach that these are unnecessary [30].

4.3.2.3 Mining, Petroleum Products, and Refining

There is great overlap in the functionality of sodium phosphates for both the mining and petroleum products and refining markets and drilling fluids and clay. In fact, the overlap is such that the extent of the use of sodium phosphates is down to subtle differences between individual customer specifications, which is beyond the scope of this book.

4.3.2.4 Plastics and Rubber

Sodium phosphates are used in the alkaline cleaning of plastic finished goods.

4.3.2.5 Pulp and Paper

Sodium phosphates are used in several ways in the pulp and paper field. STPP is used as a sequestrant in the pulp bleaching process to improve brightness; a similar functionality is used to de-ink recycled newsprint; the same property removes calcium ions in pulp that lowers the energy required in the pulping process [31]. The deposition of organic contaminants on process equipment (*pitch* and *stickies*) is detrimental to the papermaking process. Pitch refers to natural resins in virgin wood pulp, whereas stickies are adhesives or coatings from recycled paper. One approach to pitch and stickies is to add clay in a slurry to the papermaking pulp. For ease of addition, the clay is added as a slurry that requires viscosity and dispersant control; thus, STPP is added to the clay formulation. Coatings are applied to paper to improve its appearance and STPP is utilized as a dispersant in this application.

4.3.2.6 Water Treatment

Sodium phosphates (as well as phosphoric acid and tetrapotassium pyrophosphate) have an extensive range of applicability in water treatment.

4.3.2.7 Textiles

Sodium phosphates (and other phosphates, and phosphoric and phosphorous acids) have wide application in textile processing, itself a wide field. Some examples include the use of sodium phosphate as an enzyme compatible buffer [32], as a catalyst in a textile finishing composition [33], and as a sequestrant of metals that would otherwise build up and slow dyeing processes.

4.3.3 Food Uses

There are several excellent texts on the topics of food phosphates and food phosphate chemistry, including more recently Lampila and Godber [34] or going back to Waggaman [35] for the more historically minded.

The range of use is very wide both in terms of functionality and application, for example, pH adjust and buffering, sequestration and supplementation of minerals, aiding or inhibition of coagulation, and protein modification applied to products as diverse as meats, dairy products, fruit, and vegetables. Table 4.2 summarizes the functionality of sodium phosphates in food applications (but does not include other ammonium, calcium, and magnesium phosphates).

The most commonly used food phosphates are generally recognized as safe (GRAS) by the US Food and Drug Administration (FDA) under Title 21 of the Code of Federal Regulations. Food grade phosphates are also covered under Title 9 for meats; Title 21 for foods, boiler water additives, adhesives, and indirect contact; and Title 27 for alcoholic beverages. They are usually certified as kosher and halal.

TABLE 4.2
Sodium Food Phosphate Functionality

	Sodium Phosphate							
Function	**MSP**	**DSP**	**TSP**	**SALP**	**SAPP**	**TSPP**	**STPP**	**SHMP**
Acidulant	×			×	×			
Alkalinity		×	×			×	×	
Anticoagulant								×
Buffering agent	×	×	×					
Coagulant					×	×		
Dispersing agent		×			×	×	×	×
Emulsifier		×	×			×		×
Leavening agent				×	×			
Mineral supplement	×	×						
Protein modifier		×	×		×	×		
Sequestrant					×	×	×	×

Food uses are grouped into sectors, including baking and leavening; cereals; meat, poultry, and seafood; dairy and cheese; processed food; beverages; and confections.

4.3.3.1 Baking and Leavening

In common with other sectors, there is currently a drive to reduce sodium in foods that is leading to developments in baking and leavening products and a move away from sodium toward calcium and potassium. Nevertheless, sodium phosphates in the form of both SALP and SAPP remain in use.

In the baking and leavening sector, sodium phosphates are used for leavening, pH adjustment and buffering, dough conditioning, enrichment, as growth factors for yeasts, starch modification, and the manufacture of quick-cooking cereals.

In chemical leavening, a bicarbonate reacts with an acid phosphate principally generating carbon dioxide. The dough rate of reaction (DROR) measures the rate of generation of gas during mixing (2–3 min) and subsequent bench time (6–7 min). DROR varies with the type and quantity of carbonate, phosphate, moisture, temperature, and flour and other ingredients. It may be readily appreciated that different foods require specific leavening formulations, for example, bread, cake, pizza base, or pancake mix all require different amounts of raising. The timing of the raising is different, some is required immediately after mixing, some during cooking, and so on. Partially cooked or frozen goods also require different leavening. The early leavening agents of the nineteenth century mimicked yeast that generates all the carbon dioxide before baking. It was discovered that slow-acting or nowadays temperature-controlled release agents deliver the gas when it is needed. Neutralizing value (NV) is defined as the weight of sodium bicarbonate neutralized by 100 parts of the acidulant. Different baked products benefit from being both or slightly acid or slightly alkali. DROR is discussed in a little more detail in Chapter 5.

Typically, SAPP grades are available in the range SAPP-10 to SAPP-43, meaning that depending on the grade, 10%–43% of the carbon dioxide is released during the 2 min mixing stage of the test. SAPP in the 10–22 range is a slow-acting *pyro* intended for canned, refrigerated biscuit doughs, frozen batter and doughs, flour tortillas, and other formulas where the dough may be stored for up to 120 days. Minimal gas generation during storage is essential and many patents have been granted for vented package designs [36]. SAPP in the 22–28 range is considered midrange and all purpose for use in baking powders and cakes. SAPP 37 is used in a variety of institutional, bakery, and retail applications. SAPP 40 is designed for doughnut mixes and SAPP 43 is a fast-acting SAPP. As well as a leavening action, SAPP and other phosphates affect pH, elasticity, viscosity, strength, and water-holding capacity of doughs that in turn affect the form, appearance, texture, color, and bite of the final cooked product.

SALPs are resistant to cold temperatures and have DRORs in the 20%–30% range and are used in mixes for waffles, pancakes, and refrigerated biscuit doughs.

4.3.3.2 Cereals

Food phosphates decrease cooking time, aid flow through extruders for dry, extruded cereals, modify cereal color, and modify starches. (Calcium phosphates are used to fortify the calcium and phosphorus content of the food.)

DSP is added to quick-cooking cereals; it hastens cooking by increasing pH and partially gelatinizing the starch content.

DSP is added to macaroni formulations and is used in the Asian noodle market in the form of *kansui*. Kansui is a blend of sodium and potassium phosphates and carbonates and is added (at 0.1%–1.7%) to wheat flour, salt, and water to make noodles. Kansui affects texture, color, and flavor development.

Starch phosphate mono- and diesters are used widely as additives in the manufacture of desserts, sweet creams, cakes, sauces, mayonnaise, canned vegetables, and yogurts. Starch phosphates are made by esterification of corn, wheat, and other starches by phosphoric acid and sodium and potassium phosphate salts. The manufacture and application of starch phosphates is an active area of research and invention [37].

4.3.3.3 Meat, Poultry, and Seafood

Salt (NaCl) has been used as a preservative in meats since ancient times. Salt prevents spoilage and binds proteins during cooking so improving taste and texture. Question marks over the health impacts of excessive salt lead to salt replacements based on potassium chloride over 60 years ago [38]. Brines are an essential component in the processing of meat, poultry, and seafood and much study has been carried out and many patents granted in this field in this time. From a phosphate standpoint, the big discovery was that the functionality of sodium chloride was improved by the addition of various phosphates depending on the particular application.

Phosphates have wide application in meat systems, and their functionality covers color protection and color development in cured products, reduced cook-cool and thaw-drip loss, protein protection in freezing and frozen storage, inhibition of lipid oxidation, development and stabilization of emulsions, formation of myosin to a *sol* and a *gel* upon heating, enhanced succulence and flavor enhancement of the cooked product, reduced cooking time, microbial inhibition, and many others.

Consumers purchase meat products after making judgments about the quality and price of the meat. They often use color to judge the ultimate tenderness and taste of the meat product. Recognizing this, fresh pork meat, for example, is categorized in four ways, as RFN (*r*eddish pink, *f*irm, *n*on-exudative), ideal quality, pH < 6, percentage drip < 6; RSE (*r*eddish pink, *s*oft and floppy, *e*xudative), questionable quality, pH < 6, percentage drip > 6; PSE (*p*ale pinkish gray, very *s*oft and *f*loppy, very *e*xudative) quality poor, pH < 6, percentage drip > 6; and DFD (*d*ark purplish red, very *f*irm, *d*ry—free of surface fluids), quality varied, may be juicy, tender, spoils easily, pH > 6, percentage drip < 3. Immediately postmortem, there is a rapid fall of pH resulting in acidic conditions that give rise to PSE meat; antemortem stress and genetic factors are known to be contributing factors; approximately one-third of the pork flesh is subject to PSE. At slaughter, fresh pork meat contains 75% water, as the pH falls muscle fiber shrinks so also the water-holding capacity of the meat. Injection of sodium phosphate elevates and stabilizes the pH thus aiding meat succulence, reducing drip loss, and maintaining color [39].

Maintaining the freshness of fish from catch to market is an old problem. Fish is usually frozen immediately on the boat, but after delivery to the retailers, shelf life is very short. One patent teaches the use of a formulation comprising 47% STTP (with 3.7% lemon juice solids), 47% SAPP, and 6% potassium sorbate [40]. In this patent, catfish fillets were treated with a 10% solution of the formulation, breaded, and fried. A tasting panel then awarded points for flavor, texture, and odor.

Poultry is classified by age as follows: immature birds, 5–7 weeks of age; broilers and fryers, 9–12 weeks of age; roasters, 3–5 months of age; fowl, more than 10 months of age. Fowl takes longer to cook than younger poultry and the boning and recovery of cooked flesh is less efficient. One patent teaches that the addition of a 0.5% solution of STTP or SAPP reduces cooking time and improves meat recovery of cooked fowl.

The control of microbial growth in meat, poultry, and seafood products is critical to food safety. Of particular concern are *Listeria monocytogenes, Pseudomonas aeruginosa*, and *Escherichia coli.* Aqueous solutions with a pH of 7–11 consisting of STPP and optionally an organic acid are taught as effective [41].

4.4 PRODUCTION PROCESSES

Compared to the range of applications, the production processes of sodium phosphates are relatively straightforward. A sodium source, usually soda ash (Na_2CO_3) or sodium hydroxide solution, is reacted with phosphoric acid in the desired Na/P ratio and is then dried, calcined (for higher phosphates only), sized, and packed. The detail behind this simple description has been worthy of many patents and shows the development of several process technologies over the last 60 years. The difference between best practice, using best available technology, and worst has a material effect on business viability.

4.4.1 Sodium Sources

There are essentially four sodium sources for the manufacture of sodium phosphates: soda ash, sodium hydroxide, internal recycles from scrubbers and filters, and external recycles from out of specification, old, or unwanted sodium phosphate products.

Apart from TSPs where the final neutralization must be done with sodium hydroxide, the biggest driver for the choice between soda ash and sodium hydroxide is the delivered cost. The cost of sodium hydroxide is related to the Electrochemical Unit (ECU). The electrolysis of natural salt coproduces 1.1 tons of sodium hydroxide per 1.0 ton of chlorine, and an ECU represents the total cost of this production. Depending on the chlorine production technology, an ECU equates to about 3.2 MWh electricity. The cost of an ECU is about 2.5 times the electricity cost plus $200. Since the replacement of CFC refrigerants, the biggest market for chlorine is ultimately polyvinylchloride (PVC). PVC is the polymer of vinyl chloride that in turn is made from ethylene dichloride (EDC), an item of commerce. So, the achievable sales price of chlorine is driven by PVC demand and pricing, and ethylene cost with EDC having a damping effect. Together with the cost of electricity, the chlorine producer has to manage these inputs to set a price for sodium hydroxide and still make a profit. In practice, sodium hydroxide prices may vary by a factor of 10 because in many cases the customer can choose to buy soda ash instead. In the period when STPP was the main detergent builder, the volumes of either soda ash or hydroxide were sufficiently large to match the CFC and PVC demands and allowed purchasing managers real negotiating positions. With the collapse of STPP demand, this negotiating position is much weaker; nevertheless, having the option to use and purchase either soda ash or sodium hydroxide is an important commercial lever. A component of the delivered cost of the sodium source is the delivery, or logistics, cost. In practice, some phosphate producers are located too far away from one source or another for it to be realistic to use both.

Two types of soda ash are available, firstly, the product of the Solvay process and, secondly, a beneficiated, mined soda ash. The Solvay soda ash is considered more reactive than the *natural* product and is of lower bulk density; however, both types are used successfully.

Soda ash is always delivered as a powder of the specified particle size distribution. Bulk delivery is by rail car or road tanker; smaller quantities are available in big bags or sacks. Offloading to bulk storage is carried out pneumatically. Different approaches are taken thereafter. On some plants, solid soda ash is metered under weight control (this can be loss in weight belts or screw conveyors or from a smaller batching vessel on weigh cells) straight into neutralization; on others, an aqueous soda ash solution is made up and pumped to the desired location. Soda ash is maximally soluble in water at 33% concentration at 35.4°C; above this temperature solubility drops slightly, and above this concentration the monohydrate forms. If the temperature is allowed to drop, higher hydrates start to crystallize out. An alternative to either making an aqueous solution or going straight to neutralization as a solid is to make up a high sodium content soda ash slurry by mixing solid soda ash with water and, optionally, neutralization liquor. The alternative approach allows the production of 42%–65% concentration slurry.

Sodium hydroxide is usually supplied as a 50% concentration liquid. Generally speaking when considering the neutralization step, liquids are somewhat easier to handle than solids: there is no dust; there are usually less blockages with a liquid than a solid; and metering of liquids is slightly less troublesome. In terms of Na_2O delivered per road tanker, the difference is small, 38.7% for NaOH 50% solution and 41.3% for pure soda ash.

The drying, calcining, and sizing steps inevitably create dust and fume that must be filtered or scrubbed both to ensure controlled air emissions from the plant and minimize losses. Water scrubbing is used but as the phosphate concentration rises the scrub liquor must be diluted with freshwater. On a sodium phosphate plant, scrubber liquor concentration is controlled at about 10%. As the scrub liquor is diluted, the excess is pumped back to the neutralization section and bled into the neutralizer. When operating with either purified or thermal acid, there is usually little problem with impurity buildup; however, fluoride buildup can be problematic where less pure acid is used, particularly one with inadequate defluorination.

Recycling of out of specification material is potentially difficult and must be evaluated on a case by case basis. The prime consideration is that tramp material (nuts, bolts, general rubbish, etc.) must not be introduced into the process. The secondary consideration is that the rate of introduction must take account of the required

Na/P ratio of the ortholiquor coming out of the neutralizer and the capacity of the neutralizer to dissolve the recycle material.

4.4.2 Neutralization

Three approaches to neutralization are practiced that reflect downstream processes:

a. The oldest approach, with crystallization of orthophosphates downstream, is wet batch neutralization.
b. The commonest approach, with spray drying downstream, is wet either batch or continuous neutralization.
c. The third approach is *dry* neutralization; obviously this is carried out with soda ash with the objective of minimizing the amount of water that must be evaporated.

The neutralization reaction is exothermic: this must be managed in the first approach as the next step is cooling crystallization and exploited in the second and third.

Neutralization with soda ash generates carbon dioxide, and on a large plant the quantity is of sufficient interest to gas companies to warrant installation of a gas recovery unit.

In the following are more detailed descriptions of the three neutralization practices.

4.4.2.1 Neutralization for Crystallization

The crystallization plants generally carried out batch neutralization in two or three neutralization vessels. Figure 4.6 shows the batch neutralization and crystallization steps using soda ash for the main neutralization with sodium hydroxide available optionally for the production of TSP. If the neutralization was carried out with sodium hydroxide, the slurry vessel would be redundant. The operation could be carried out under computer control but more usually would be manually controlled.

Filtrate from the centrifuge and filter washes is pumped into the agitated slurry vessel. Water is pumped in to a specified weight. Finally, soda ash is added via a metering screw from the soda ash storage until the specified weight is reached.

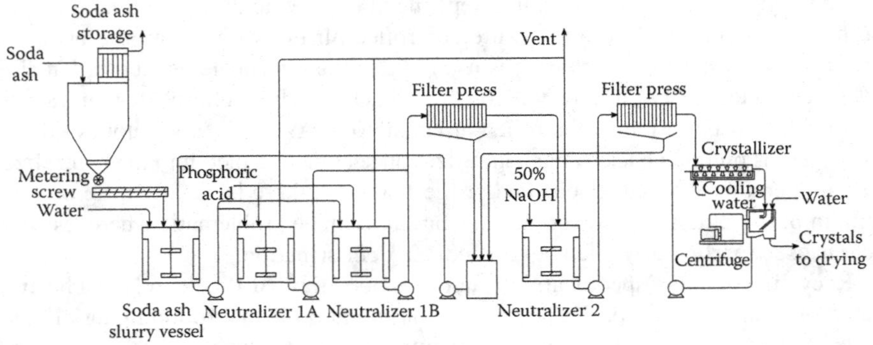

FIGURE 4.6 Sodium phosphate batch neutralization and crystallization flow sheet.

Agitation is maintained until the main neutralization vessel is ready to receive the next batch of soda ash slurry. Once empty, the first neutralization vessel receives the soda ash slurry while maintaining agitation. Phosphoric acid is metered into the neutralization vessel to 90% of the desired Na/P ratio; steam and carbon dioxide evolve and are vented. Once the neutralization is complete, the batch is pumped through a plate filter to the second neutralizer where sodium hydroxide is added to bring the batch up to the intended Na/P ratio. The batch is sampled and tested for pH, temperature, and density from which data Na/P is verified. The batch is then pumped to the crystallizer via a second filter. Both filters are water washed to recover dilute sodium phosphate liquor; the washes are collected together with the crystal mother liquor and wash from the centrifuge and pumped to the soda ash slurry vessel. In the crystallizer, the batch is cooled and the sodium orthophosphate crystallizes out. On older plants, the crystals and mother liquor were transferred manually to a centrifuge to separate the mother liquor out on the first spin. Then a little water was added for crystal washing. Finally, the crystals were transferred for drying, sizing, and packing.

4.4.2.2 Neutralization for Spray Drying

While spray driers are used on small plants and development units and are available down to bench scale, their application on sodium phosphate plants tends to be at full industrial scale in the range 25,000–125,000 tons per year. Consequently, although at the lower end of this scale there are many plants that operate batch neutralization, many operate continuously. Modern plants are instrumented and under computer control.

Figure 4.7 shows a continuous neutralization unit. The following description assumes that although there is a facility to run partially or completely with soda ash the operation described is making an ortholiquor that will ultimately be calcined to STPP from sodium hydroxide. The scale of operation is that required to make 50,000 tons STPP per year.

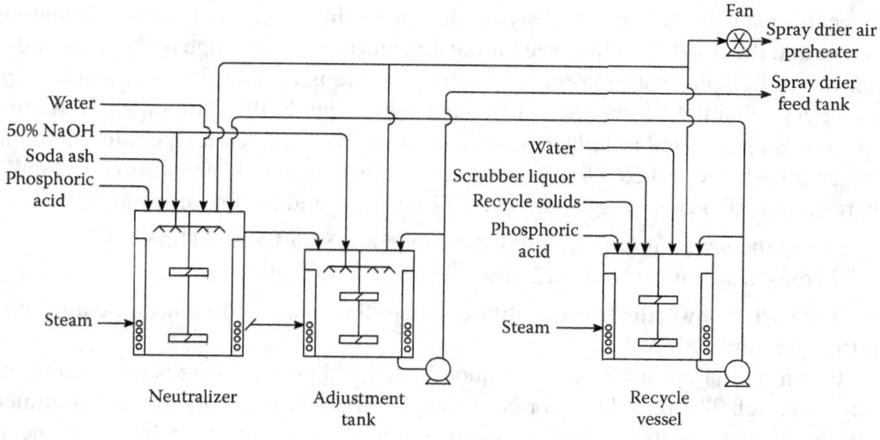

FIGURE 4.7 Sodium phosphate continuous neutralization flow sheet.

Phosphoric acid (85% H_3PO_4) is pumped at a controlled rate to the neutralization vessel via a mass flowmeter. The neutralization vessel is a 16 m^3, baffled, agitated 316 L stainless steel vessel fitted with steam coils. When running with sodium hydroxide, the coils are only required to aid start-up—bringing the reaction mass up to temperature—or to maintain temperature during temporary shutdowns, thus avoiding blockages. (When operating with soda ash, the first stage of neutralization is often run at 85°C at a pH of 6.5–6.7; therefore, there is little driving force from the sodium hydroxide that completes the neutralization to raise the temperature near to the ortholiquor boiling point of 108°C. Therefore, the steam coils are required in this circumstance.)

Sodium hydroxide (50% NaOH) is also pumped at a controlled rate to the neutralization vessel via a mass flowmeter. The soda stream is distributed in the neutralizer such that it sprays uniformly on the liquor surface. The ultimate Na/P ratio of the ortholiquor is 1.67; however, the sodium hydroxide flow rates are split 90% to the neutralizer and 10% to the subsequent adjustment vessel. The rates are led by the acid flow that is set to achieve the desired overall plant rate.

Thus, assuming annual operating hours of 8000 and neglecting efficiency losses (e.g., unrecovered powder losses from air filters), the feed rates are calculated as follows:

Hourly rate of STPP = 50,000/8,000 = 6250 kg/h
Therefore,
Dry orthosolids feed to calciner = 6250 × 404/368 = 6860 kg/h
STPP contains 57.88% P_2O_5; therefore,
Feed acid mass flow = 6250 × 0.5788 = 3617.5 kg/h P_2O_5
As 85% H_3PO_4 acid = 3617.5/0.6158 = 5874 kg/h
For a feed acid temperature of 50°C, specific gravity is 1.664; therefore,
Feed acid volumetric flow rate = 3.53 m^3/h
STPP contains 42.12% Na_2O; therefore,
The total 50% NaOH mass flow = 5874 × 0.4212/0.3874 = 6795 kg/h 50% NaOH
Thus, total mass flow into the neutralizer = 5,874 + 6,795 = 12,669 kg/h

To minimize the drying duty of the spray drier, the ortholiquor is held at its boiling point of 108°C for sufficient time to allow the ortholiquor to concentrate. The target concentration is in the range 59.5% (which is on the intersection of the 100°C solubility line and the Na/P = 5:3 line, see Figure 4.8) and 62%. This range is chosen in order that the ortholiquor entering the spray drier nozzle is at least 5°C below the point where crystals might form and cause blockages. Equally the ortholiquor feed temperature is maximized to reduce drier load; therefore, the feed temperature is in the range 95°C–103°C. At 59.5% concentration and between 95°C and 100°C, normally there are no blockages; edging toward 62% operator find a point of inconvenience.

Taking the lower figure, the spray drier feed = 6,860/0.595 = 11,537 kg/h
Therefore, steam evolved = 12,669 − 11,537 = 1,132 kg/h

1132 kg/h is a worthwhile quantity of steam that is used to preheat the air going to the spray drier air heater.

Returning to Figure 4.7, recycle liquor from the plant scrubbers is returned to the recycle vessel. The recycle liquor Na/P ratio is checked regularly and programmed into the plant computer so that the fresh acid and soda feeds are adjusted to incorporate a small stream of liquor into the neutralizer. Periodically, solid sodium

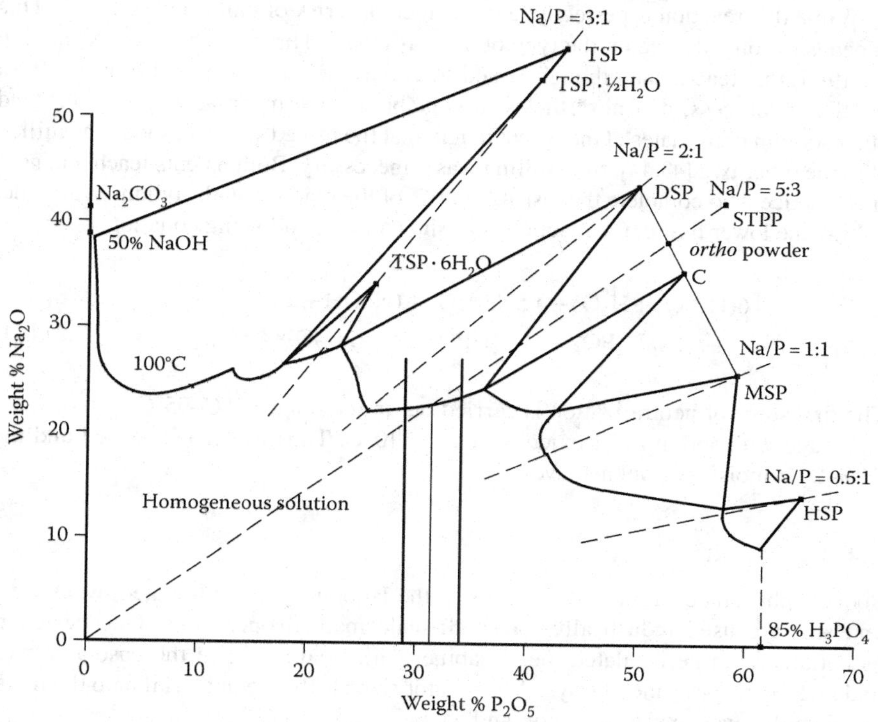

FIGURE 4.8 The Na_2O–P_2O_5–H_2O phase diagram showing the operating zone for producing STPP.

phosphate–recycled materials are tipped into the recycle vessel. To ensure that these materials are dissolved completely, the liquor pH is reduced to less than pH 4 by the addition of phosphoric acid.

Larger-scale plants that do not operate a continuous neutralization section usually run a semi-batch operation with three or more neutralizers.

4.4.2.3 *Dry* Neutralization

Dry neutralization is carried out primarily with solid soda ash. The product is a slurry with a water content in the range 12%–20%. As such it is unsuitable for spray drying. Nevertheless, this composition is readily dried in heated mixers, belt driers, or rotary drier/calciners. The obvious advantage of this process is that less drying is required; however, there is less scope to alter the physical form of the dried product from this starting point.

The reaction to produce feed material for conversion to STPP is shown in the following equation:

$$6(H_3PO_4 + H_2O) + 5Na_2CO_3 \rightarrow 2NaH_2PO_4 + 4Na_2HPO_4 + 11H_2O + 5CO_2 \quad (4.10)$$

where (H_3PO_4 + H_2O) approximates 85% acid.

While this reaction is perfectly feasible, there is a risk of unreacted soda ash. This depends to some degree on the type of soda ash used. Three patents were granted in the 1980s that teach that although mined soda ash is less reactive than soda ash from the Solvay process, it is nevertheless acceptable for the manufacture of STPP (and other sodium phosphates). One patent taught that the mined soda ash should be milled [42]; the other two [43,44], that milling was unnecessary. Both patents teach and general practice is to complete the last 10%–20% of the reaction with sodium hydroxide. Taking the lower figure, the reaction is as shown in the following equation:

$$6(H_3PO_4 + H_2O) + 4.5Na_2CO_3 + (NaOH + 2.22H_2O) \rightarrow 2NaH_2PO_4 + 4Na_2HPO_4 + 12.72H_2O + 4.5CO_2 \quad (4.11)$$

The first stage of neutralization is carried out in the range 65°C–95°C, and the second stage with sodium hydroxide at 105°C–110°C. Thus, water is released and the 12%–20% moisture level achieved.

4.4.3 Drying

Sodium phosphates are dried in ovens in the laboratory. Steam-heated ovens were and still are used industrially for small-scale manual operations. One design is essentially as large insulated metal cabinet with steam coils in the base and back and racking to carry metal trays. An operator shovels the wet material onto the trays, loads up the oven, closes the door, and waits.

The drying oven is made continuous by use of a moving belt. In a moving-belt oven, a wet material is dropped onto a moving belt that carries the material through the oven where moisture is removed. Some ovens operate under slight vacuum and are heated in different ways. A further enhancement is to turn the moving belt into a fluid bed. These ovens are widely used across the chemical, pharmaceutical, and food-processing industries.

A variant of the moving-belt concept is the drum drier. A typical drum drier configuration has two counterrotating stainless steel drums side by side. The drums are steam heated (or oil heated for higher temperatures). Wet material, usually a liquor, is distributed on the drums forming a thin layer. Water is evaporated and conducted away via a vent hood above the drums. A scraper separates the dried material from the drum that then drops into a hopper and passes on for further processing. Drum driers on phosphate applications have proved to have very high maintenance costs. Figure 2.4 shows drum driers operating on sodium ortholiquor.

In the 1950s, the Marchon factory at Whitehaven, which became Albright & Wilson (A&W), used drum driers in their STPP process. When in the 1960s rapid expansion of STPP production was considered, executives visited plants in the United States that were using Bowen spray driers. William Spencer Bowen, a citizen of the United States and resident of New York and New Jersey, was granted several patents between 1922 and 1948, most of which related to drying (including an open steam box belt drier [45]) and in particular spray drying. In due course, A&W installed three large Bowen Engineering spray driers (10 m diameter) for the manufacture of

FIGURE 4.9 Bottom cone of an STPP spray drier. (Courtesy of Solvay.)

STPP in Whitehaven. Figure 4.9 shows a photograph of the base of one of the spray driers. Bowen became part of the Niro organization in 1981, a Danish company established by Johan Ernst Nyrop in 1933 that also developed spray and other driers, which in turn became part of what is now the GEA group. A&W went on to install both Niro and Anhydro (a Danish competitor of Niro and now part of the SPX group) spray driers on STPP and other sodium as well as calcium and potassium phosphate plants around the world. Other companies have produced spray driers for use in the STPP process, and one of these, Deutsche Babcock, installed 26 complete STPP plants worldwide, including their own spray drier design.

There is an extensive body of literature on spray drying [46] and industrial drying [47] so the following is only the briefest overview of spray drying technology that has been used principally on sodium phosphates plants around the world.

The spray drying section of a plant comprises the drying chamber, a cyclone separation system after the drier, a fan to pull air through the system, a scrubbing unit, and an air heater. The spray is created either by a rotating disk atomizer or by a high-pressure pump and spray nozzles in the drying chamber. Depending on the duty and scale of operation, the cyclone and scrubber are combined into a cyclone reverse jet filter unit.

Figure 4.10 shows a flow sheet of a rotating disk atomizer spray drier typical of those supplied by both Bowen Engineering and Niro for STPP manufacture. Sodium ortholiquor, from the neutralization section, at 60% concentration and 95°C is pumped to the atomizer inlet. The atomizer is driven by a variable speed inverter drive and is rotating at 6,000–15,000 rpm. The ortholiquor is sprayed into

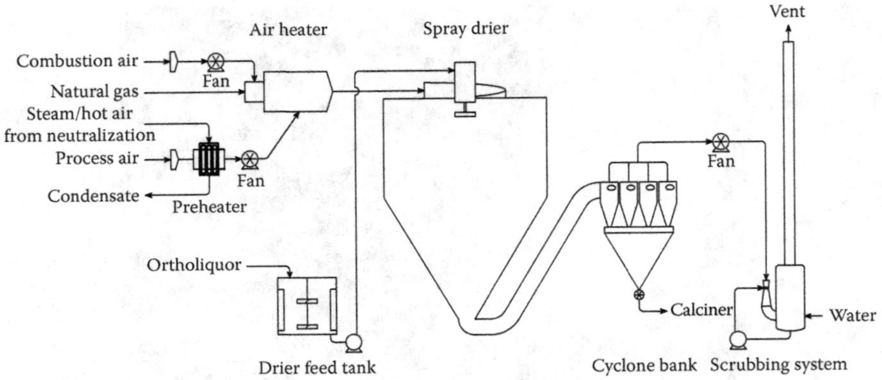

FIGURE 4.10 Typical Bowen Niro spray drier flow sheet for STPP production.

the chamber forming droplets. Droplet size control is achieved by varying the atomizer speed. The droplets are dried in a stream of hot air forming spherical particles of orthobead. The size and structure of the orthobead are linked to drying temperature (and therefore rate), droplet size, feed concentration, and chemical composition. In this configuration, all the air, water vapor, and dry orthobead pass out of the base of the drying chamber through a large duct that is directed into a quartet of cyclones. Cyclones are used for larger-scale disengagement, from single through to a bank of six, depending on throughput and space. In the cyclones, the orthobead is disengaged and is discharged through a rotary valve to a conveyor to the next stage. The air is drawn through the cyclones by the main fan that exhausts to a venturi scrubber. Input air is heated directly by natural gas in the air heater. There are many different designs of air heater: some as indicated here are essentially a simple cylindrical chamber with a gas burner head, and others are complex constructions that are designed to provide a uniform temperature distribution across the inlet duct to the spray drier. In this drier design, the hot air enters the drier through a curved duct that creates a swirling air flow.

Figure 4.11 shows a slightly different configuration. Here 75% of the orthobead is captured and discharged from the drying chamber through a rotary valve. The air flow and remaining 25% of finer solids pass out of the side of the drying chamber to a single cyclone. The finer material is discharged through a rotary valve and is mixed with the material from the drying chamber.

Figure 4.12 shows a high-pressure nozzle spray drier typical of an Anhydro design. The drying chamber is shorter than that shown in Figures 4.10 and 4.11 as the air flow within the drier follows a different path. Hot air is swirled into the chamber around the spray zone, passes downward, then returns and exits via vents in the chamber roof and flows to a cyclone. The ortholiquor is fed to several nozzles located centrally at the top of the drying chamber. Droplet size is varied by nozzle shape, orifice size, and pump pressure. Pump pressures are typically in the range 25–75 bar. The droplets are dried in the hot air and fall into a fluidized bed at the base of the drying chamber. Dried orthobead passes from the fluid bed to downstream processing. Vapor from the top of the cyclone is scrubbed.

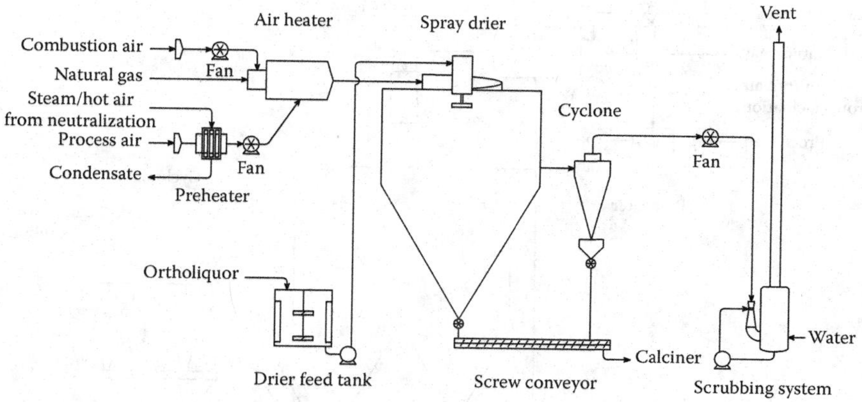

FIGURE 4.11 Typical Niro spray drier flow sheet for sodium phosphates production.

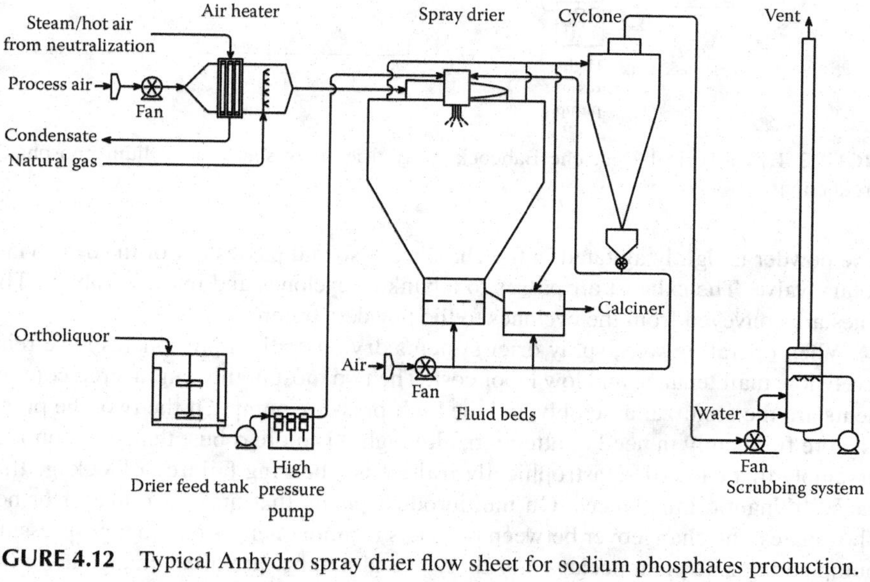

FIGURE 4.12 Typical Anhydro spray drier flow sheet for sodium phosphates production.

Fines from the base of the cyclone are pneumatically conveyed to the drying chamber inlet and recycled.

Figure 4.13 shows a high-pressure nozzle spray drier typical of a Deutsche Babcock design. Hot air is ducted to the top of the drying chamber and enters through a central duct passing straight down the chamber. A number of spray nozzles are arranged around the periphery of the chamber roof and direct the ortholiquor spray into the path of the hot air. The droplets are dried and fall to the base of the chamber. The air, water vapor, and fine material exhausts from a central location in the bottom cone of the drier. A top hat obstructs powder from falling directly into the exhaust duct.

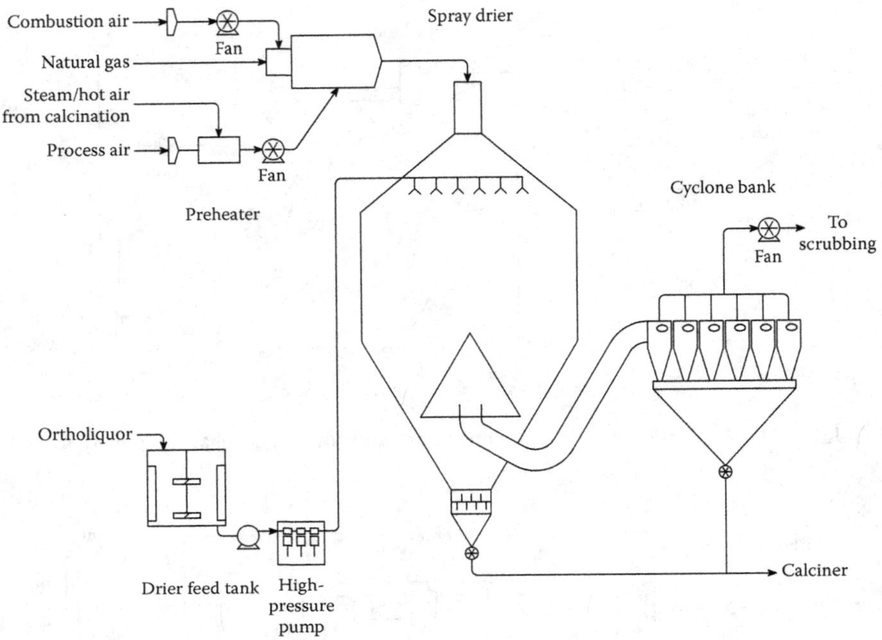

FIGURE 4.13 Typical Deutsche Babcock spray drier flow sheet for sodium phosphates production.

The powder is lightly agitated in the chamber base and passes out of the drier via a rotary valve. The exhaust air passes to a bank of cyclones and into a scrubber. The fines are conveyed from the cyclones to the powder stream.

When operating well, spray drier systems drying sodium phosphates have relatively low maintenance and low labor cost. The two most highly engineered components are the atomizer assembly and the high-pressure pump. Of the two, the pump is more frequently in need of attention, although it is more robust than the atomizer assembly that can fail catastrophically if there is a bearing failure or blockage that causes dynamic imbalances. On multiproduct plants that make a range of orthophosphates, the changeover between products is more onerous on the high-pressure pump and nozzle plant because nozzle type and number are often changed. On the atomizer plant, the only change is the wheel speed. Finding the optimum settings for each product and grades within a product range can take a considerable time in commissioning even if the product has been trialed at the spray drier manufacturers pilot facilities.

Returning to the calculation earlier, the spray drier ortholiquor feed is 11,537 kg/h, at a concentration of 59.5% and temperature of 95°C.

Fundamentally, the spray drier is evaporating water, and its efficiency is maximized by operating with the largest possible temperature difference between inlet and outlet. The outlet temperature is set at 165°C to avoid condensation that in turn would promote blockages. The inlet temperature is limited to 500°C by the materials of construction of the spray drier inlet and the need to avoid creation of glassy phosphates that would then condense and cause blockages.

Allowing orthobead product moisture of 0.5% and taking specific heats of 2.65 kJ/kg K for ortholiquor and 1.13 kJ/kg K for orthosolids,

Sensible heat in liquor feed at 95°C = 11,537/3,600 × 95 × (4.23 − (6,860 × 100/11,537) × 0.0233) = 866 kW

Orthobead moisture = 6860 × 0.005/0.995 = 34.5 kg/h

Water for evaporation = 11,537 − 6,860 − 34.5 = 4,642.5 kg/h

Latent heat required = 4642.5/3,600 × 2,501 = 3225 kW

Heat of crystallization = 6860 × 70.3/3600 = 134 kW

Sensible heat in orthobead at exit at 165°C = 6860/3600 × 165 × 1.13 = 355 kW

Sensible heat in exit vapor at 165°C = 4642.5/3600 × 165 × 2.5399 = 540 kW

Therefore,

Useful heat required = 3225 + 134 + 355 + 540 − 866 = 3388 kW

Losses @ 5% = 3388/0.95 × 0.05 = 178 kW

Therefore,

Total heat required = 3388 + 178 = 3566 kW

4.4.4 Calcining

On most plants, the dried orthobead and the fines pass into a screw conveyor that enters a rotary calciner (kiln). On some plants, a mixer is incorporated into this transfer step, and a double shafted device is sometimes used. Sometimes a spray is incorporated into this mixer; this spray has two functions. Firstly, a small flow of ortholiquor, or water, is sprayed onto the mixed surface that helps attach fine material to the coarser orthobead, which in turn reduces fines losses to the scrubbing system. Secondly, the spray can provide a capacity boost if the spray drier is at full capacity because the calciner is able to carry out drying as well as calcining duties. Heating air passing through the calciner flows to a cyclone, with the fines returning to the inlet mixer/conveyor.

The calciners on sodium phosphates duties are typically gas-fired rotating tubes made from either a high-temperature carbon steel or stainless steel mounted on rollers at a slight inclination to the horizontal. Usually, but not always, the calciner has several banks of internal fittings that serve to move the powder inside the tube in different directions. For example, at the entrance to the calciner where the feed material may still be slightly moist and sticky, there is often a short section of helical blades that push the powder along the tube. There is always a temperature gradient along the calciner axis and the process may require the powder to sojourn at a particular point corresponding to a particular temperature needed for a phase transformation. To achieve a local sojourn, back lifter flights are fitted that fling the powder backward. In other zones in the calciner, the heat transfer between hot air and particle may be particularly important and here simple lifters are incorporated that pick up the powder and drop it between the 11 and 1 o'clock positions of the calciner rotation. This has the effect of creating a powder curtain and maximizes gas to solid heat transfer. For materials that have a tendency to agglomerate and form lumps, chains are hung inside the calciner.

When making STPP, the overall residence time in the calciner is about 90 min. The speed of a particle through the calciner is a function of the rotational speed, the calciner diameter, and the maximum angle of repose. Therefore, if the calciner were a simple tube, the particle would have a certain residence time. To increase residence time, the calciner is always fitted with an end plate with a circular hole; the height of the weir thus created increases residence time. These weirs are sometimes fitted within the calciner as well, usually only one (which would be central) or two or occasionally three on the largest calciner, and are known as dam rings.

Calciner sizes, for industrial-scale production of sodium phosphates, range from 1.5 m diameter by 14 m long through to 4 m diameter by 30 m long. There does not appear to be an accepted precise sizing formula; a comparison of both specific diameter and volume varies significantly between many examples. Clearly, for the production of STPP, which requires the greatest heat input, a theoretical sizing might be arrived at for a simple shell. This is then complicated by the provision of different flights, dam rings, and rotational speed. The industry appears to have arrived at sizings that work and can be adapted. To give the reader some guidance, the following sizes are known to be satisfactory: 25,000 tons per year, 2.2 m diameter by 18 m long; 50,000 tons per year, 3 m diameter by 23 m long; 90,000 tons per year, 3.6 m diameter by 27 m long.

The calciners are mounted on cast iron rings that run on rollers. These rollers are lubricated by water, grease, or graphite; the first is unreliable and expensive, the second is satisfactory only if managed well, and the third is effective and requires little maintenance management. The calciner is turned by a geared drive ring linked to a gearbox and electric motor. Ideally, the motor is inverter controlled that allows rotation speed variation. A safety feature not always incorporated is a backup drive; the calciner usually runs sufficiently hot to *banana* if it is not continually rotated. If the primary drive breaks down, the backup needs to be running within 30 min. For the same reason, some care is needed when altering or adding dam rings as this increases the load on the calciner. For smaller calciners, this is not problematic but for larger calciners the manufacturer should be consulted. Typical rotation speeds are in the range 1.5–4 rpm. Calciner inclinations vary but 1.25° is not unusual.

The calcined powder passes out of the kiln via an end hood and often into a simple hammer mill that serves to break up any large lumps that have formed. The powder then passes into a cooler. Historically, two types of cooler have been used, either air- or water-cooled rotary cylinders. The water-cooled devices are generally smaller and utilize cooling water. Air cooling offers the opportunity for heat recovery; however, this benefit is somewhat negated by the electrical load supplying the fan that drives the cooling air. More recently, a water-cooled bunker has been utilized to cool the powder. The bunker is fitted with many flat plate cooling panels, supplied with cooling water; the powder falls into the bunker and is steadily withdrawn from the bottom outlet. This technology has many advantages over the traditional approach.

Coolers are usually located on the ground floor. On all plants, the cooled powder is conveyed to high level by a bucket elevator and then goes through a series of sieves, mills, and recycles to achieve the desired particle size distribution. Most sodium phosphates are relatively soft and easily milled; consequently, quite a wide range of mills have been used on these plants.

Generally, a multiproduct sodium phosphates plant is operated to produce a maximum of two products at a time, powder and granular. Mills and sieves are set up (as is Na/P ratio in the neutralization section) for a campaign and the products are packed off into either sacks, big bags, or bulk. For the larger plants, form–fill–seal packing machines are used, which, once set up, are effective and accurate for filling 25 kg sacks.

The neutralizer–spray drier–calciner–cooler–separation plant covers most sodium phosphates manufacturing processes and with appropriate materials of construction and minor adaptation can and is used for other phosphate manufacture.

4.4.5 *Hexameta*

In the context of sodium phosphates, the manufacture of sodium hexametaphosphate is unique. As for all sodium phosphate processes, the first step is neutralization. For many hexameta plants, an ortholiquor of the appropriate Na/P ratio is made and then pumped to the furnace. On some plants, an ortho paste is made, with the desired Na/P ratio and conveyed into the furnace by screw, and the advantage of the paste is its significantly lower water content [48]. The first stage of the hexameta furnace process is the evaporation of all free water; therefore, a low water content feed consumes less energy. The second reason is that water vapor contributes to the high corrosivity of the vapor phase in the furnace.

There are as many different hexameta furnace types as there are producers. Essentially, the furnace is a brick-lined structure, often encased with a carbon steel shell and insulated. There are at least two layers of bricks in the walls and roof and, in many examples, the floor (one East German design uses cooled stainless steel for the floor); the bricks are usually of high zircon content. At one end, the furnace has one or more gas burners, sometimes co- and sometimes countercurrent to the hexameta flow. The ortho material drops in a moving pool of hexameta molten glass and is rapidly converted. The glass overflows and leaves the furnace. Radiative heating provides a significant proportion of the heat energy. In order to maximize this, some furnaces are designed with curved internal roof profiles that are thought to reflect heat back to the surface of the molten pool. Some designs are constantly cylindrical; others have different roof curvature and distance from the pool at different locations along the axis of the furnace. The A&W furnace at Widnes was a simple box. Dimensionally, the furnaces are 1.5–3 m wide by 4–8 m long. Operating temperature is around 900°C.

The molten glass usually passes down a water-cooled trough (cooled for corrosion protection) and into a cooling device. Several different devices are used including water-cooled drums, plates, or belts. The common feature is a water-cooled metal surface on which the molten glass is quenched to solid glass and is then scrapped off the surface. The broken hexameta glass is conveyed to milling. Three product forms are sold: powder, granular, and glass. The former is thoroughly milled, the latter only lightly milled. A sieving process ensures the correct particle size distribution going to the final product storages. Hexameta is very hygroscopic, as a result the overall residence time in the plant is minimized, the product is covered as much as possible, and occasionally the packing off area is air-conditioned to lower ambient humidity.

Hot gases from the furnace pass to a stack. It is well recognized that these gases have great potential for heat recovery. Many have attempted to do so but struggled with corrosion issues. The most obvious use of the waste heat is in preheating the combustion air for the furnace.

REFERENCES

1. B. Wendrow and K. A. Kobe, The alkali orthophosphates. Phase equilibria in aqueous solution, *Chem. Rev.*, 54(6), 891–924, December 1954.
2. R. N. Bell, Hydrates of trisodium orthophosphate, *Ind. Eng. Chem.*, 41(12), 2901–2905, December 1949.
3. E. Tillmanns and W. H. Baur, Stoichiometry of trisodium orthophosphate dodecahydrate, *Inorg. Chem.*, 9(8), 1957–1958, August 1970.
4. R. M. Lauck, R. E. Vanstrom, and J. W. Tucker, Novel complex sodium aluminum orthophosphate reaction products and water insoluble reactions thereof, US Patent 3097949, July 1963.
5. R. E. Benjamin and T. E. Edging, Process for preparing alkali metal aluminum phosphate, US Patent 4260591, April 1981.
6. G. A. McDonald, Complex alkali metal-aluminum and alkali metal-iron acid phosphates, US Patent 2550490, April 1951.
7. L. D. Mathias, Chlorine-containing compound, US Patent 1555474, September 1925.
8. C. Y. Shen, Process for producing chlorinated trisodium phosphate, US Patent 4402926, September 1983.
9. L. Steinbrecher and J. F. Hazel, Thermal gravimetric studies of orthophosphates, *Inorg. Nucl. Chem. Lett.*, 4(10), 559–562, October 1968.
10. L. A. Kramer and L. E. Netherton, Stabilized low reaction rate sodium acid pyrophosphate, US Patent 2844437, July 1958.
11. R. H. Tieckelmann, Alkali metal acid pyrophosphate leavening acid compositions and methods for producing the same, US Patent 4804553, February 1989.
12. J. R. Van Wazer, *Phosphorus and Its Compounds Chemistry*, vol. 1, Interscience Publishers, New York, 1958.
13. A. D. F. Toy, *The Chemistry of Phosphorus*, Pergamon Press, New York, 1975.
14. J. D. McGilvery and A. E. Scott, The role of water in the formation of sodium triphosphate by calcination, *Can. J. Chem.*, 32(12), 1100–1111, December 1954.
15. J. W. Edwards, Method of producing phosphates, US Patent 2916354, December 1959.
16. T. Graham, Researches on the arsenates, phosphates and modifications of phosphoric acid, *Phil. Trans. Roy. Soc.*, 123, 253–284, 1833.
17. L. S. Henson, Beverages containing mixed sodium-potassium polyphosphates, US Patent 6440482, August 2002.
18. K. Dietz, Process for the production of industrial slurries of reduced moisture content in relation to viscosity, US Patent 2709661, May 1955.
19. R. P. Bright and D. D. Double, Chemically bonded ceramic armor materials, European Patent 0299253, January 1989.
20. C. M. Jones, Chromia-magnesia ramming cement, US Patent 4507395, March 1985.
21. W.-C. Wei and S.-J. Lu, Aqueous colloidal dispersions of sub-micrometer alumina particles, US Patent 5518660, May 1996.
22. J. E. Neely and J. R. Martinet, Refractory, US Patent 3278320, October 1966.
23. J. E. Neely and M. L. Mayberry, Refractory shape, US Patent 3392037, July 1968.
24. W. Feldenheimer, Treatment of clay, US Patent 1438588, December 1922.

25. M. J. Garska, C. R. L. Golley, R. J. Pruett, and J. Yuan, Integrated process for simultaneous beneficiation, leaching, and dewatering of kaolin clay suspension, US Patent 7122080, October 2006.
26. S. A. Bolkan, G. Byrnes, and S. Dunn, Alkali metal cleaner with zinc phosphate anticorrosion system, US Patent 5712236, January 1998.
27. S. Ikeda, Y. Matsuura, and K. Yasuhara, Alkaline tin-plate degreasing detergent, US Patent 4756846, July 1988.
28. M. Maruta and K. K. Kaisha, Neutral tin electroplating baths, US Patent 4329207, May 1982.
29. O. A. Ashiru and S. J. Blunden, Electroplating, US Patent 5378346, January 1995.
30. N. A. Fotinos and G. D. Kent, Non-accelerated iron phosphating, US Patent 5073196, December 1991.
31. P. Engstrand, L.-A. Hammar, M. Htun, B. Sjogren, and B. Svensson, A method of producing cellulosic pulp, European Patent 0500674, September 1992.
32. G. M. Boston, A. K. Clarkson, D. K. Collier, P. T. Graycar, M. M. Hartzell, Y.-L. Hsieh, and A. E. Larenas, Enzyme treatment to enhance wettability and absorbency of textiles, European Patent 088531129, November 2006.
33. W. M. Scheper, R. R. Gardner, M. R. Sivik, and V. M. Arredondo, Textile finishing composition and methods for using same, US Patent 6989035, January 2006.
34. L. Lampila and J. Godber, Food phosphates, in *Food Additives*, vol. 116, J. H. Thorngate, S. Salminen, L. Branen, and M. Davidson, Eds., CRC Press, Boston, MA, 2001.
35. W. H. Waggaman, *Phosphoric Acid, Phosphates, and Phosphatic Fertilizers*, Reinhold Pub. Corp., New York, 1952.
36. J. C. McDilda and M. J. Rice, Container for refrigerated dough and method of forming a refrigerated dough product, US Patent 5084284, January 1992.
37. W. Bindzus, P. A. Altieri, J. J. Kasica, and P. T. Trzasko, Starch phosphate ester composition, process and method of use in food, US Patent 6428836, August 2002.
38. E. D. Davy, Salt substitute, US Patent 2471144, May 1949.
39. R. G. Kauffman, M. L. Greaser, E. Pospiech, and R. L. Russell, Method of improving the water-holding capacity, color, and organoleptic properties of beef, pork, and poultry, US Patent 6020012, February 2000.
40. E. Brotsky and W. E. Swartz, Increased shelf life for refrigerated fish, US Patent 4937092, June 1990.
41. L. S. Henson, R. V. Manley, and K. J. Fennewald, Use of antimicrobial polyphosphates in food processing, US Patent 6509050, January 2003.
42. E. J. Powers, Preparation of sodium orthophosphate mixtures, US Patent 4224294, September 1980.
43. N. E. Stahlheber and J. E. Lyon, Continuous process for preparing sodium orthophosphate slurries from natural soda ash and orthophosphoric acid, US Patent 4661331, April 1987.
44. N. E. Stahlheber and J. E. Lyon, Continuous process for preparing sodium orthophosphate slurries from natural soda ash orthophosphoric acid, US Patent 4853200, August 1989.
45. W. S. Bowen, Open steam box drier, US Patent 2360100, October 1944.
46. K. Masters, *Spray Drying Handbook*, 5th edn. Longman Scientific & Technical, Harlow, U.K.; Wiley, New York, 1991.
47. A. S. Mujumdar, *Handbook of Industrial Drying*, 3rd edn., CRC Press, Boca Raton, FL, 2006.
48. H. E. Buckholtz, Production of hexametaphosphates from other phosphates, US Patent 49976380, March 1991.

5 Calcium Phosphates

5.1 INTRODUCTION

Of the alkaline earth phosphates (Mg, Ca, Sr, Ba), the calcium phosphates are produced in the largest volumes and have received the most academic study, particularly in the field of dentistry and general physiology [1]. From an industrial standpoint, much of the academic study is informative rather than definitive, for example, when considering tooth decay or the deposition of plaque, the investigator is working in a pH range of 5–8 and molar concentrations in the range 0.001–0.1 M; by contrast, industrial processes operate at molar concentrations between 1 and 4 M and the full pH range.

Calcium orthophosphates of high purity are produced for bakery, dental, nutritional, or pharmaceutical applications; careful attention is paid to the impurities in the calcium source as well as the acid. The total worldwide production of these calcium phosphates is around 150,000 tons/year, of which dental grade dicalcium phosphate dihydrate (DCPD) is the largest proportion. On the other hand, animal feed grade dicalcium phosphate (DCP) production worldwide is an order of magnitude larger at approximately 4,000,000 tons. Animal feed DCP is made either by acidulating phosphate rock with hydrochloric acid or by utilizing wet process phosphoric acid (WPA), to which is added lime or limestone (CaO or $CaCO_3$). The phosphoric acid component is self-evidently less pure than thermal or purified acid; similarly, the calcium source requirements are less stringent. Nevertheless, the principles of the chemistry of the production of calcium phosphates are the same regardless of the purity of the final product, and although animal feed DCP is not produced from purified or thermal phosphoric acid, it is addressed in this chapter because of its commercial significance.

The higher calcium phosphates (pyrophosphate and polyphosphate) are produced commercially but in relatively low volumes. In general, manufacturing processes for the higher phosphates are based on the heat treatment of orthophosphates.

5.2 CHEMISTRY OF CALCIUM ORTHOPHOSPHATES

The chemistry of calcium orthophosphates of commercial interest is best described with reference to the $CaO–H_2O–P_2O_5$ phase diagram as depicted in Figure 5.1. The diagram is drawn from the data of Elmore and Farr [2] and has been studied extensively; the recent study has made use of more precise equipment giving more accurate results [3]; however, for industrial practice, the older data are satisfactory. At low ion concentrations typical of *in vivo* studies, the behavior of calcium phosphates is both complex and difficult to observe with precision; however, this is of no concern for a manufacturing process that operates at far higher concentrations for reasons of economic equipment sizing.

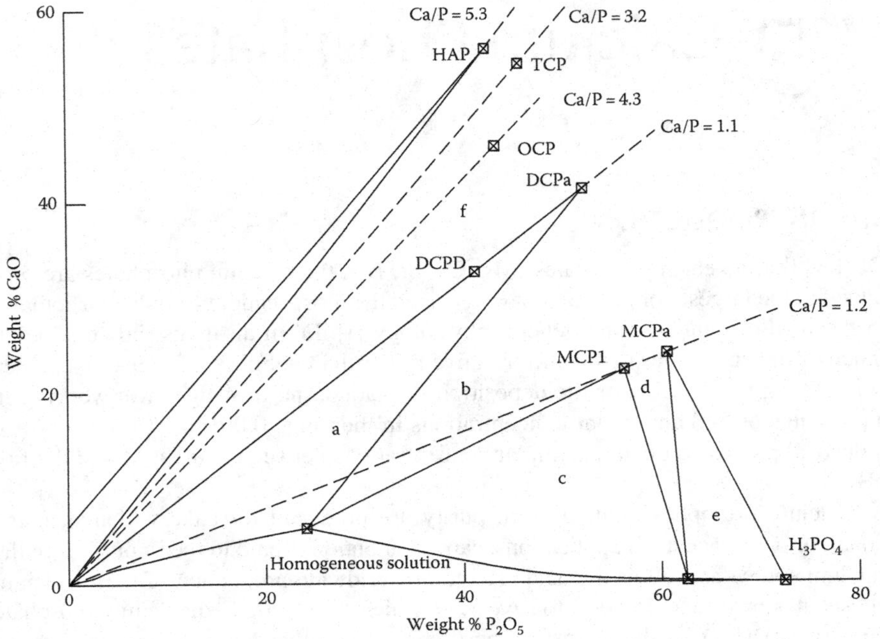

FIGURE 5.1 The $CaO–P_2O_5–H_2O$ phase diagram. (Adapted with permission from Elmore, K.L. and Farr, T.D., Equilibrium in the system calcium oxide–phosphorus pentoxide–water, *Ind. Eng. Chem.*, 32(4), 580–586. Copyright 1940 American Chemical Society.)

There are four distinct groups of calcium orthophosphates: the monocalciums comprising anhydrous monocalcium phosphate (MCPa, $Ca(H_2PO_4)_2$) and monocalcium phosphate monohydrate (MCP1, $Ca(H_2PO_4)_2 \cdot H_2O$, brushite); the dicalciums comprising DCPD ($CaHPO_4 \cdot 2H_2O$, monetite) and the anhydrous salt (DCPa, $CaHPO_4$); one of variable composition (a Ca/P mole ratio in the range 1.40–1.75) collectively, and generally incorrectly, known as tricalcium phosphate (TCP, $Ca_3(PO_4)_2$) although, in fact, it is often predominantly hydroxylapatite (HAP, $Ca_5(OH)(PO_4)_3$); and others of little commercial interest including octacalcium phosphate (OCP, $Ca_8H_2(PO_4)_6 \cdot 5H_2O$). All of these compounds are obtainable by precipitation from an aqueous reaction mixture of phosphoric acid and hydrated lime although commercial practice does not always do so. The fixed points on the phase diagram are these pure products; Table 5.1 sets these out with their Ca/P (molar) and CaO/P_2O_5 (weight) ratios and CaO, P_2O_5, and H_2O weight percentages. These are easily calculated from molecular weights. OCP ($Ca_8H_2(PO_4)_6 \cdot 5H_2O$) is included for reference but is an undesired product; manufacturing processes are designed to avoid making this product. Tetracalcium phosphate, hilgenstockite (TTCP, $Ca_4P_2O_9$), is included in Table 5.1 but not in Figure 5.1.

Referring to the diagram, which is representative of the system at 25°C, the system remains in solution in acidic conditions and low CaO concentrations. There is a triple point where MCP1, DCPa, and the saturated solution are in equilibrium. Elmore and Farr [2] measured this point at 24.10% P_2O_5 and 5.785% CaO at 25°C;

TABLE 5.1
Summary of Calcium Phosphate Compositions

Name	Formula	MW	Ca/P	CaO/P_2O_5	%CaO	$\%P_2O_5$	$\%H_2O$
MCPa	$Ca(H_2PO_4)_2$	234.1	0.50	0.40	23.96	60.65	15.39
MCP1	$Ca(H_2PO_4)_2 \cdot H_2O$	252.1	0.50	0.40	22.25	56.31	21.44
DCPa	$CaHPO_4$	136.1	1.00	0.79	41.22	52.16	6.62
DCPD	$CaHPO_4 \cdot 2H_2O$	172.1	1.00	0.79	32.59	41.24	26.17
TCP	$Ca_3(PO_4)_2$	310.2	1.50	1.19	54.24	45.76	0.00
HAP	$Ca_5(OH)(PO_4)_3$	502.3	1.67	1.32	55.82	42.39	1.79
OCP	$Ca_8H_2(PO_4)_6 \cdot 5H_2O$	982.6	1.33	1.05	45.66	43.34	11.00
TTCP	$Ca_4P_2O_9$	366.3	2.00	1.58	61.24	38.76	0.00

this point moves along an invariant curve with increasing temperature; at 100°C, it is 40.15% P_2O_5 and 5.529% CaO.

In the area (a) bounded by the saturated solution line, the line from the triple point to the DCPa point and the Ca/P = 1:1 line, DCPa is the stable saturating solid. Adjacent to this area, as P_2O_5 concentration increases (b), solid MCP1 and DCPa coexist in saturated solution. The Ca/P = 1:2 line cuts through this area; if solid MCP1, which lies on this line, is mixed with water, some will dissolve and be converted to DCPa (and phosphoric acid); eventually, as more water is added, the conversion will be complete. As a result, dry, neutral commercial grades of MCP1 will always contain a proportion of DCPa.

This conversion is described in the following equation:

$$Ca(H_2PO_4)_2 \cdot H_2O + xH_2O \rightleftharpoons CaHPO_4 + H_3PO_4 + (x+1)H_2O \tag{5.1}$$

This equation illustrates the incongruent dissolution of monocalcium phosphates, that is, dissolution is accompanied by reaction, which also applies to the DCPs.

In the adjacent area (c), bounded by the saturation line, the line from the triple point to the MCP1 point and down to a point at 62.30% P_2O_5 and 0.2% CaO, MCP1 is the stable saturating solid. Both the monohydrate and anhydrous monocalcium salts exist in solution in the area (d) bounded by the Ca/P = 1:2 and pure salt to 62.30% P_2O_5 and 0.2% CaO lines; both also exhibit positive temperature coefficients of solubility. In the adjacent area (e), bounded by the MCPa to 62.30% P_2O_5 and 0.2% CaO line and the MCPa to pure H_3PO_4 line, MCPa alone exists in solution.

Above the Ca/P = 1:1 line at higher Ca/P ratios (f), DCPD decomposes into a mixture with free phosphoric acid and HAP. DCPD also decomposes to DCPa with moisture, higher temperatures, and fluoride ions and is unstable above 36°C–40°C. These properties are significant when considering processes to manufacture DCPD for formulation into fluoride toothpastes. If the DCPD contains any free lime, there is a risk of conversion to HAP and the product will turn rocklike; if it converts to DCPa, the toothpaste becomes gritty. Consequently, for dental formulations, DCPD is manufactured on the acid side of neutral to avoid HAP formation and incorporates stabilizers to mitigate conversion to DCPa.

TCPs lie in a narrow band above the Ca/P = 3:2 line. Normally, TCPs are manufactured by adding acid to slaked lime, $Ca(OH)_2$. Clearly, it is possible to start from acid and continue to add lime although the calcium phosphate must undergo several changes including transforming from DCPD to the intermediate OCP before arriving at a TCP.

While the processes to manufacture the calcium orthophosphates are accurately described as precipitation reactions, they may also be described as reactive crystallizations. Both the product purity and its physical form are fundamental aspects of the product specification. Both of these attributes may be altered by postreaction processes, for example, through washing or milling; however, the bar is set through the control of the crystallization reaction; no amount of washing will remove unreacted lime from the core of a DCPD crystal. Several factors control crystal growth: supersaturation, pH, ionic strength, temperature, Ca/P ratio, and additive concentration. In the industrial reactor, whether batch or continuous, these factors are in turn influenced by the accuracy, consistency, and rate of reactant addition, agitation, and heating or cooling of the reaction mass. Final crystal size also depends on the reactor residence time and agitation; these factors are largely set when the plant is designed and built, in that reactor and agitator shape and size cannot easily be altered; limited adjustment is possible by changing the overall plant rate or the rotation speed of the agitator.

Of these factors, temperature measurement is the easiest and most accurate, followed by liquid flow rate so reaction temperature and heat loss/gain and acid feed rates are reasonably accurate. pH is relatively easy to measure, but pH probes require frequent maintenance and recalibration and on an industrial reactor may, in practice, only be accurate to ±0.25 pH. The rate and quantity of calcium addition, whether in the form of a $Ca(OH)_2$ or $CaCO_3$ slurry, are difficult to measure accurately. As a result, some processes use pH measurement as the control measure, controlling the addition of the calcium source based on the quantity or rate of acid against a fixed pH set point. BK Ladenburg was granted a patent [4] for the use of conductivity measurement as an improved control of calcium phosphate production. Given the inaccuracies, plant control is supplemented by laboratory analysis and as steady and consistent an operation as possible.

A number of authors have set out the governing equations and propose models that can assist the design of an industrial calcium phosphate process [5–8]. Neglecting additives and considering a pure calcium phosphate system, there are eight reactions that are shown in Equations 5.2 through 5.9 together with their associated reaction equilibrium constants:

$$H_3PO_4(aq) \rightleftharpoons H_2PO_4^- + H^+ \quad K_1 = \frac{\gamma_{H_2PO_4^-}\gamma_{H^+}\left[H_2PO_4^-\right]\left[H^+\right]}{\left[H_3PO_4\right]_{aq}} = 10^{-2.148} \tag{5.2}$$

$$H_2PO_4^- \rightleftharpoons HPO_4^{2-} + H^+ \quad K_2 = \frac{\gamma_{HPO_4^{2-}}\gamma_{H^+}\left[HPO_4^{2-}\right]\left[H^+\right]}{\gamma_{H_2PO_4^-}\left[H_2PO_4^-\right]} = 10^{-7.199} \tag{5.3}$$

$$HPO_4^{2-} \rightleftharpoons PO_4^{3-} + H^+ \quad K_3 = \frac{\gamma_{PO_4^{3-}}\gamma_{H^+}\left[PO_4^{3-}\right]\left[H^+\right]}{\gamma_{HPO_4^{2-}}\left[HPO_4^{2-}\right]} = 10^{-12.35} \tag{5.4}$$

$$Ca^{2+} + PO_4^{3-} \rightleftharpoons CaPO_4^- \quad K_4 = \frac{\gamma_{CaPO_4^-}\left[CaPO_4^-\right]}{\gamma_{Ca^{2+}}\gamma_{PO_4^{3-}}\left[Ca^{2+}\right]\left[PO_4^{3-}\right]} = 10^{6.4} \tag{5.5}$$

$$Ca^{2+} + OH^- \rightleftharpoons CaOH^+ \quad K_5 = \frac{\gamma_{CaOH^+}\left[CaOH^+\right]}{\gamma_{Ca^{2+}}\gamma_{OH^-}\left[Ca^{2+}\right]\left[OH^-\right]} = 10^{1.3} \tag{5.6}$$

$$Ca^{2+} + HPO_4^{2-} \rightleftharpoons CaHPO_4 \quad K_6 = \frac{\left[CaHPO_4\right]_{aq}}{\gamma_{Ca^{2+}}\gamma_{HPO_4^{2-}}\left[Ca^{2+}\right]\left[HPO_4^{2-}\right]} = 10^{2.74} \tag{5.7}$$

$$Ca^{2+} + H_2PO_4^- \rightleftharpoons CaH_2PO_4^+ \quad K_7 = \frac{\gamma_{CaH_2PO_4^+}\left[CaH_2PO_4^+\right]}{\gamma_{Ca^{2+}}\gamma_{H_2PO_4^-}\left[Ca^{2+}\right]\left[H_2PO_4^-\right]} = 10^{1.4} \tag{5.8}$$

$$H_2O \rightleftharpoons OH^- + H^+ \quad K_8 = K_W = \gamma_{OH^-}\gamma_{H^+}\left[OH^-\right]\left[H^+\right] = 10^{-13.997} \tag{5.9}$$

where

[ ] represents the molar concentration

γ_i represents the activity coefficient and is used to modify the molar concentrations to account for their deviation from ideality

K_1–K_8 are equilibrium constants and may be found in the literature [9], for example, $K_1 = 10^{-2.148}$

Different authors have suggested that the activity coefficient, γ, may be calculated from a number of different equations [10]; the extended form of the Debye–Hückel equation proposed by Davis [11] is one of these:

$$\log \gamma_{\pm} = -A\left|Z_+Z_-\right|\frac{\sqrt{I}}{1+\sqrt{I}} - 0.3I \tag{5.10}$$

where

Z is the species charge

I is the ionic strength

The coefficient A depends on temperature [8] and is calculated from

$$A = 0.486 - 6.07\times10^{-4}T + 6.43\times10^{-6}T^2 \tag{5.11}$$

The material balance equations are

$$\left[\mathrm{Ca}\right]_T = \left[\mathrm{Ca^{2+}}\right] + \left[\mathrm{CaPO_4^-}\right] + \left[\mathrm{CaOH^+}\right] + \left[\mathrm{CaHPO_4}\right] + \left[\mathrm{CaH_2PO_4^+}\right] \quad (5.12)$$

$$\begin{aligned}\left[\mathrm{P}\right]_T = &\left[\mathrm{H_3PO_4}\right]_{aq} + \left[\mathrm{H_2PO_4^-}\right] + \left[\mathrm{HPO_4^{2-}}\right] + \left[\mathrm{PO_4^{3-}}\right] + \left[\mathrm{CaPO_4^-}\right] \\ &+ \left[\mathrm{CaHPO_4}\right] + \left[\mathrm{CaH_2PO_4^+}\right]\end{aligned} \quad (5.13)$$

$[Ca]_T$ and $[P]_T$ are the known initial concentrations of calcium and phosphorus, and pH may be measured; alternatively pH, $[Ca]_T$, and $[P]_T$ may be specified. Either way, a system of 10 equations and 10 unknowns is now set and does submit to treatment either by software (relatively straightforward, e.g., using MATLAB®, TK Solver, or Microsoft Excel) or by hand (not straightforward).

Equations 5.2 through 5.13 describe a calcium phosphate system at equilibrium and unsaturated. As the product species concentration exceeds the solubility product, K_{sp}, the product begins to precipitate, for example, if the objective was to precipitate DCPD, this would start to occur when

$$\left[\mathrm{Ca^{2+}}\right]\left[\mathrm{HPO_4^{2-}}\right] \geq K_{sp\mathrm{DCPD}} \quad (5.14)$$

or for HAP when

$$\left[\mathrm{Ca^{2+}}\right]^5\left[\mathrm{PO_4^{3-}}\right]^3\left[\mathrm{OH^-}\right] \geq K_{sp\mathrm{HAP}} \quad (5.15)$$

For a supersaturated system, the material balance equations (5.12 and 5.13) are modified by adding the precipitated solid phase concentration to both equations.

Once crystal nucleation has occurred, crystal growth rate is dependent on supersaturation, β; for DCPD,

$$\beta = \frac{\gamma_{\mathrm{Ca^{2+}}}\gamma_{\mathrm{HPO_4^{2-}}}\left[\mathrm{Ca^{2+}}\right]\left[\mathrm{HPO_4^{2-}}\right]}{K_{sp}} \quad (5.16)$$

Crystal growth rate may be expressed as

$$G_i = k_{gi}(\beta - 1)^n \quad (5.17)$$

where

G_i is the growth rate in a specific direction (radial or axial)

k_{gi} is a growth rate constant

One paper published the following values for DCPD [8]:

$$\begin{aligned}G_x &= 1\times10^{-11}(\beta-1)^{1.01} \\ G_r &= 7\times10^{-13}(\beta-1)^{0.98}\end{aligned} \quad (5.18)$$

In an industrial setting, crystals do both break and agglomerate; thus, the growth rates in (5.17) should be viewed as indicative; nevertheless, they do inform decisions on residence time.

Different species precipitate at different pH, and a number of solubility isotherm diagrams are available in the literature [3,12,13]. The order of precipitation has been studied extensively; however, most of the literature refers to more dilute systems than industrial practice so these may only be indicative. Crystal growth is facilitated by the presence of seed crystals; an operating plant will tend to make one product; therefore, even if a batch plant is being operated, it is likely to leave deposits of the final product that will act as seeds for the next batch. Consequently, even if under laboratory conditions an undesired product, for example, HAP, begins to precipitate before the desired product, say, DCPD, on the plant, so long as the Ca/P ratios are correct, and the temperature, and adequate agitation avoids any local Ca/P imbalances, DCPD seed crystals will aid the system in growing the desired product.

Reaction temperature is an important consideration. As stated earlier, DCPD is considered unstable above 36°C–40°C, and given that the correct Ca/P ratio is maintained, it will start to convert to DCPa above this temperature range. DCPa growth rates improve above 60°C [11]. TCPs are produced at similar temperatures to DCPa, that is, greater than 60°C. MCP1 and MCPa are produced at temperatures greater than 100°C.

5.3 CHEMISTRY OF CALCIUM PYROPHOSPHATES AND POLYPHOSPHATES

The onset of the conversion of MCPa to calcium acid pyrophosphate (CAPP, $CaH_2P_2O_7$) occurs at 220°C [14]; an earlier patent of 1911 points the way [15]. A Monsanto patent [16] observes that simply heating MCPa results in an impure product and proposes the addition of MCPa to an excess of phosphoric acid at 200°C–220°C. CAPP is used in bakery applications.

When DCPD is dehydrated thermally, amorphous calcium pyrophosphate (CPP, $Ca_2P_2O_7$) is formed at 360°C–450°C. With continued heating, amorphous CPP is converted exothermically to γ-$Ca_2P_2O_7$ at 530°C. A process to make γ-$Ca_2P_2O_7$ for use as luminescent phosphors describes heating DCPa to 500°C–600°C [17]. Upon further heating, the γ-form is converted to the β-form at 750°C and to the α-form at 1171°C–1179°C, which melts at 1352°C [18]. A method to make a pure, spherical form of γ-$Ca_2P_2O_7$ commences by dissolving DCPa in nitric acid, then spray drying the solution to form predominantly DCPa but also MCPa in spherical form, and then heating the powder to 700°C for 4 h [19]. A process to make γ- and β-forms of CPP as dentifrice abrasives commences with the production of DCPD, which is heated to 100°C to form DCPa, which in turn is heated to 450°C–475°C to form γ-$Ca_2P_2O_7$ and further heated to 650°C–800°C to form β-$Ca_2P_2O_7$ [20].

Hydrated forms of CPP are made by the reaction of an aqueous solution of calcium chloride with tetrapotassium pyrophosphate or by the reaction of $Ca(OH)_2$ with an aqueous solution of solid pyrophosphoric acid ($\geq$95% $H_4P_2O_7$). If the latter reaction is held at below 40°C, the product is $Ca_2P_2O_7 \cdot 4H_2O$; between 45°C and below 65°C,

$Ca_2P_2O_7 \cdot 2H_2O$; and between 80°C and below 100°C, $Ca_2P_2O_7$ [21]. CPP dihydrate is also made in the body, forming crystals and causing pain in joints.

Calcium polyphosphates may be represented by the general formula $(CaO)_m(P_2O_5)_n \cdot xH_2O$ where the m/n ratio may vary from approximately 1.66–2.00 (in other words, a Ca/P of 0.83–1.00) and x from 0 to 8 [22]. Earlier publications suggest a value for x up to 30. A 1958 patent to Victor Chemicals [22] describes the production of calcium tripolyphosphate octahydrate, $Ca_5P_6O_{20} \cdot 8H_2O$, by the reaction of a 9.7% solution of sodium tripolyphosphate with a 4.7% solution of calcium chloride at 70°C–75°C, followed immediately by neutralization with dry $Ca(OH)_2$ and filtration. The octahydrate dehydrates to the tetrahydrate at 110°C and to the anhydrous salt at approximately 420°C. Calcium tripolyphosphate may be made by the transformation of MCP1 with microwave radiation [23]; this process is not yet commercialized, and given its liveliness, the product is most likely the anhydrous form. Calcium tripolyphosphate is formed as an unhelpful scale when low-concentration sodium tripolyphosphate is used as a detergent builder with hard waters [24].

Calcium tripolyphosphate may be used in a number of applications. It acts as a stabilizer for DCPD in toothpaste formulations and is an effective anticorrosive pigment for paint [25]; however, to date, calcium tripolyphosphate has not been produced in large commercial quantities.

5.4 APPLICATIONS

5.4.1 Bakery Applications

Both sodium and calcium phosphates are used heavily in bakery applications as chemical leavening agents, especially different grades of sodium acid pyrophosphate (SAPP) and monocalcium phosphates. As well as chemical leavening, uses include pH adjustment and buffering, enrichment, growth factors for yeasts, starch modification, and the manufacture of quick cook meals. The function of these phosphates in chemical leavening is to act as baking acid in reaction with sodium bicarbonate and so release carbon dioxide in the dough or batter. Sodium bicarbonate, or *common baking soda*, is still the most used bicarbonate; however, increasing health concerns about sodium levels in the diet are leading to a move to potassium or ammonium bicarbonate and an emphasis on the calcium rather than sodium phosphate salts. Two product qualities specific to leavening agents are neutralizing strength or neutralizing value (NV) and dough rate of reaction (DROR).

NV is a measure of the acid required within a bakery formulation and is expressed as parts by weight of the acid salt required to neutralize exactly 100 parts of sodium bicarbonate. Equations 5.19 and 5.20 represent the final reactions between sodium bicarbonate and MCP1 and SAPP, respectively:

$$3Ca(H_2PO_4)_2 \cdot H_2O + 8NaHCO_3 \rightarrow 8CO_2 + Ca_3(PO_4)_2 + 4Na_2HPO_4 + 11H_2O \quad (5.19)$$

$$Na_2H_2P_2O_7 + 2NaHCO_3 \rightarrow 2CO_2 + 2Na_2HPO_4 + H_2O \quad (5.20)$$

According to these equations, the theoretical neutralizing strength of MCP1 and SAPP is 88.9 and 60.9, respectively. Commercial baking acids have lower than theoretical neutralizing strengths, for example, MCP1 always contains up to 10% DCPa, so a commercial MCP1 may have an NV of 80.0. The desired end pH ranges from slightly acid to slightly alkali; for example, chocolate cakes are made in the pH range 7.1–8.0. NV is determined by simple titration. For example, 0.8401 g of the sample acidulant, 20 g NaCl, 5 mL of a 25% sodium citrate solution, and 25 mL distilled water are blended in a 375 mL white casserole. Immediately after 30 s of vigorous stirring, 120 mL of 0.10 normal NaOH is added and the mixture heated to boiling. After 7 min of boiling and removed from heating, 0.05 mL phenolphthalein indicator is added. The solution is titrated with 0.2 normal HCl until the pink color disappears. The solution is boiled for 1 min. NV is calculated from the following [26]:

$$NV = \text{mL}\,0.10\,\text{M}\,\text{NaOH} - 2(\text{mL}\,0.2\,\text{M}\,\text{HCl}) \tag{5.21}$$

DROR measures the rate of generation of carbon dioxide during the mixing, usually 2–3 min, and the subsequent bench time of a simple biscuit dough formulation at 27°C and at specific times up to 60 min. Different products require different gas generation profiles: gas quantity, release rate, and timing are all relevant depending on the product. For example, pancake dough will require a completely different DROR than refrigerated cookie dough.

The procedure for determining DROR is broadly similar throughout the industry and based on Parks [27]; however, the precise configuration of the DROR apparatus and the procedural details vary depending on the manufacturer. A typical procedure begins with preparation of a 5:3 ratio flour and water mix; a 50 g sample is placed in a mixing vessel and mixed for 1 min. To this mix is added the rest of the dough ingredients, including the leavening agents. Both vary, but commonly the ingredients and quantities are sodium bicarbonate ~1.1 g, salt ~0.1 g, sugar ~1.5 g, dry milk ~4.5 g, shortening (hydrogenated vegetable oil) ~5.5 g, and leavening acid ~1.3 g. The mixing vessel is held at 27°C in a water bath. The vessel is sealed and vents into equipment that will measure the quantity of carbon dioxide released. The simplest measuring device is a Chittick apparatus, as shown in Figure 5.2. The Chittick apparatus comprises the vent from the mixing vessel to the top of a measuring burette; a leveling/venting stopcock is located on a tee on the vent line, a measuring burette, and a leveling bulb. The bottom of the measuring burette is connected via rubber tubing to the leveling bulb, which is itself vented to atmosphere. The leveling bulb is easily moved up and down so as to match the liquid levels in the burette. Colored water is poured into the leveling bulb, through the flexible tube and into the measuring burette. Before commencing the DROR procedure, the leveling stopcock is opened and a liquid level established. With the system now sealed, the leveling bulb is lowered a little to create a slight negative pressure and the test commences. As the leavening acid reacts with the sodium bicarbonate, carbon dioxide begins to be released, changing the liquid levels in the burette and leveling bulb. The leveling bulb is moved every minute to reestablish

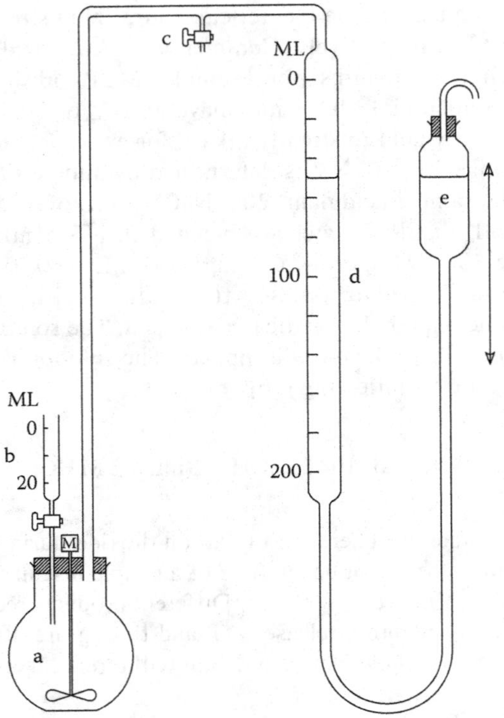

FIGURE 5.2 Chittick apparatus: (a) mixing vessel, (b) filling burette, (c) vent, (d) measuring burette, and (e) levelling bulb.

levels, and the volume reading noted. After the initial gas generation, the period between readings is extended as the generation rate levels out. Small corrections for ambient temperature and atmospheric pressure are made to the readings to give a reasonably accurate measurement of gas quantity and evolution rate. More complicated automated equipment is used, some with electronic sensors to send signals to data loggers; however, the reaction rate is sensitive to pressure, and if the headspace pressure rises—and adjustment of the manual Chittick apparatus avoids this—the reaction is suppressed.

Batter rates of reaction (BRORs) are determined at different temperatures using a batter formulation. BROR gives a better indication of how a dough or cake mix will perform during the cooking time.

Figure 5.3 shows typical DROR profiles for different leavening acids.

5.4.1.1 Application Properties

MCPa is fast reacting and hygroscopic. MCPa is made as a product itself, as an intermediate that is partially or wholly converted to CAPP and as a coated product. The coating is a combination of potassium and aluminum phosphates and delays the

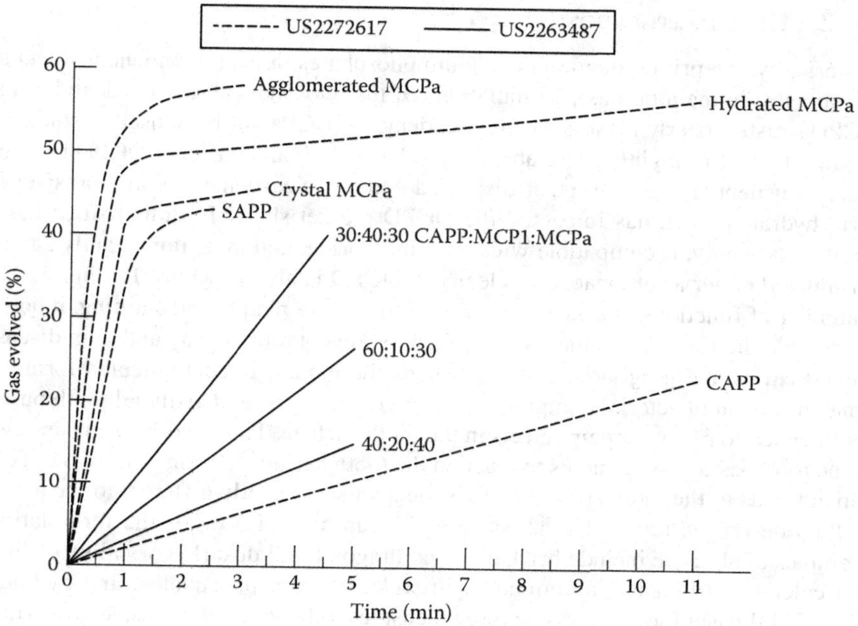

FIGURE 5.3 Typical DROR profiles for different leavening acids.

leavening reaction; it is useful when a fast reaction is required, but a long shelf life is also needed. It is often used in leavening blends for dry mixes or by itself in baking powders.

Dry mixes are often formulated by blending corporations and sold either as, for example, cake mix in the supermarket or on an industrial scale to a cake maker. In either case, either milk or water is added and the mix cooked.

MCP1 is a very fast acting agent, and its primary function is to create a large number of gas cells during mixing, which later serve as nuclei for expansion during heating. Up to 60% of the available carbon dioxide is released during mixing with the balance only when heating temperatures exceed 60°C. One of MCP1's many applications in different bakery products is the inhibition of ropiness in bread, which is caused primarily by the bacterium *Bacillus subtilis* as well as other *Bacillus* species. Active bacilli are destroyed by normal baking temperatures; however, internal loaf temperatures may not be sufficient to kill its spores. As the loaf cools, it reaches temperatures ideal for the spores to grow. The spores secrete enzymes that break down the bread and produce extracellular slimy polysaccharides that form fine threads, hence the term *ropiness*.

DCPD has low water solubility and decomposes at baking temperatures to MCP, HAP, and free phosphoric acid. Its application is best suited for cake mixes with long bake time and high pH such as chocolate cakes.

CAPP was developed for bakery applications as an alternate to SAPP.

5.4.2 Dental Applications

Historically, the prime function of calcium phosphates in dental hygiene is to act as the abrasive in the toothpaste formulation. Of the phosphate abrasives, dental grade DCPD satisfies nearly most of the market demand. DCPa has been used in smoker's toothpaste as it is slightly more abrasive, as has CAPP and CPP. As DCPD became more prevalent in the market, it displaced sodium bicarbonate as an abrasive; in turn, hydrated silica has impacted the DCPD market share. Hydrated silica has a similar abrasivity, is compatible with sodium fluoride, and most importantly can be formulated either as an opaque or clear gel; DCPD is always white. Toothpaste has a number of functions: it acts as an abrasive to remove plaque and staining, it delivers soluble fluoride ions to the tooth surface to prevent tooth decay and gum disease (dental caries and gingivitis), and it *freshens* the mouth. Recent patented formulations include antibacterial components to tackle gingivitis and artificial tooth builders intended to fill and repair defects in the tooth surface. These work by causing two of the toothpaste's components to react on the tooth surface to form amorphous TCP, which bonds to the tooth structure. This mechanism is subtly different to the action of fluoride on the tooth. The abrasive makes up around 50% of the formulation; examples of abrasive include brick dust and dragon tooth dust (historically), sodium and calcium carbonates, aluminum hydroxide, calcium phosphates, and hydrated silica. Toothpaste formulators use three measures to benchmark abrasive properties: Mohs hardness number, Perspex abrasion value [28], and RDA value. RDA is short for radioactive dentin abrasion test, a recognized method for assessment of dentifrice abrasivity set out as part of ISO 11609:2010 and adopted by the American Dental Association. In this procedure, extracted human teeth are irradiated with a neutron flux and subjected to a standard brushing regime. The radioactive phosphorus 32 removed from the dentin in the roots is used as an index of the abrasion of the dentifrice tested [28]. Clearly, a tooth abrasive that is too abrasive will wear away the teeth, which is undesirable; one that is too soft will not be sufficiently abrasive, and similarly one that dissolves easily in water will be of little utility.

The delivery of fluoride ions is achieved by including either sodium fluoride or monosodium fluorophosphate (SMFP) in the formulation; sufficient is added to yield 1000–1500 ppm soluble fluoride ion. Sodium fluoride is cheaper and easier to make than SMFP; however, it reacts with DCPD forming insoluble calcium fluoride and fluorapatite. Consequently, the free fluoride concentration is reduced. This starts to happen as soon as the formulation is mixed and continues, while the toothpaste is in storage. This property is known as *fluoride stability*. There are two accelerated aging tests for fluoride stability that are intended to be equivalent to 2 years' storage: the short test, during which the toothpaste is held at 60°C for 5 days, and the long test, during which the toothpaste is held at 49°C for 21 days. At the conclusion of these periods, the toothpaste is analyzed for soluble fluoride concentration. In toothpaste formulations where DCPD is the abrasive, fluoride stability is addressed in its manufacture (usually with the use of magnesium that competes with the calcium ions for the fluoride ions, so reducing the formation of calcium fluoride and fluorapatite) and by the use of SMFP.

Another important aspect of dental grade DCPD is its *hydration stability*. DCPD is only slightly soluble in water; however, in water, it does nevertheless break down to

DCPa and ultimately HAP. In practice, the breakdown to DCPa is of prime concern, and so another aspect of dental grade DCPD manufacture is the inclusion of a stabilizer that will slow this deterioration down. Different approaches are used, but a commonality is the incorporation of pyrophosphate on the crystal surface. The degree of stabilization is determined by what is known as the *glycerol test*. DCPD is mixed with glycerol and water to give a mobile paste. The paste is then held for 30 min at 100°C in a water bath. The paste passes the test if on stirring it is found to have an unchanged consistency with no sign of hardness. The assessment of unchanged consistency is somewhat subjective; there is no mechanical or electronic measurement.

5.4.3 Nutritional Applications: Calcium-Fortified Beverages, Calcium Food Supplements

Calcium, as the most abundant mineral in the body, is an essential component of the diet, particularly for children and young adults, pregnant women, and the elderly when bone breakdown exceeds formation and may lead to osteoporosis. The recommended dietary allowances for women of 50 years and above and men of 70 years and above, 1200 mg/day, are only slightly lower than the 9–18 years age group [29]. Milk, yogurt, and cheese are rich natural sources of calcium; however, for a number of reasons, these foods are not always taken in sufficient quantity to meet the dietary needs. One common approach to this need is to offer calcium tablets; another is to fortify foods and drinks with calcium. Calcium carbonate is inexpensive and a common agent with high calcium content; it does though have some limitations. Different calcium phosphates offer an alternative: TCP is used in orange juice, and a blend of MCP1 and MCPa is used to fortify clear fruit juices [30]. Regardless of which calcium fortifier is used, it is useless if it unfavorably affects the beverage or food; consequently, pH alteration, texture or *bite*, and taste are all tested when developing the additive.

5.4.4 Pharmaceutical Applications: Excipients, Biocement, Synthetic Bone Ash

Drug delivery is an interesting and diverse subject. Over 80% of all drug delivery is by mouth via liquid suspensions, syrups, capsules, and tablets. Specific grades of DCPD, DCPa, and TCP are used as tablet excipients. An excipient is usually pharmacologically inactive; it acts as both the tablet builder and the structure that carries the active ingredient. One obvious exception to this is the use of coated TCP [31] in tablets given for the prevention and treatment of osteoporosis where the TCP is both active ingredient and excipient. Several considerations are required when designing a tablet and include the following:

a. The drug dose, the rate of delivery, and the location of delivery (e.g., mouth, stomach, or small intestine).
b. The tablet size, shape, taste, color, and strength—these properties determine the ease of use of the tablet (and thus, in part, the likelihood of it being taken as prescribed) and its ability to withstand packaging, distribution, and handling.

c. The ease and reliability of tablet manufacture. Compressed tablet machines may produce a million tablets an hour so the blend of active ingredient and excipients that is fed to the machine must be of uniform constituency, with good flow characteristics and not requiring excessive pressure to achieve a satisfactory tablet. Once formed, the tablet must be ejected easily from the machine without sticking or breaking.

The principle of compressed tablet manufacture has not been altered since Thomas Brockedon was granted a patent for his machine, comprising a single die, in 1843. Modern machines are automated, computer-controlled, rotary devices with many dies each with an upper and lower punch although manual, single die machines are still sold to the artisan market. Each die/punch combination is arranged on a rotating, circular table. As the table rotates, each die/punch enters different areas where cams move the punches up and down depending on the stage of tablet production. Initially, the lower punch is lowered in the die creating a space for the powder to fill, in the fill zone. In the fill zone, the powder formulation passes from a hopper above the machine into a holder on the surface of the table. Within the holder, moving arms sweep powder into the die hole. As the die moves out of this zone, excess powder is swept from the die surface by the sweeping arms of the powder holder, and the lower punch drops slightly. In the next zone, the upper punch comes down and compresses the powder into a tablet. Finally, the lower punch rises up and ejects the formed tablet, and the cycle begins again. Figure 5.4 depicts both the tabletting sequence of the dies and punches and a plan view of a tabletting machine.

To address the design considerations, the tablet formulation as fed to the tabletting machine comprises several constituents: lubricants, diluents, binders, and disintegrants. Lubricants prevent the tablets adhering to the punches, are typically less than 1% of the formulation, and are often stearates. Diluents bulk up the tablet to a practical size both for manufacture and use particularly when the active ingredient dose is small. Examples of diluents include calcium carbonate, microcrystalline cellulose [32], and di- and tricalcium phosphates. Binders are necessary when the other constituents lack sufficient cohesiveness; examples include starch and sugars such as glucose and dextrose. Disintegrants are added to govern the rate of disintegration of the tablet and so ensure that the active ingredient is delivered at the desired rate and in the desired place in the gastrointestinal tract. Examples of disintegrants include starches, sodium bicarbonate (effervescent disintegration), modified starches (sodium carboxymethylcellulose), and cross-linked polyvinylpyrrolidone. Post tabletting processes such as coating also govern the drug delivery rate. Often, a constituent will have more than one function, microcrystalline cellulose might act as both diluent and disintegrants, and granular forms of DCPa act as diluents requiring no binder.

Prior to tabletting, the tablet formulation undergoes one of three processes: dry granulation, wet granulation, or direct compression. Dry granulation processes include mixing the constituents and processing the mix in other ways such as milling and screening. Wet granulation processes are more extensive and include wet mixing, extrusion, spheronization, drying, and milling, all with the goal of producing a formulation with good flow and tabletting characteristics. Direct compression requires that after simple blending, the formulation is ready to go direct to the

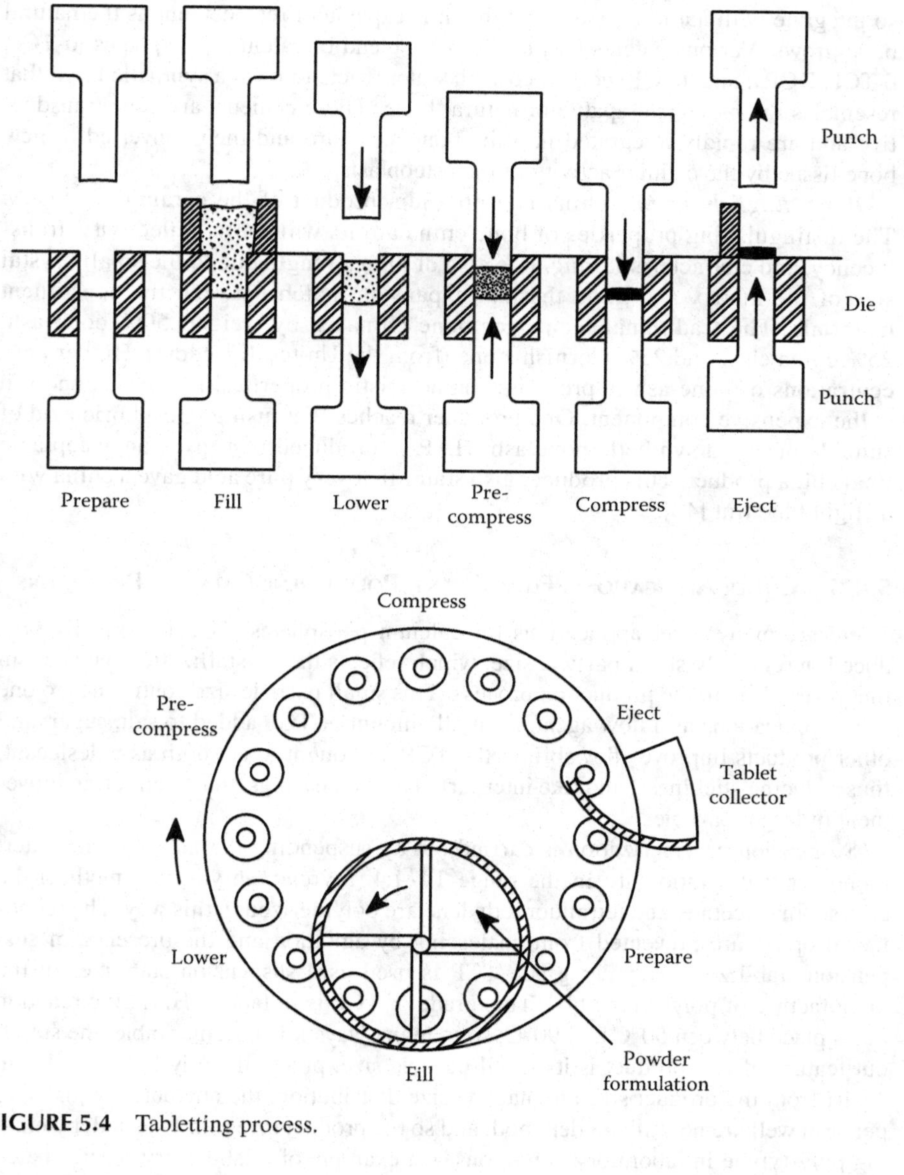

FIGURE 5.4 Tabletting process.

tabletting machine. Direct compression is obviously the simplest process, and so excipient producers aim to design their products for this use. Both DCPa and TCP powder grades are suitable excipients but require binders. Small modifications to the processes that produce DCPD, DCPa, and TCP have allowed suppliers to offer these chemicals in a form suitable for direct compression [31,33–36].

Biocement is used in orthopedic surgery to fix metal prostheses such as hip replacements and bone screws when repairing complex fractures. Biocement is also used in the dental field [37,38]. Biocement must be physiologically acceptable and so integrate with natural bone tissue that it is capable of replacement as the natural bone grows. Various patents [39] teach that a blend of calcium phosphates (α-TCP, β-TCP, DCPa, and TTCP) combined with water reacts to form a synthetic HAP that resembles the biological apatite of natural bone. These cements are osteotransductive and are rapidly integrated into the bone structure and then converted to new bone tissue by the cellular activity of the osteoblasts.

Bone, *English*, or *fine* china is a premium product of the ceramics industry. The distinguishing properties of bone china are its whiteness, reflectivity, translucency, and characteristic *ring*. It is also of high strength due to the small crystal size of the matrix and good thermal expansion match between the constituent materials. The traditional recipe for bone china is, by weight, 50% bone ash, 25% china clay, and 25% Cornish stone (from the United Kingdom). Higher concentrations of bone ash improve its characteristic properties; however, bone ash is the expensive component. One producer teaches that, using phosphoric acid of suitable purity, a synthetic bone ash, HAP, is produced that gives an acceptable final china product. This producer also states that very pure acid gave a china with a slight blue tint [40].

5.4.5 Other Applications: Flow Agent, Polystyrene Catalyst, Phosphors

There are many other applications for calcium phosphates. TCP is typically produced in relatively small particle size, which reflects the crystallization mechanism that takes place in the production process. This small particle size contributes to one of its applications as a flow agent. A small amount of TCP added to salt, sugar, and other products improves flowability; the TCP is thought to act both as a desiccant, thus reducing stickiness, and like interparticlar ball bearings, thus facilitating movement of larger particles.

Suspension polymerization is carried out by suspending the monomer in water; monomer water ratios are in the range 1:1–1:4. Styrene, acrylic and methacrylic esters, vinyl acetate, and tetrafluoroethylene are polymerized in this way. The monomer droplets are prevented from coalescing by agitation and the presence of suspension stabilizers. Polymer grade TCP is used as a suspension stabilizer in the manufacture of polystyrene [41]. This grade of TCP is in fact an HAP; the reaction takes place between 50°C and 90°C, so in many ways it is unremarkable; the standout feature of this product is its small particle size, predominantly less than 1 μm. Apart from the product size and narrow size distribution, the attributes required to perform well are not fully understood, and so the product is regularly tested, by making polystyrene in laboratory conditions. An example of a laboratory reactor batch is 358 g water, 495 g styrene, 1.54 g TCP, 0.9 g benzoyl peroxide, and 2.5 g 0.5% Nacconol solution, and the test is to demonstrate that the polystyrene resulting from this reaction has particle sizes in the desired range.

Phosphors are solid materials that emit light when exposed to electromagnetic radiation. Phosphors are used in fluorescent lamps and electroluminescent displays. There are hundreds of different phosphors that produce different colors. Calcium phosphates provide a structure for dopants such as rare earth metals in the manufacture of phosphors or the starting molecule for the manufacture of calcium halophosphates, which are phosphors in their own right. In general, halophosphates are represented by the formula $3M_3(PO_4)_2 \cdot M'L_2$ where M and M′ are alkaline metals and L is a halogen. A common calcium halophosphate has the formula $3Ca_3(PO_4)_2 \cdot Ca(F,Cl)_2$ and is doped with antinomy and manganese. The manufacturing process involves dry mixing a calcium phosphate and other constituents then firing the mixture under nitrogen at 1200°C–1500°C [42]. The form of the calcium phosphate source influences the final product quality; one process takes DCPa, dissolves it in nitric acid, and then spray dries the liquor to produce a spherical calcium phosphate powder suitable for the manufacture of pure gamma CPP, which is then further processed [19].

5.4.6 Animal Feed Calcium Phosphates

Just like humans, livestock have daily dietary requirements of many elements including calcium and phosphorus. Deficiency of these elements leads to disease, weakness, and poor growth. Calcium and phosphorus come from the food the livestock consume. Phosphorus is locked up in vegetable matter as phytic acid, also known as myoinositol hexakisphosphate. Phytic acid chelates (phytates) incorporate essential nutritional metals such as magnesium, iron, and zinc as well as trace elements such as copper and molybdenum. The metabolization of phytic acid and phytates requires the enzyme phytase. Mammals do not produce phytase although phytase-producing bacteria build up in the rumen of ruminant or polygastric animals such as cows and some naturally occurs with the feed itself. As a result, absorption of these elements especially in nonruminants is poor. Consequently, these elements are excreted with the animal feces, which without further treatment lead to environmental challenge such as eutrophication. Farmers address these challenges by adding three types of supplement to the diet: calcium phosphates, both MCP and DCP are used; mineral supplements; and phytase. Animal feed calcium phosphates have high levels of available phosphorus compared to vegetable matter. One measure of the effectiveness of phytase supplements is the level of phosphorus in the animal feces. The difference in phosphorus concentrations in different livestock feces with and without phytase feed supplement is striking with a 70% reduction of P loss for some animals. Phytase is manufactured from modified yeasts or fungi [43].

The animal feed calcium phosphate market is highly competitive. Not only is there competition between manufacturers but also the product competes in part with feed crops and phytase; as a result, production costs are kept to a minimum. Production costs for any product are usually minimized; however, in this case, it means that the final product could not carry the cost of starting from purified or thermal phosphoric acid. Three processes are common: acidification of animal bones with hydrochloric acid (HCl), neutralization of wet process phosphoric acid with lime, and acidification of phosphate rock with HCl and subsequent neutralization with lime. The first process is very old and was the subject of investigation in Europe as to its safety with respect to bovine spongiform encephalopathy (BSE) in Europe; this process accounts for 15% of the market in Europe. The latter processes, utilizing either WPA or rock, require some purification steps to meet animal feed standards that avoid harm. Fluoride and heavy metal levels are the main concern. On the other hand, there are some metals present in the product that are helpful to the animal and that are not present in calcium phosphates made from purified acid.

5.5 PRODUCTION PROCESSES

5.5.1 Calcium Sources and Processing

The calcium component of calcium phosphates is introduced in several different forms: as quicklime, CaO, usually referred to as lime; as hydrated or slaked lime, $Ca(OH)_2$, the term hydrated usually implying the powder and slaked, a slurry in water; as raw or mined calcium carbonate (chalk); as precipitated calcium carbonate; or as phosphate rock, in other words a combination of but primarily HAP. The last form, phosphate rock, is only of relevance here when considering animal feed DCP; while it is possible to make very pure calcium phosphates utilizing the calcium content of the rock, it is not practiced by the major world producers of dental, food, and pharmaceutical grade calcium phosphates.

Four considerations drive the choice of calcium source: price, availability and reliability of supply, purity, and physical condition of the material. For example, in this context, the purest form of calcium is precipitated calcium carbonate. This material is available as a free-flowing powder and therefore requires minimal processing. It is also the most expensive form of calcium, and the resultant carbon dioxide evolution requires careful management. Quarried calcium carbonate is not usually pure enough for direct use (it is after all usually sent to be converted to lime); however, there are some deposits around the world of suitable quality. At the other end of the scale, a cheap, poor-quality, impure lime is readily available in varying sizes from powder to big rocks.

It is feasible to process a poor-quality lime on the plant; whether the additional capital and operating cost of doing so is justified is a separate question. In the first instance, the lime is comminuted in a ball or hammer mill to achieve a maximum particle size of less than 25 mm. In this comminution process, a larger inert rock, sufficiently hard to withstand the mill, is separated as waste. At this size, it is relatively straightforward to transfer the lime by belt, screw conveyor, or pneumatically. Lime delivered in by rail or road containers is usually conveyed to storage silos pneumatically.

This material is then available for slaking. There are three types of slaking equipment: a slurry slaker that uses a water/solid ratio of up to seven, a paste slaker that uses a water/solid ratio of up to 2.5, and a ball mill slaker. The ball mill slaker is more expensive than the slurry or paste slaker and is intended for higher throughputs than usually required for the type of calcium phosphate plant under consideration.

The slurry and paste slacker have respective strengths and weaknesses. The slurry slaker produces slacked lime slurry ready for the plant; however, the slaking temperature, which affects the lime reactivity, is lower than the paste slaker. The paste slaker operates at a higher slaking temperature, producing a more reactive paste, which must then be diluted for use on the plant. The higher reactivity in the paste slacker leads to shorter residence time and so smaller equipment volumes. Both slakers remove inert grit using gravity separation and rotating rakes or screws.

Finally, the slaked lime slurry is often passed through a polishing filter prior to reaction with phosphoric acid. The ease of use of the polishing filter and the optimum slurry concentration are determined by operating practice and, depending on lime quality, lie in the range 16%–20% calcium hydroxide.

Given sufficient process time, a poor-quality lime can be made ready for processing to a calcium phosphate; at this stage, as much as 10%–30% of the starting material has been separated as inert waste. Trace impurities remain, which may incorporate within the phosphate crystal or lie on the crystal surface in the mother liquor. If the impurity is in the mother liquor and the phosphate is washed at some stage, for example, during filtration, then ultimately the lime is acceptable. If on the other hand the impurity lies within the crystal or there is no washing step, then the lime is only acceptable if the final product meets its impurity specification.

5.5.2 Monocalcium Phosphate Processes

The earliest patented process for the manufacture of monocalcium phosphate was granted to Ebenezer Horsford in 1856 [44]. MCP processes have developed since then although in one location in the United Kingdom even up to the 1970s, the process was very simple and manual; operators mixed acid into a pile of lime with shovels and allowed the resultant mass to mature over a number of days, after which it was milled and packed. Today, there are broadly two processes: a dry mixer process where preheated acid is intimately mixed with solid calcium hydroxide and the resultant powder post processed and a spray drier process wherein a suspension of DCP in dilute phosphoric acid is spray dried and either packed, after a short maturing period for sale, or post processed. The mixer process was first patented in 1935 [45]; subsequent patents have claimed invention in the process conditions and additions rather than the basic mixing plant. The spray drier process was first patented in 1923 [46], and over the next 10 years, others followed [47]. Spray drying is an efficient drying process and typically produces a spherical product of reasonably uniform particle size. The particle shape often imparts useful product qualities such as superior flowability when compared to otherwise identical material dried by other processes and subsequently milled. If the product from the spray drier is directly saleable without further processing, such as screening, milling, and classification, then this process is advantageous. In order to feed the spray drier with a material that will both pump and form droplets in the drier, the feed liquor has the equivalent of a water phase in the 40%–60% range. If the product out of the drier still requires further processing, such as milling, then the advantages of spray drying disappear, in the case of monocalcium phosphates, because the solid content of the dry mix process is so much higher and therefore requires much less energy to achieve the dryness specification. In fact, with careful process management, most of the drying energy required in the dry mix process is provided by the heat of dilution of the phosphoric acid and the heat of reaction of the reactants.

The most important parameters of the control of all calcium phosphate processes, whether mono, di, or tri, are the Ca/P ratio and reaction temperature. All reactions are carried out at atmospheric pressure, so this parameter is of secondary importance. Given an appropriate Ca/P ratio, a suitable process plant is capable of

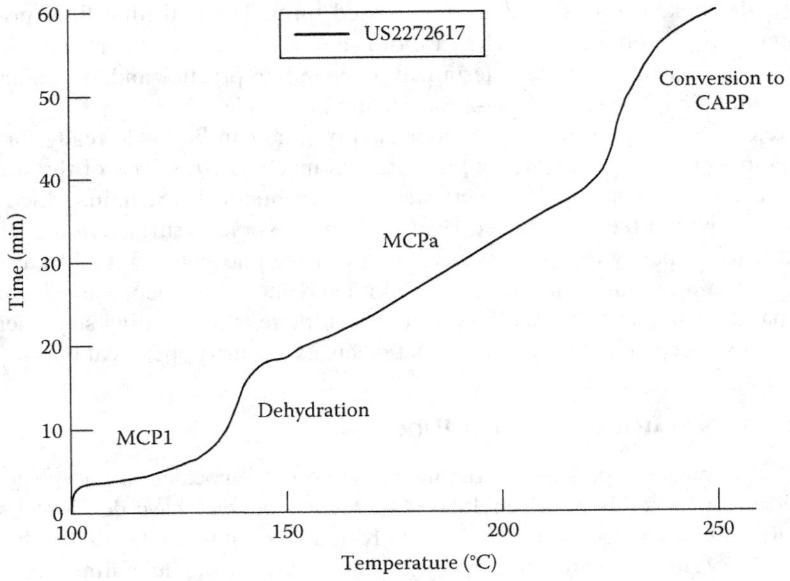

FIGURE 5.5 Development of MCP1–CAPP.

producing MCP1, MCPa, and CAPP, depending on the temperature at which the reaction mass is held. Figure 5.5, from US Patent 2272617, shows the development of monocalcium phosphate from the hydrated to the anhydrous form and subsequently the pyrophosphate with temperature and time.

On these plants, it is almost impossible to eliminate all moisture completely. The product emerging from any drier will contain up to 0.5% water, and unless it is immediately packed and sealed, the product is exposed to air and the prevailing humidity levels. Consequently, according to Equation 5.1, even the driest MCP1 will begin to convert to DCP and free phosphoric acid. In turn, the acid, being itself hygroscopic, will attract more water, and the product degradation continues. Eventually, equilibrium is set up; however, in practice, this means that the MCP product quickly becomes sticky, lumpy, and difficult to handle if it is exposed to moist air for anything other than short periods. As a result of this unhelpful property, several approaches are adopted to ameliorate the problem. Firstly, the operating Ca/P molar ratio is always greater than stoichiometric, that is, a lime excess to ensure all the acid is reacted; a typical target is a Ca/P of 1:1.8, equating to a 10% Ca excess. After this, the most common and perhaps simplest approach to this problem, particularly when making MCP1, is to make MCPa and then hydrate it with a water spray [48]; other older patented processes taught the addition of conditioning agents including calcium and other alkaline metal hyperphosphates [49] or magnesium carbonate [50].

Coated MCPa was patented in 1939 [51] and was developed as an alternative to SAPPs. SAPP is offered in different grades with different DRORs, which are much slower and generally more useful than the fast rates of MCP; however, as a pyrophosphate, it sometimes imparts an undesirable, astringent *pyro* taste to bakery products. MCPa does not give a *pyro* taste. Coated MCPa delays and slows the leavening

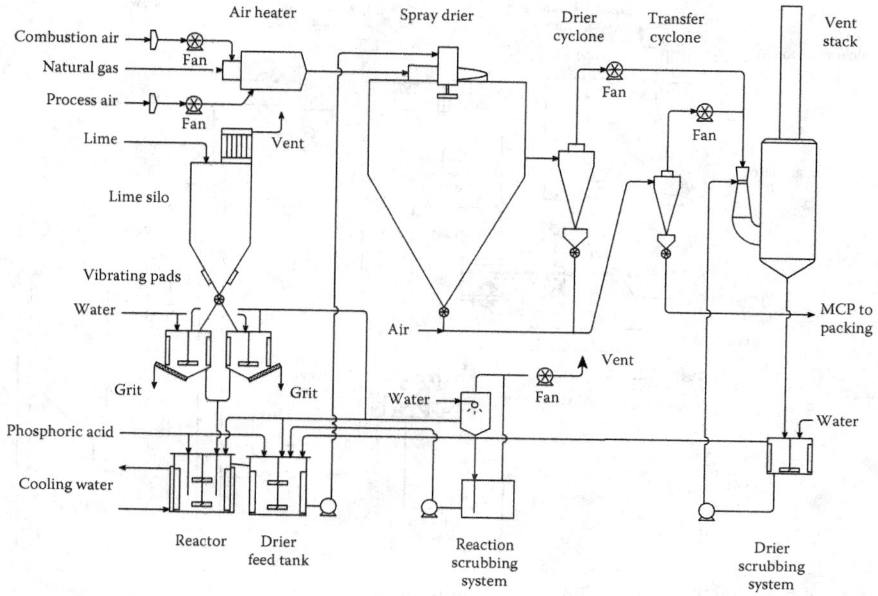

FIGURE 5.6 MCP spray drier process flow sheet.

reaction and has increased moisture toleration. More recently, food legislators have required reductions in sodium levels in processed food; calcium is an obvious, safe alternative. The SAPP family is not replaceable with one calcium phosphate; however, a calcium phosphate blend does permit the configuration of products that match NV and DROR; the developments of the art taught in the older patents have led to the latest patented products [52].

Two MCP plants are described later with reference to the flow sheets of Figures 5.6 and 5.7. The flow sheets are illustrative and intended to show some of the different unit operations that both are feasible and are used on operating plants around the world. Figure 5.6 shows a spray drier plant that will only make either MCP1 or MCPa. Figure 5.7 shows a dry mix plant capable of making the full MCP/CAPP range.

5.5.2.1 MCP Spray Drier Process

In Figure 5.6, the lime is delivered to a storage silo. The size of the silo is set with reference to the production rate and supply reliability; a capacity in the range 50–200 m^3 is normal for the scale of these plants. The silo is usually made from carbon steel although concrete is sometimes specified for larger sizes and occasionally stainless steel for smaller sizes or where it might be within a process building rather than outside in a tank farm. The lime is charged into the silo either pneumatically or mechanically via a bucket elevator or other suitable devices. Inevitably, the charging process creates dust, and so the silo is vented via an air filter; a reverse jet filter is shown in the flow sheet. The design of the vent arrangements depends on the charging method; a pneumatic conveying system obviously delivers a lot of air with the

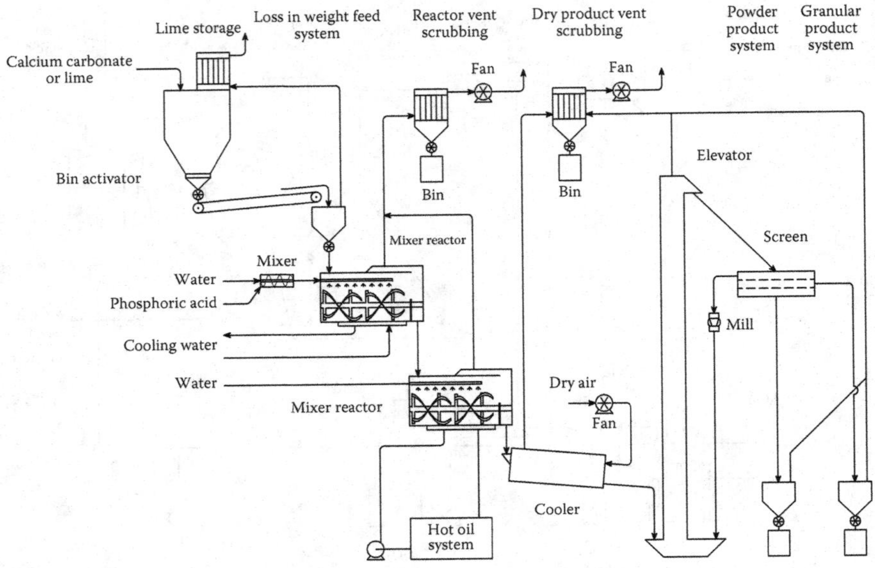

FIGURE 5.7 MCP mixer reactor process flow sheet.

lime, which must be managed. The lime does not always move freely, and it can bridge in the silo; therefore, the lower cone is often specified at 60° or greater and, as shown, is fitted with flow enhancing vibration pads. The lime discharges, batchwise, into two slaking tanks via a rotary valve. The slaking tanks are baffled and fitted with agitators and made from carbon steel. Water is added to achieve the desired concentration; 16%–20% calcium hydroxide is a typical, manageable concentration, verified by regular sampling and a combination of weigh measurement of the tanks and flow measurement of the water. The tank sizes will depend on the required residence time for the lime quality that is envisaged. A screw conveyor is shown at the tank base to remove inert insoluble waste.

Slaked lime is transferred to the reactor, an agitated, baffled reactor fitted with a cooling water jacket; phosphoric acid is co-fed. As shown, the reactor operates continuously with the product overflowing into the spray drier feed tank; batch operation works as well. Reaction temperature is controlled below 95°C by a combination of cooling and control of the feeds. The slurry is pumped via a centrifugal pump to the spray drier atomizer; as depicted, this is a centrifugal atomizer, and the spray is generated as the atomizer spins; the speed of rotation is variable and controlled with an electrical inverter drive, depending on the size and type of atomizer in the 6,000–15,000 rpm range. (The alternative nozzle spray drier requires a high-pressure delivery pump.)

Figure 5.8 shows the relevant part of the $CaO–H_2O–P_2O_5$ phase diagram for the manufacture of MCP. The initial wet part of this process takes place just below 100°C; therefore, the solution curve, below which everything is in solution, is drawn from the solubility data at 100°C [2], which makes a small difference to the phase diagram calculations that follow. The target product line is drawn from the origin

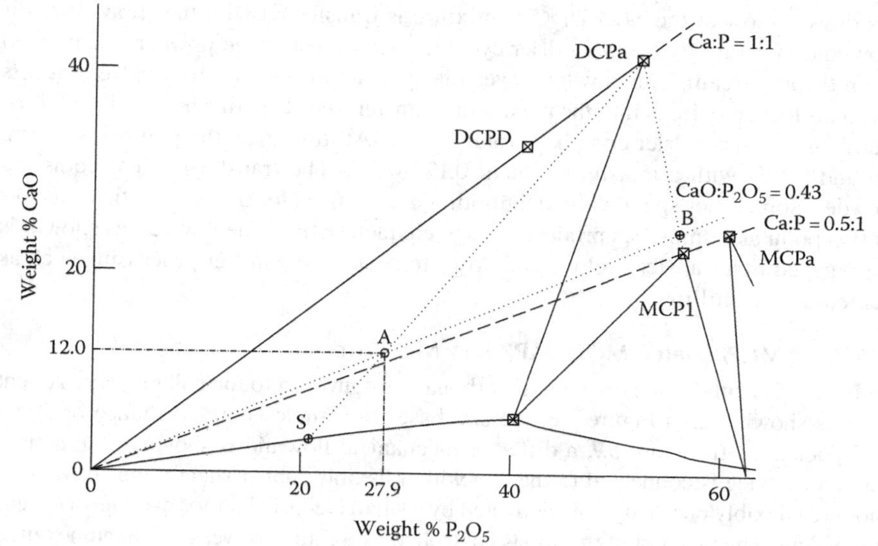

FIGURE 5.8 MCP production phase diagram. (Adapted with permission from Elmore, K.L. and Farr, T.D., Equilibrium in the system calcium oxide–phosphorus pentoxide–water, *Ind. Eng. Chem.*, 32(4), 580–586. Copyright 1940 American Chemical Society.)

with a gradient of 0.43, representing the weight percentage CaO/P_2O_5 ratio of 0.43. This line intersects the MCP1 to DCPa line at point B, which defines the ratio of DCPa and MCP1 in the final product, at 100% dryness. The proportion of DCPa in this example is 9.48% by weight. Assuming a calcium hydroxide slurry feed with a 12% CaO concentration, we can draw a horizontal line to the point A where it intersects the target product line. The ratio of the distance from the origin, O, to this point and from this point to the final product point B, OA/OB, gives the percentage of water that must be evaporated, 49.9%, and therefore the spray drier duty but not the slurry content of the feed to the spray drier. The actual solid content of the spray drier feed can be calculated from the phase diagram by drawing a line from DCPa to A and then extending this line to the solution boundary at S; the solid weight percentage is calculated from the ratio SA/SDCPa, 22.8%. Point A, representing the feed slurry, is clearly within the DCPa/DCPD area of the phase diagram. The reaction temperature tells us that the product of the reaction is DCPa; therefore, this material is a mixture of DCPa and free phosphoric acid. As the slurry droplets are flung from the atomizer, they undergo a reaction to MCP. By limiting the drying temperatures to no more than 120°C, the final dry product is MCP1. The initial DCPa reaction and the subsequent spray drying reaction are shown approximately in the following equations:

$$6Ca(OH)_2 + 11H_3PO_4 + 70H_2O \rightarrow 6CaHPO_4 + 5H_3PO_4 + 82H_2O \quad (5.22)$$

$$6CaHPO_4 + 5H_3PO_4 + 82H_2O \rightarrow 5Ca(H_2PO_4)_2 \cdot H_2O + CaHPO_4 + 77H_2O\uparrow \quad (5.23)$$

As drawn, most of the MCP1/DCPa mixture is transferred with the spray drier air flow and moisture to the spray drier cyclone. Here, most of the powder is separated from the air stream. The powder leaves the cyclone through a rotary valve that acts as an air lock and drops into the pneumatic transfer line. This line also picks up powder from the spray drier cone base rotary valve. At this stage, the powder is warm, around 90°C, with a moisture content 0.1%–0.2%. The transfer process cools the powder. Sometimes specifically dehumidified air is used for the pneumatic transport at this point as it improves product storage characteristics. The powder, as shown, is disengaged in a transfer cyclone and drops to packing or further processing such as screening and milling.

5.5.2.2 MCP/Coated MCP/CAPP Dry Mix Process

In Figure 5.7, either lime or calcium carbonate is conveyed to the calcium silo. A vent filter is shown. As in Figure 5.6, the silo base cone angle is 60° to reduce bridging and poor flow. In Figure 5.7, a different mechanical flow aid is shown, a bin activator. This device is connected to the silo with a flexible rubber sheath; the activator is mounted flexibly on springs and activated by a small electrical motor driving an eccentric weight. The flow out of the silo is controlled by a rotary valve. The calcium source is conveyed by belt to the reactor mixer feed bin. Either the belt or the bin, or both, are mounted on weigh cells and thus are able to meter the weight of material flowing to the reactor. Together with laboratory analysis of the calcium content of the feed material, this loss in weight system allows the control of the calcium feed rate. With accurate metering of the phosphoric acid feed, the control of the CaO/P_2O_5 ratio is feasible.

When making MCP1, this way, the first mixer reactor is set up to make MCPa. This follows the discovery that making MCP1 directly produces a product with greater tendency to stickiness and caking in storage, whereas hydrating MCPa produces a superior product [48]. The feed control is set up to achieve a Ca/P molar ratio of between 1:2 and 1:8, as discussed earlier, and the reactor temperature set at 140°C. This temperature is achieved through a combination of using the heat of dilution of the phosphoric acid and heat of reaction. At this temperature, much of the water of dilution and reaction is driven off as steam. If calcium carbonate is used as a dry feed, then carbon dioxide is also driven off in the reaction. A careful design of the venting of the reactor is essential to avoid losing a lot of material to the vent filtration system. As depicted in Figure 5.7, the phosphoric acid, normally 85% H_3PO_4, is mixed in a static mixer with water and then sprayed onto the reaction mass in the mixer reactor. The mixer reactor is drawn as a continuous ribbon blade mixer; however, several different types are used including kneaders and plow share mixers. Both batch and continuous operations are possible.

One of several quality and specification measures for many phosphates is loss on ignition (LOI). The LOI test is carried out by placing a dry sample in an oven at 800°C for 30 min. The sample is weighed before and after the test, and the difference is the LOI. The test is quick and simple, and although not always conclusive, it can be helpful in production management as well as quality assurance (QA). Preparing a sample for the LOI test includes ensuring that it is dry; this can be done at the same time as the dryness test, which usually requires that the sample is held at 60°C–75°C for 30–60 min for both production and QA.

Referring to Figure 5.5, it is clear that MCP1 is dehydrated to MCPa after 20 min at between 145°C and 150°C; therefore, the first mixer reactor is sized for a residence time of this order. The now predominantly MCPa is transferred into the second mixer reactor, and water is sprayed on the MCPa, 5%–10%. The water spray leads to cooling by evaporation and heating by hydration; the hydration reaction is rapid at these temperatures. The net effect is that hydrated MCP, with an LOI of typically 18%–20%, passes to the cooler at approximately 75°C where it is cooled with dry air to ambient temperatures. The dry air ensures that free moisture is reduced to less than 0.5%.

The dry powder flows by gravity into an elevator and onto a rotating screen. The purpose of the screen is a three-way separation of oversize material, which overflows to a mill, granular material, and powder. The screen as shown has two decks, the top deck with a mesh size of 150 μm and the lower deck with a mesh size of 75 μm, typical of the sizes that differentiate powder and granular MCP products. Different and more complex screen and mill combinations are possible; however, here the mill envisaged is a fine impact mill. The powder and granular material pass to the final product storages and are then sent for packing.

The coated MCPa is MCPa with a thin coating of alkali metal metaphosphatic salts. This coating slows the rate of reaction although not sufficiently for the coated MCPa to be a direct competitor to various SAPP leavening agents. The coated MCPa is made by heat treating MCPa that has been made with an enriched food grade phosphoric acid; less than 1% aluminum, potassium, iron, and sodium are added with sulfuric acid. The MCPa is made in the first mixer reactor at a temperature of 160°C–170°C and is then heat treated in the second mixer reactor at up to 230°C, so as to avoid conversion to CAPP. The coated MCPa is cooled and screened and milled as required. As with the production of MCP1 earlier, the temperature in the first mixer reactor is governed by the amount of water added to phosphoric acid to generate heat of dilution and the rate of addition to the reactor of the reactants. The temperature for heat treatment is achieved by the use of hot oil in a jacket around the mixer reactor. Industrial hot oil systems are used to achieve temperatures over 300°C and are either electrically heated or fuel fired (oil, gas) and have the advantage of being low pressure, compared to steam, and the disadvantage of flammability.

CAPP is made by heat treating MCPa in the second mixer reactor. A number of patents teach [53,54] that a blend of CAPP with MCPa and or MCP1 gives a range of calcium-based leavening agents that are a useful alternative to sodium-based equivalents. Figure 5.3 shows the gas evolution rates for different calcium phosphates [55] and three CAPP/MCP1/MCPa blends [53].

5.5.3 Dicalcium Phosphate Processes for Food, Pharmaceutical, Dental, and High-Purity Uses

Both DCPa and DCPD may be made on a wide range of equipment, so long as the Ca/P ratio is correct and, for DCPD, the temperature is maintained low enough during the reaction and drying stages to avoid making DCPa. Much industrial DCP process practice is wet and batch; however, both a dry mix process and at least one continuous wet process are also practiced. For the wet processes, many different types of equipment are used: there are several examples of different filter types, as well as centrifuges; equally, different types of driers are used after the filtration step; the same is true of the milling and separation stages. From the process chemistry standpoint, the manufacture of stabilized DCPD for the dental market is approached in almost as many ways as there are plants making this product. The common theme is that both magnesium and pyrophosphate components are added. Magnesium is added as solid magnesium oxide, dimagnesium phosphate trihydrate (DMP3), or trimagnesium phosphate octahydrate (TMP8). The oxide is only added early in the reaction, at a relatively low pH, typically less than 2.4 [56]. Arguably, this is an example of process intensification in that the magnesium oxide reacts to form magnesium phosphates at the same time as the calcium phosphate is formed. DMP and TMP may be added early and late or dry blended prior to packing. Although DMP and TMP are items of commerce, DCPD producers tend to make these products themselves on small units for exclusive use on the DCPD plant.

Both for profit and environmental reasons, manufacturers aim to minimize waste and energy consumption. The dry mixer plant inherently obviates P_2O_5 losses to the mother liquor in the main reaction and minimizes energy consumption for drying but gives up any meaningful control of product morphology and must be milled and screened. As is common with most phosphate plants, the milling and screening of dry powders produce dusts that must be filtered and recycled to reduce product loss. Given that the wet processes lose P_2O_5 to the mother liquor, indeed, some purposely operate at pH levels that necessitate unreacted P_2O_5 in the mother liquor, and the filtration step always loses the smallest crystals to this stream; it is essential to put in place process steps that will recover this P_2O_5. The P_2O_5 recovery is usually achieved by neutralizing the mother liquor with lime to the extent that HAP is formed; the HAP is then recycled to the reaction stage.

Figure 5.9 shows a stabilized DCPD process based on lime. The plant produces a single product for the dental market. Lime from the lime silo is transported into a paste slaker via a weigh belt. A weigh belt, or *loss in weight* belt, is fitted with strain gauges and can measure the mass flow of the solid material it is carrying. In the slaker, the lime is wetted by recycled water from the HAP settler and conveyed along the slaker by rotating paddles that also mix the slaked lime into a paste. The paste passes over an internal weir, which governs retention time. Just before falling from the slaker, a rake removes larger pieces of inert material such as small rocks. The slaked lime falls into a rotating screen designed to remove inert grit such as sand. The slaked lime paste is transferred to the slaked lime storage where more recycle water is added to achieve the desired concentration. The slaked lime is sometimes pumped through a polishing filter (to remove residual grit) to both the reactors and the mother liquor neutralizer. Three reactors are shown in Figure 5.9, R1, R2, and R3; all are jacketed, baffled, and fitted with agitators and constructed from 316L stainless steel; as drawn, the reactors can operate either continuously or batchwise; continuous operation is described and allows smaller equipment for the same output. The function of reactor R1 is to accept all the phosphoric acid for the reaction, dilute the acid to the required strength, and receive only as much calcium as will allow the reaction mass to remain completely liquid, in other words, just below the onset of

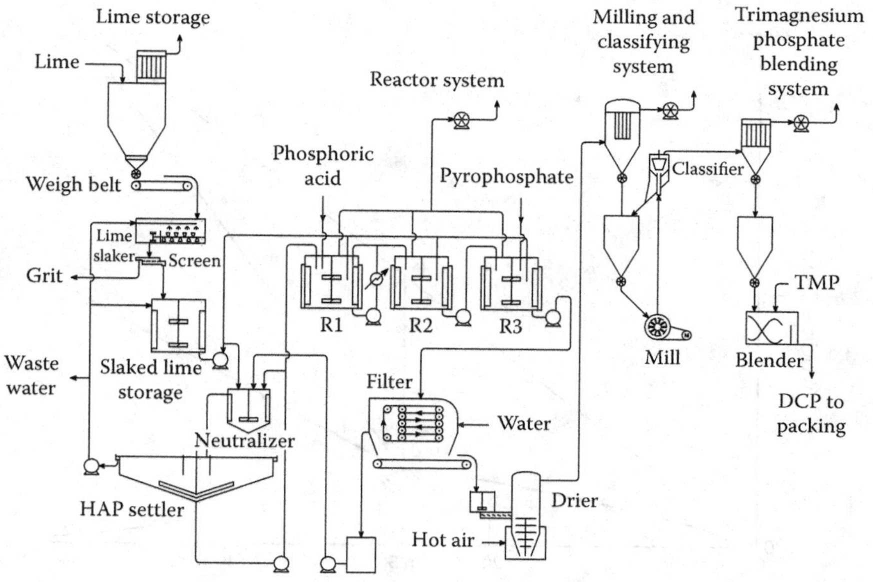

FIGURE 5.9 Stabilized DCPD process flow sheet.

precipitation. The operating point for R1 is in the shaded area of the phase diagram of Figure 5.10. (The solution curve is drawn from the solubility data at 40°C [2].) In this area, where the lime concentration is just below 6% CaO, no precipitation has occurred, and the heat of dilution as well as some heat of reaction is easily removed in a water-cooled condenser. In operating practice, there is a risk of precipitation and consequent scaling on the exchanger surfaces; the operating conditions in terms of temperature and pressure are not particularly onerous; therefore, a simple plate heat exchanger is a good choice here. There are three calcium contributions: fine DCPD crystals that bypass the filtration stage and are converted to HAP in the neutralizer, lime that is added to the neutralizer to make this conversion, and lime added directly to R1. The reaction liquor is pumped forward to reactor R2 with some material recycling to aid temperature control in R1 and the rest going forward. Most of the slaked lime is added in R2, and DCPD starts to precipitate. Reaction temperature is controlled at 36°C with cooling water in the reactor jackets; for practical purposes much below this temperature, the reaction starts to proceed slowly, much above, and DCPa begins to form; a practical operating temperature range of 30°C–40°C is satisfactory. During the hottest months in warmer climates, cooling water temperatures are sometimes too high to be of use; one solution at these times is to inject chilled water into the cooling water circuit. The reaction proceeds up the dotted line on the phase diagram (Figure 5.10) from S toward D. At D, the mother liquor is theoretically pH 7; however, as this point is approached, at pH > 6.5, the likelihood of making HAP increases. Therefore, the normal practice is to operate a little way below D, in an acidic zone of the phase diagram and so avoid making HAP. Doing this has the added advantage of leaving some free acid to attack the smallest crystals

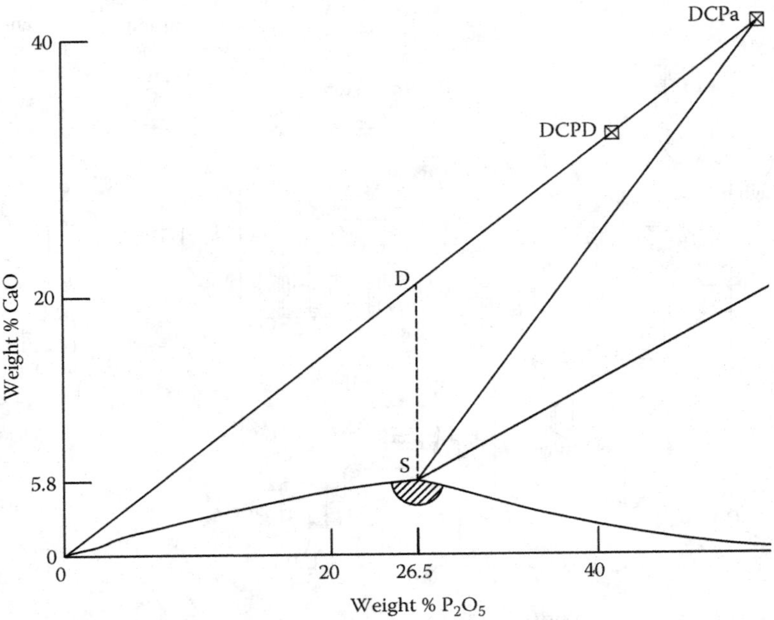

FIGURE 5.10 DCPD production phase diagram. (Adapted with permission from Elmore, K.L. and Farr, T.D., Equilibrium in the system calcium oxide–phosphorus pentoxide–water, *Ind. Eng. Chem.*, 32(4), 580–586. Copyright 1940 American Chemical Society.)

remembering that the next stage after reaction is filtration and generally larger crystals filter better than fine crystals.

Some producers have suggested operating as low as pH 4–5; others aim to complete batch reactions at pH 7. Both academic study and analysis of the solubility equations suggest that maximum crystal growth rates occur above pH 6. One design objective is to achieve a crystal size that will filter well; maximizing the crystal growth rate minimizes reactor residence time and therefore size and cost. The conditions for maximum crystal growth can also lead to uncontrolled nucleation and lots of tiny crystals and also the inclusion of impurities in the crystal structure. Consequently, one operating approach is to run R2 at pH 4–5 at a moderate crystal growth rate and then increase the pH in R3 to maximize growth rate on established crystals. The three reactors would then run under the following regime: R1 pH < 2.2–2.4, R2 pH 4–5, and R3 pH 6–6.5.

So in R3, the DCPD crystal is allowed to mature; any fine adjustment of the Ca/P ratio is carried out, and pyrophosphate is added that has the effect of lowering the pH slightly. Usually, sodium tetrapyrophosphate is the material of choice at 1%–2% concentration with respect to DCPD. The pyrophosphate is added either as a metered solution or directly as powder.

Some care is necessary in the design of the reactors particularly to ensure good mixing at all stages and the introduction of feeds into well-mixed zones. The reasons for this should now be obvious: in R1, good mixing ensures a uniform and therefore consistently measurable temperature in the reactor, which is essential for the control of the cooling water flows. In R2, good mixing ensures that local concentrations either of Ca- or P-rich material are minimized, which is necessary again for temperature control as well as ensuring only DCPD is precipitated. In R3, the reaction mass must be kept well mixed as the last of the lime and the pyrophosphate is added, as well as avoiding undermixing and risking unhelpful agglomeration of crystals or overmixing and some crystal breakdown. Pitched turbine agitators at 0.3–0.5 D turning at 60–90 rpm have been satisfactory.

The reaction slurry is pumped to filtration. Vacuum drum filters, pressure plate filters, and centrifuges are or have all been used at this stage. A pressure filter with a continuous filter cloth that has all of the attributes of a plate filter is depicted. This type of filter is manufactured by Outotec Larox in Finland. The continuous filter cloth separates a number of filter pressure plates. At the beginning of the filtration cycle, the plates are pressed together holding the cloth and the slurry is pumped into the space between the plates and the cloth and some mother liquor passes through the cloth to drain. A high-pressure pump then raises the pressure in the plate void squeezing out the liquor through the cloth. Clean water is then pumped into plate void and again squeezed through the cloth. The plates are then hydraulically separated, and the continuous cloth belt is progressed through the filter over rollers; as the cloth travels over the rollers, the dry filter cake falls into a hopper and onto a conveying belt and is transferred to drying. This type of filter achieves pressures far greater than those achieved by vacuum (on drum or standard continuous belt filters) but less than a tube filter, and moving the cloth over a series of rollers removes most of the filter cake (which can be a problem with standard plate filters).

The clean water wash makes up the overall process water balance. The water could be introduced anywhere, but doing so here allows the DCPD crystals to be washed of impurities in the mother liquor. The total amount of water in the wet part of the process is governed by the concentration needed to achieve trouble-free slurry pumping (slaked lime and reaction slurry). The amount of water added needs to balance that evaporated in drying, which is affected by filtration performance; that needed to wash the crystals effectively, which is governed by the filter plate void volume; and that needed to keep the concentration of impurities to a level, which will not impact crystal growth, final product quality, or specification. For example, food grade PWA may contain sulfate in hundreds of parts per million; at thousands of parts per million, sulfate influences the crystal habit of DCPD. Similarly, food grade PWA is usually below 1 ppm lead, a heavy metal, and DCPD specifications require similar purity levels. Even though the acid entering the process, or the calcium source, is sufficiently pure on a once through basis, with a recycle and no *blow down*, it is obvious that impurities could build up to unacceptable levels.

As depicted, the dry filter cake, typically 20%–25% water from this type of filter, is transferred to a flash drier. The filter cake falls into a small agitated drier feed tank and is then screwed into the drier body. Hot air, heated by gas, steam, electrical elements, or a combination of these, is blown into the drier base and passes up the body. An agitator breaks up the filter cake as it enters the drier and dry particles are conveyed up to a cyclone filter. A fan exhausts clean air to atmosphere.

Drier temperatures are the minimum that avoids condensation in the transfer line and cyclone and does not induce conversion to DCPa; typically, inlet temperature is 120°C–140°C and outlet temperature 95°C–105°C.

The dry powder passes to the premill storage; this provides some buffer capacity and ensures a constant feed to the mill. With a Mohs hardness of 2.5, DCPD is not a hard material and is easily milled; often in fact, the milling is deagglomeration rather than particle size reduction. Older plants use large roller mills; other plants use different types of air mill. Air mills are only really necessary for the production of very fine material, in this context less than 10 μm, and are expensive to operate because of the large amount of air that is required. For example, an air mill processing 1200 kg/h powder may require 3500 m^3/h air at 7 bar pressure; the air compressor to deliver this air may require a 350 kW drive. Roller mills use much less air; nevertheless, the majority of the electrical load is used to drive the fans that convey the powder through the mill. An impact mill is depicted in Figure 5.9, which has a relatively low electrical load. Milled powder is conveyed by the air movement of the fan action of the mill itself to a turbo classifier. This unit combines the function of a cyclone and a powered impellor to give a precise particle size cutoff that can be varied with impellor speed; it is a compact alternative to a screen and is less prone to blockage [57]. For dental grade material, the cut is set at 45 μm; any material above this size is returned to the premill storage for further milling, and everything below this size and therefore within specification is disengaged in the preblending cyclone filter and drops into the preblending storage. DCPD passes into a ribbon blender, either continuously or on a batch basis, and is blended with dry TMP8; the product is then conveyed to the packing area. Usually, the final product is packed in 25 kg sacks or 1000 kg big bags.

Meanwhile, the mother liquor from the filter, essentially acidic water with fine DCPD particles, is pumped to the neutralizer where it is mixed with slaked lime and some recycle material from the base of the settler. In the neutralizer, the pH is raised to pH 9–10 to convert all the acid to HAP. The recycle is used to provide nucleation sites for the growth of HAP crystals, which will then settle more easily in the HAP settler. When operating well, the HAP separates and recycle is pumped forward to R1. Clear liquor is used to slake the lime, and a proportion is pumped away to site effluent treatment. The relative cost of lime and phosphoric acid means that it is more economic to overdose the lime and ensures that the P_2O_5 is recycled as HAP. Environmental regulations often have a low level of permitted P_2O_5 release to effluent streams outside site. A DCPD plant is often on a multiproduct site that produces an acidic effluent; as a result, an excess lime in the waste stream is not wasted in that it reduces the lime loading at the site effluent treatment plant.

There are many options to this flow sheet. As stated in the introduction, a dry mixer reaction section is feasible. The use of calcium carbonate is practiced, in which case care must be taken to vent evolved carbon dioxide safely. The reaction slurry could be pumped to a spray drier and onto final blending, thus avoiding filtration and drying. Lastly, a much more complex milling and screening arrangement is possible to produce granular and powder products.

DCPa is manufactured on the same equipment simply by adjusting the reaction conditions and ensuring that the reaction temperatures are in the range of 70°C–90°C.

An alternative approach is to dehydrate DCPD in a drier; the main issue is ensuring sufficient residence time (at least 30 min at up to 150°C) to achieve dehydration [34].

5.5.4 Tricalcium Phosphate Processes for Food, Pharmaceutical, Dental, and High-Purity Uses

The majority of plants around the world that produce what is generally called *tricalcium phosphate* are, in fact, manufacturing HAP materials (i.e., HAP and DCPa mixtures with a similar Ca/P ratio) and do so via a batch process. There are examples of continuous processes that bring advantages in terms of process control and equipment size. True TCPs (α- and β-forms being the most common) are manufactured by sintering a calcium phosphate of the correct Ca/P ratio that has been made via the aqueous precipitation route.

The key difference between making DCP and HAP, apart from the Ca/P ratio, is that in the latter case, the reaction always starts with a calcium hydroxide slurry to which phosphoric acid is added. The sense of this is most easily explained with reference to the phase diagram (Figure 5.11). If, for example, the reactor, as for DCP, was charged with a solution containing 17.2% P_2O_5 and 4.4% CaO, at A in Figure 5.11, and then more hydrated lime (at a slurry concentration of 22.7% CaO, equivalent to 30% $Ca(OH)_2$) was added to bring the reaction mass up to 22.7% CaO, at B, on the HAP line, it is likely that the resultant reaction mass would contain predominantly DCPD with some OCP, HAP, and free lime. Possibly, over a long time and with good agitation, eventually, the calcium ions would redistribute to form a high HAP content. If, on the other hand, the reaction commences with a 22.7% slurry in the reactor, at C in Figure 5.11, as acid is added, the reaction moves along the dotted

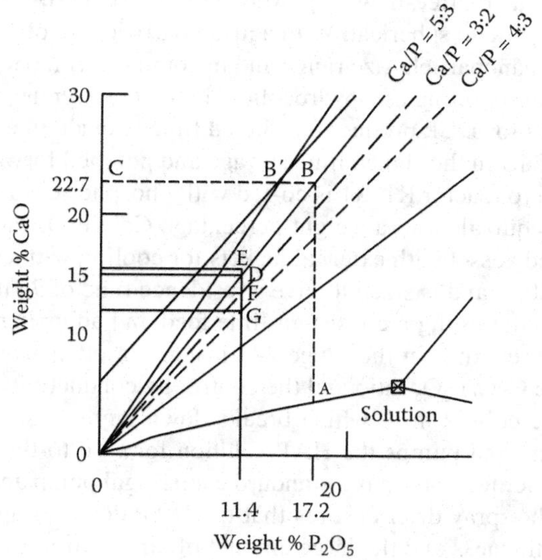

FIGURE 5.11 TCP production phase diagram. (Adapted with permission from Elmore, K.L. and Farr, T.D., Equilibrium in the system calcium oxide–phosphorus pentoxide–water, *Ind. Eng. Chem.*, 32(4), 580–586. Copyright 1940 American Chemical Society.)

line to B′ when HAP starts to precipitate and the reaction is concluded at B. Some plants operate with more dilute lime slurries, for example, at 15% CaO to as low as 8% CaO; this is for several reasons; most usually the lime system is easier to operate at this level and also that the calcium phosphate crystal growth rates are slower at lower ion concentrations and therefore produce better crystal form. Referring again to Figure 5.11 and operating with a lime slurry with a 15% CaO set point, it is clear that to meet the HAP line at D requires a 11.4% P_2O_5 acid solution. Given that the lime slurry requires the mixing of a solid with a liquid and that the solid may have slightly varying amount of CaO, it is possible that the lime concentration varies ±0.5% so although the control target is at D in Figure 5.11, the average operating point might be up to E. Good agitation ensures that the reaction mass is well mixed; however, some concentration of the acid feed is inevitable at the point of addition. If this concentration is short lived, then there will be no problem; however, if it is not or if there are dead zones in the reactor, it is possible that calcium phosphates other than the desired HAP will coprecipitate as impurities, represented at F and G. If this happens, it can also lead to unhelpful viscosity changes with consequent plant damage.

While important if operating a continuous plant, knowing when the reaction to HAP is complete is critical to the control of a batch plant. Usually pH is measured, and around pH 7.5 is taken as the point at which the reaction is complete; however, at this point, the pH is falling rapidly, and a small difference in pH can indicate a large difference in product purity. For example, in Figure 5.12 [4] at P1, HAP and some CaO are present, at P2 HAP, CaO, and β-TCP. US5180564 teaches that conductivity is a more useful parameter than pH to control the reaction.

Figure 5.13 shows a plant to make polymer grade HAP. The goal is to produce HAP in the form of a dry, free-flowing powder with a particle size range of 0.5–5 μm with each particle close to spherical. With a mean particle size of 1 μm, this product is at the top of the nanoparticle size range and in common with one important property of nanoparticles is acting as a hydrocolloid in the polystyrene process.

As in Figure 5.9 for DCP production, slaked lime is made in a paste slaker and then held at 15% CaO in the slaked lime storage and pumped forward continuously under flow control to reactor R1. R1 is co-fed with phosphoric acid also under flow control at a ratio equivalent to a weight percentage CaO/P_2O_5 ratio of 1.32. R1 is an agitated, baffled vessel with external jackets for cooling with cooling water. R1 is maintained at 80°C and is sized to give a residence time of 75 min. The process is controlled by maintaining a consistent lime feed and adjusting the acid feed to keep conductivity readings in the range 2–3 ms/cm; in other words, the set point is governed by the CaO/P_2O_5 ratio and the control by conductivity. The HAP solution flows into the colloid mill, which breaks down any oversize material to the required size range and pumps the HAP solution forward to the spray drier feed tank. The spray drier feed pump is a standard centrifugal pump and feeds the spray drier atomizer. The spray drier ensures that the HAP does not agglomerate; however, the product fineness and the large amount of air required for spray drying do make the exhaust air filtration challenging. Usually, a cyclone or set of cyclones follow a spray drier; however, for this level of product fineness, a baghouse with reverse jet filters is required. The product flows by gravity from the spray drier and

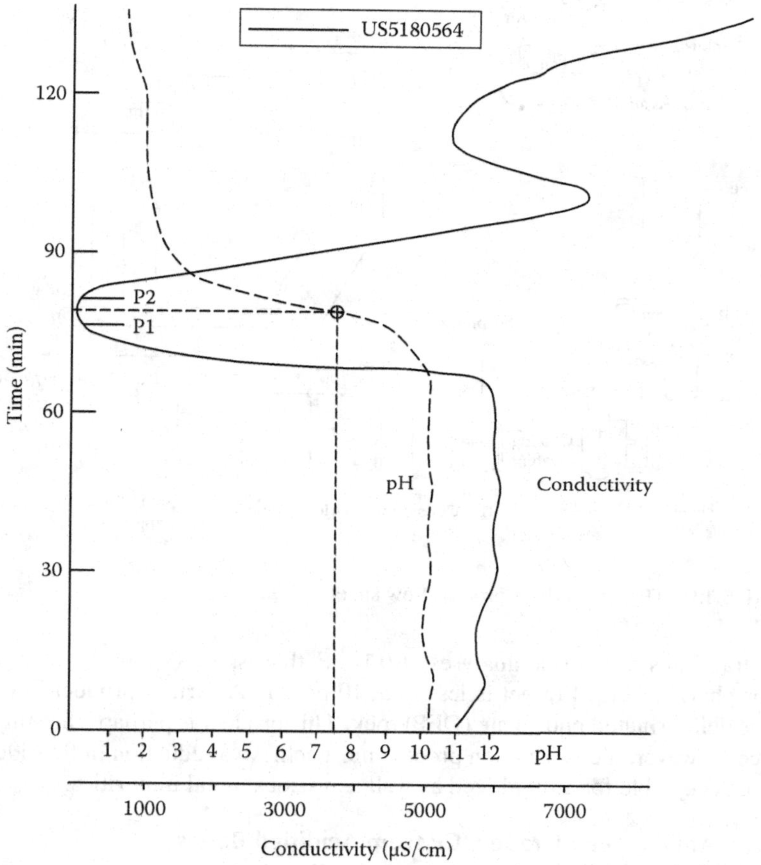

FIGURE 5.12 TCP production pH and conductivity diagram.

filters, via rotary valves into a tube conveyor, which transfers the product to packing with no further treatment.

5.5.5 Animal Feed Calcium Phosphates

The predominant animal feed calcium phosphate is DCP. Different producers supply different blends of DCP, which comprise both the anhydrous and hydrated forms, DCPa and DCPD. MCP is also sold on its own or in a blend. All these inorganic calcium phosphates have high phosphorus availability in the region of 80% with some products slightly higher than others. The chemical principles of Ca/P ratio and temperature are exactly the same for animal feed grade products as for dental or pharmaceutical grade calcium phosphates. The main differences between animal feed and the higher-purity calcium phosphates are that the specifications for impurities and product size are less stringent for the animal feed products. One example of this difference both of specification and mind-set in the producers of the different products is that the animal feed community seeks to achieve high P_2O_5/F ratios,

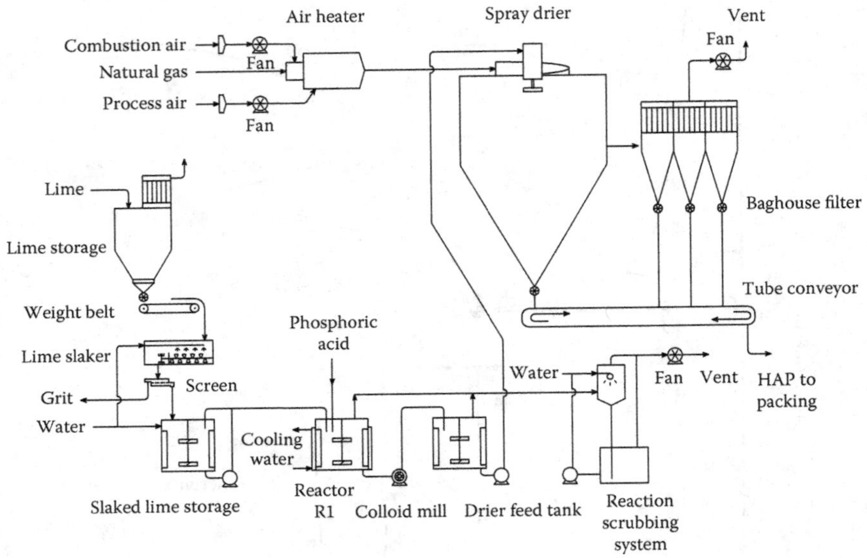

FIGURE 5.13 TCP spray drier process flow sheet.

which translates to fluoride analyses at 0.1% F, that is, 1000 ppm; by contrast, the food or pharmaceutical target is less than 10 ppm F. A further product in this category is defluorinated phosphate (DFP) rock. This product is perhaps the simplest to produce; however, the production process must achieve a reduction in fluoride levels to those acceptable for animal feed as well as various metal impurities.

5.5.5.1 Animal Feed Grade DCP from Acidified Bones

Animal bone is recovered from slaughter houses and butchers for recycling. In some parts of the world, legislation requires that some bone is incinerated and not used as animal feed so as to avoid the possibility of causing BSE. A typical process to make DCP from animal bone commences by crushing the bones and then degreasing them in hot water at 80°C–85°C. The bone chips are mixed with weak hydrochloric acid over several days in a series of tanks of increasing HCl concentration; the range is 2%–5% HCl. The resultant leachate is a mixture of calcium chloride and MCP and is separated from the remaining solids, which are processed further to gelatin. The leachate is further reacted with lime to precipitate DCP. This DCP is separated from the calcium chloride liquor by filtration and water washing and then dried, milled, and packed. The animal feed DCP from this process is particularly low in fluoride and other undesirable constituents.

5.5.5.2 Defluorinated Phosphate Rock

The opportunity to utilize, and the requirement to defluorinate, phosphate rock as an animal feed supplement was recognized in the 1930s. One early patent [58] describes the calcination of phosphate rock in a vertical shaft kiln at 1400°C. Milled rock is mixed with silica and fed into the top of the kiln; milled coal is fed into the middle of the kiln and provides the heat input. Moist combustion air is blown up through the kiln from the base. The reaction temperature is just below that which would start to produce phosphorus. The presence of water is essential to liberate hydrogen fluoride and related gases, which are captured in an absorber. Over the years, the process has developed as taught in many patents, for example, [59–61]. The main features of the process are as follows:

a. Milled phosphate rock, whether fluorapatite or a mixture of fluoride bearing rocks, is mixed with phosphoric acid, usually WPA, sodium carbonate, and water to satisfy three ratios given in Equations 5.24 through 5.26. 5.24 and 5.5 are mole ratios, and 5.26 is a weight ratio. Pure fluorapatite contains 3.7% F; up to this level, there is often sufficient silica in the rock to satisfy Equation 5.25. At higher fluoride levels, there may be insufficient silica present to satisfy Equation 5.25; in this case, silica is added. In some versions of this process, ammonia is added to reduce the buildup of solids in the calciner, the so-called fused ring and ball materials [61]:

$$\frac{CaO + MgO + Na_2O + K_2O - SO_3 - F_2}{P_2O_5 - Fe_2O_3 - Al_2O_3} = 2.5 - 3.8 \tag{5.24}$$

$$\frac{CaO + Na_2O - 3P_2O_5}{SiO_2} = 1.6 - 2.0 \tag{5.25}$$

$$\frac{CaO}{P_2O_5} = 1.010 - 1.058 \tag{5.26}$$

b. The rock, acid, soda, and lime mix is dried, screened, and milled to ensure a uniform particle size and then fed to a calciner operating at 980°C–1350°C. Some processes utilize a rotary calciner, and others a fluid bed calciner; the more simple processes combine the drying and calcining steps in a single rotary calciner. The prime function of the acid, soda, and silica addition is to facilitate the defluorination process. If the ratios are not followed, there is a risk of fusion of the mix. The calcium addition to trim the Ca/P ratio contributes to maximizing the available P_2O_5 in the final product.
c. Quenching the calcined material to less than 100°C ensures that most of the product is the more P_2O_5 available α-TCP.
d. After separating out fine solids, the gas from the calciner is predominantly hydrogen fluoride and the combustion products supplying the heat. The calciner off-gas is scrubbed in weak hydrofluoric acid, which can itself be concentrated and sold as a product.

Figure 5.14 shows a flow sheet adapted from US Patent 4101636 [60] for the production of DFP. Phosphate rock is received into the rock storage and is conveyed to the mill system where the rock is comminuted to 75 μm. The rock is conveyed to the mixing vessel and mixed with sodium carbonate, water, and phosphoric acid/lime slurry from the acid slurry mixer. The rock slurry flows to the granulator and,

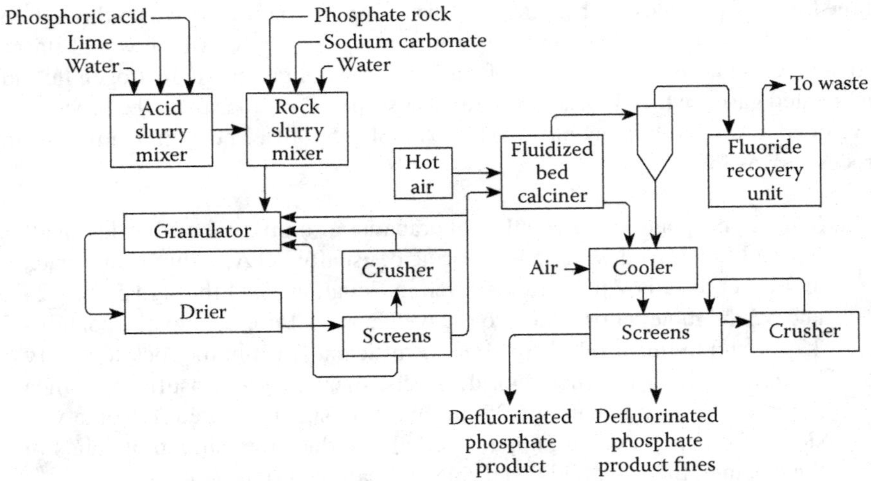

FIGURE 5.14 Defluorinated phosphate rock production block diagram.

combining with fines from the screens and crushed oversize, forms granules that are fed to the drier. The drier is gas or oil fired with an average air temperature of 450°C and outlet temperature of 90°C. The dried product is elevated to screens. The screens separate out the calciner feed in the size range 1–2.5 mm; oversize passes to a small crusher mill, the comminuted material is recycled to the granulator; screen fines are recycled to the granulator. The fluid bed calciner receives the dried granules and heats them to 980°C–1350°C with a residence time of 2–4 h; the hot air providing the heat also introduces water vapor as a combustion product, and the water vapor facilitates defluorination. The calcined granules pass to an air cooler where the product is cooled to less than 100°C. The cooled granules are then screened and milled; the target size is 1–1.5 mm with up to 5% fine material at 75 μm.

5.5.5.3 Animal Feed Grade DCP from WPA

Many plants produce animal feed DCP from WPA around the world. It is the most obvious route given the preponderance of WPA plants for fertilizer manufacture. The first step is to reduce the fluoride levels in the acid so that the final product will meet specification. Producers are concerned to achieve high P_2O_5/F ratios. The defluorination of WPA is carried out in a variety of ways either through the addition of activated silica in conjunction with concentration or through the addition of sodium or potassium salts to form sodium or potassium hexafluorosilicate precipitates, which are then separated from the acid. Defluorination is covered in more detail in Chapter 2. The objective is to reduce fluoride levels to 0.1%–0.2% F. Depending on the rock source, the acid may still contain undesirable elements such as heavy metals like arsenic, lead, and cadmium, at unacceptable concentrations. Heavy metals are removed by precipitation, commonly as sulfides. To accomplish this, WPA is reacted with either sodium bisulfide or hydrogen sulfide. Other processes use reduction by the addition of iron followed by chelation with various agents. Heavy metal removal, particularly arsenic, is also covered in detail in Chapter 2. Instead of defluorinating the WPA and removing heavy metals, the DCP producer may simply source WPA from a supplier prepared to supply at the required specification. Chemically, the process to make animal feed DCP is very similar to the process to make dental or pharmaceutical grade DCP; usually, for cost reasons, limestone is used as the calcium source rather than lime or precipitated calcium carbonate. One process pioneered in the 1950s by Texas City Chemicals carries out the neutralization in two steps, the first step to the monocalcium phosphate solution stage that allows the removal of principal metal phosphate impurities, silicates, and gypsum and the second to complete the neutralization to DCP [62].

5.5.5.4 Animal Feed Grade MCP and DCP Using Hydrochloric Acid

MCP and DCP are made by a small number of producers via the hydrochloric acid route and represent a significant proportion of total world production. The HCl route is economically viable when the producer has access to low-cost or by-product HCl. Chlorine producers are usually based close to a brine resource and utilize the chlorine in the production of a number of chlorine products. The production and sales imbalances often result in excess HCl, which is then available as an inexpensive feed to this route. Many processes produce HCl as a by-product. Of particular relevance is the production of sulfate of potash (SOP), K_2SO_4, from KCl or muriate of potash. While KCl is a major component of NPK fertilizers, SOP is seen as a higher-purity and more efficacious product particularly for higher-value liquid fertilizers. The oldest and most common process to manufacture SOP is the Mannheim process where KCl is reacted with sulfuric acid in a cast iron brick-lined reactor; a rotating internal rake mixes and transfers the reactants. The reaction proceeds chemically in two steps, the first an exothermic reaction and the second an endothermic reaction, which are given in Equations 5.27 and 5.28. The heat for the second reaction is provided by burning gas or oil. The Mannheim process is high cost due to the expense of the equipment needed to cope with the process conditions and the energy needs; other lower-cost processes are taught [63]:

$$KCl + H_2SO_4 \rightarrow KHSO_4 + HCl \tag{5.27}$$

$$KCl + KHSO_4 \rightarrow K_2SO_4 + HCl \tag{5.28}$$

The HCl is released from the reactor as a gas and is water scrubbed to hydrochloric acid. This acid is then available for DCP manufacture. As described in 5.5.5.1, a weak HCl is used to acidulate animal bones to make DCP: this process substitutes phosphate rock for the bone and often utilizes more concentrated acid (20%–30% HCl compared to 2%–5%).

As with the process to make DCP from WPA, the process utilizing HCl and phosphate rock requires the removal of undesirable metals and fluoride. Broadly speaking, two slightly different processes have developed since the 1960s from roots in the 1930s [64] and the earliest patents now over 100 years old [65]. Both processes remove impurities by precipitation. Some patents teach processes that work in the laboratory but require complicated and expensive equipment at plant scale [66]. The main difference between the two practiced processes is that the first accepts some P_2O_5 loss concurrent with removal of unwanted metals [67], whereas the second strives for a higher P_2O_5 efficiency through the use of solvent extraction [68] and coproduction of PWA. Figure 5.15 shows the block diagrams of both routes. In the initial reaction between phosphate rock and HCl, the first process

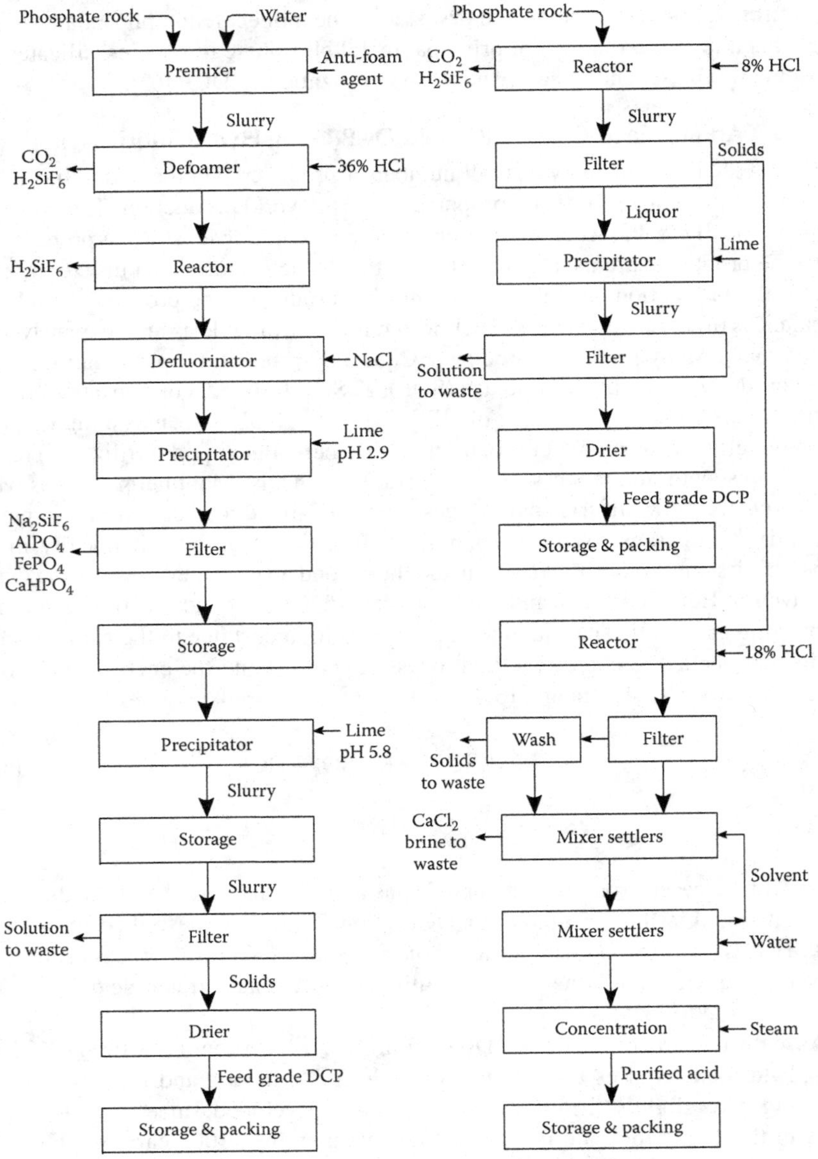

FIGURE 5.15 Animal feed DCP from HCl production block diagrams.

calls for concentrated acid, dissolves the rock completely, and then carries out two neutralizations, the first to remove aluminum and iron phosphate and the fluoride not removed during the initial reaction and the second to complete the neutralization to DCP. The second process uses weak acid (8% HCl) and carries out a partial dissolution in the first reactor; the solids and the supernatant liquor are separated; the solids are water washed and the wash added to the liquor, which is then neutralized to DCP. The washed solids are dissolved in 18% HCl, and the resultant liquor separated from insoluble matter. The insoluble matter is water washed and discarded; the wash is added to the dissolution liquor and mixed with solvent to extract the phosphoric acid. The extraction takes place in a number of mixer settlers. The calcium chloride remains in the aqueous phase and is duly discarded. The acid is re-extracted into water, together with some HCl, and sent for concentration. During concentration, the HCl is separated and recycled and a purified phosphoric acid results.

5.6 ECONOMICS

The economics of the high-purity calcium phosphate products is dominated by the cost of the P_2O_5 component, which is always either purified or thermal acid. The precise figure depends on the product being made and the price of each cost component; typically, the acid cost represents 60%–80% of the total product cost. Utilities represent a relatively minor proportion of the product cost. For the smaller-volume plants, the proportion of maintenance, labor, and depreciation costs is relatively high; this is often the case across many industries. Product packaging might add $50/ton and logistics $100/ton depending on the requirements and location of the customer.

Applying these economics to plant design, it is clear that the first priority is to design a process that will maximize P_2O_5 efficiency. In other words, the process should put as much in specification product into the sales packages as possible.

REFERENCES

1. L. Wang and G. H. Nancollas, Calcium orthophosphates: Crystallization and dissolution, *Chem. Rev.*, 108(11), 4628–4669, November 2008.
2. K. L. Elmore and T. D. Farr, Equilibrium in the system calcium oxide–phosphorus pentoxide–water, *Ind. Eng. Chem.*, 32(4), 580–586, April 1940.
3. H.-B. Pan and B. W. Darvell, Calcium phosphate solubility: The need for re-evaluation, *Cryst. Growth Des.*, 9(2), 639–645, February 2009.
4. A. Maurer-Rothmann, K. Merkenich, and F. Wahl, Process for the production of an aqueous suspension of hydroxylapatite, US Patent 5180564, January 1993.
5. J. N. Butler, *Ionic Equilibrium: Solubility and pH Calculations*, 2nd Rev edn., Wiley-Blackwell, New York, 1998.
6. T. P. Feenstra, A note on the calculation of concentrations in the case of many simultaneous equilibria, *J. Chem. Educ.*, 56(2), 104, February 1979.
7. C. Oliveira, Dicalcium phosphate dihydrate precipitation: Characterization and crystal growth, *Chem. Eng. Res. Des.*, 85(A12), 1655–1661, December 2007.
8. A. Tadayyon, S. M. Arifuzzaman, and S. Rohani, Reactive crystallization of brushite under steady state and transient conditions: Modeling and experiment, *Ind. Eng. Chem. Res.*, 42(26), 6774–6785, December 2003.

9. D. R. Lide and T. J. Bruno, *CRC Handbook of Chemistry and Physics 2012–2013*, CRC Press, Boca Raton, FL, 2012.
10. M. Abutayeh and S. W. Campbell, Predicting the citrate soluble loss of the dihydrate process, *Ind. Eng. Chem. Res.*, 48(18), 8670–8677, September 2009.
11. M. Frèche and J. C. Heughebaert, Calcium phosphate precipitation in the 60–80°C range, *J. Cryst. Growth*, 94(4), 947–954, April 1989.
12. W. E. Brown and L. C. Chow, Combinations of sparingly soluble calcium phosphates in slurries and pastes as mineralizers and cements, US Patent 4612053, September 1986.
13. G. Vereecke and J. Lemaître, Calculation of the solubility diagrams in the system $Ca(OH)_2$–H_3PO_4–KOH–HNO_3–CO_2–H_2O, *J. Cryst. Growth*, 104(4), 820–832, September 1990.
14. E. Cox and R. H. Kean, Calcium acid pyrophosphate composition and method of production, US Patent 2272617, February 1942.
15. W. B. Brown and R. A. Holbrook, Baking Preparation, US Patent 990699, April 1911.
16. D. R. Gard and B. B. Heidolph, Leavening composition and process of making, US Patent 5409724, April 1995.
17. J. A. G. Bruce, H. M. Levy, and P. Oborn, Phosphate and phosphors prepared there from, US Patent 3635660, January 1972.
18. A. D. F. Toy, *The Chemistry of Phosphorus*, Pergamon Press, New York, 1975.
19. C. M. Forster and S. A. Meilicke, Synthesis of gamma calcium pyrophosphate, US Patent 5667761, September 1997.
20. E. J. Griffith and W. C. McDaniel, Calcium phosphates, US Patent 4721615, January 1988.
21. C. Y. Shen, Production of alkaline earth metal pyrophosphates, US Patent 3407035, October 1968.
22. R. N. Bell and L. E. Netherton, Calcium polyphosphate and method of producing the same, US Patent 2852341, September 1958.
23. J. P. Godber and H. Jenkins, Condensation polymerization of phosphorus containing compounds, US Patent 5951831, September 1999.
24. F. Gauthier, J. Shulman, B. Weinstein, A. Keenan, and Y. Duccini, Scale inhibitors, US Patent 6395185, May 2002.
25. V. F. Vetere, M. C. Deyá, R. Romagnoli, and B. del Amo, Calcium tripolyphosphate: An anticorrosive pigment for paint, *J. Coatings Technol.*, 73(917), 57–63, June 2001.
26. T. E. Edging, Leavening acid composition, US Patent 4500557, February 1985.
27. J. R. Parks, A. R. Handleman, J. C. Barnett, and F. H. Wright, Methods for measuring reactivity of chemical leavening systems. I. dough rate of reaction, *Cereal Chem.*, 37, 503–518, July 1960.
28. J. R. Newton, J. P. Quinn, and P. W. Stanier, Precipitated silicas, US Patent 5098695, March 1992.
29. Dietary Supplement Fact Sheet: Calcium—Health Professional Fact Sheet. [Online]. Available: http://ods.od.nih.gov/factsheets/Calcium-HealthProfessional/ (accessed April 16, 2013).
30. J. Godber and A. Shaheed, Calcium fortification substance for clear beverages, US Patent 0274264, November 2008.
31. W. Camarco, L. R. Hendricks, and J. M. Jobbins, Directly compressible tricalcium phosphate, US Patent 7226620, June 2007.
32. E. A. Hunter, B. E. Sherwood, and J. H. Staniforth, Pharmaceutical excipient having improved compressibility, US Patent 5585115, December 1996.
33. G. Brachtel, G. Raab, and D. Schober, Process for the preparation of coarse-particle dicalcium phosphate dihydrate, US Patent 4755367, July 1988.
34. C. A. Ertell and C. G. Gustafson, Granular anhydrous dicalcium phosphate compositions suitable for direct compression tableting, US Patent 4707361, November 1987.

35. T. W. Gerard, A process of compacting fine particles of calcium phosphate and the compacts obtained thereby, European Patent 0054333, April 1986.
36. J. Godber and L. Leite, Hydroxyapatite calcium phosphates, their method of preparation and their applications, US Patent 013044, May 2009.
37. M. Jarcho, Hydroxylapatite ceramic, US Patent 4097935, July 1978.
38. W. E. Brown and L. C. Chow, Dental resptorative cement pastes, US Patent 4518430, May 1985.
39. F. C. M. Driessens and R. Wenz, Tricalcium phosphate-containing biocement pastes comprising cohesion promoters, US Patent 6206957, May 2001.
40. G. D. Irvine, Synthetic bone ash, US Patent 4274879, June 1981.
41. R. E. Vanstrom and F. McCollough, Polymer suspension stabilizer, US Patent 3387925, June 1968.
42. H. D. Layman and R. E. Taylor, Production of calcium halophosphate, US Patent 3655576, April 1972.
43. T. Brugger, T. Friedrich, S. Haefner, A. Knietsch, E. Scholten, and O. Zelder, Phytase, US Patent 0081331, March 2009.
44. E. N. Horsford, Improvement in preparing phosphoric acid as a substitute for other solid acids, US Patent 14722, April 1856.
45. B. D. Saklatwalla, H. E. Dunn, and A. E. Marshall, Process of making monocalcium phosphate, US Patent 2012436, August 1935.
46. C. T. Whittier, Process of preparing dry granular calcium acid phosphate, US Patent 1442318, January 1923.
47. J. N. Carothers and P. Logue, Method for the production of dried acid calcium phosphate, US Patent 1818114, August 1931.
48. W. J. Balfanz, R. E. Benjamin, and T. E. Edging, Method for preparing monocalcium phosphate compositions with reduced caking tendencies, US Patent 3954939, May 1976.
49. N. E. Stahlheber, Processes and products for conditioning caking salts, US Patent 3367883, September 1968.
50. F. R. Deutman, Free flowing monocalcium phosphate, US Patent 1913796, June 1933.
51. J. R. Schlaeger, Heat-treated monocalcium phosphate, US Patent 2160232, May 1939.
52. F. H. Y. Chung and T. E. Edging, Leavening acid composition, US Patent 6080441, June 2000.
53. E. Cox and R. H. Kean, Leavening agent, US Patent 2263487, November 1941.
54. F. H. Y. Chung, Leavening acid composition produced by heating monocalcium phosphate at elevated temperatures, US Patent 5925397, July 1999.
55. R. H. Kean, E. Cox, and W. K. Enos, Calcium acid pyrophosphate composition and method of production, US Patent 2272617, February 1942.
56. H. Buhl, F.-J. Dany, and J. Holz, Process for the preparation of dicalcium phosphate, US Patent 5024825, June 1991.
57. M. Hosokawa, F. Nakagawa, and T. Yokoyama, Powder classifier, US Patent 3670886, June 1972.
58. T. F. Baily, Method of calcining phosphate bearing materials, US Patent 2121776, June 1938.
59. A. E. Henderson, Method of defluorinating phosphates, US Patent 4405575, September 1983.
60. M. E. Clark and H. V. Larson, Defluorinated phosphate rock process using lime, US Patent 4101636, July 1978.
61. J. R. Gruber, E. A. Gudath, D. H. Michalski, R. R. Riddle, and R. R. Stana, Defluorination kiln restriction control agent and method, US Patent 4716026, December 1987.
62. Making DCP, *Phosphorus Potassium*, 1996:36–44, February 1996.

63. J. B. Sardisco, Production of potassium sulfate and hydrogen chloride, US Patent 4045543, August 1977.
64. W. R. Seyfried, Process for treatment of phosphatic solutions, US Patent 2164627, July 1939.
65. E. Bergmann, Method of producing dicalcium phosphate, US Patent 852372, April 1907.
66. E. A. Fallin, Production and purification of phosphatic materials, US Patent 3161466, December 1964.
67. P. R. Cutter, Method for preparing feed grade dicalcium phosphate, US Patent 3391993, July 1968.
68. C. Fink and D. Loewy, Processes for the manufacture of feed-grade dicalcium phosphate and phosphoric acid, US Patent 3988420, October 1976.

6 Other Phosphates

6.1 INTRODUCTION

Most sodium and calcium phosphates are made on plants sized in ten thousands of tons per year output with the largest individual sodium tripolyphosphate (STPP) plants at about 100,000-ton scale. The next most significant industrial phosphates in terms of volume are ammonium and potassium phosphates. Typically, these plant sizes are in the 10,000–20,000 tons per year range. Aluminum, magnesium, and lithium phosphates are manufactured on a smaller scale in hundreds or low thousands of tons per year. Other phosphates are made but at the kilogram scale (e.g., phosphates of Ba, Cr, Cu, Fe, Pb, Mn, Sn, and Zn and the pyrophosphates of Cu, Mg, Mn, Sn, and Zn were once made in Albright & Wilson's Stratford works many years ago, in small quantities, by the *alchemists*).

Generally, the manufacturing process steps for these phosphates are very similar to those of the sodium and calcium phosphates and include neutralization, crystallization, spray and other types of drying, calcination in wet and dry kilns, and milling and sizing.

6.2 ALUMINUM PHOSPHATES

The chemistry of the pure aluminum orthophosphates is best described with reference to the Al_2O_3–P_2O_5–H_2O phase diagram (see Figure 6.1). Figure 6.1 is similar in form to other metal oxide–P_2O_5–H_2O phase diagrams with a number of identifiable crystal forms, areas where these crystals exist in solution, and an area of homogeneous solution. The phase equilibria of aluminum orthophosphates in aqueous solutions were studied comprehensively by d'Yvoire in the 1960s. More recent studies [1,2] have added to knowledge about different forms and the thermal evolution of aluminum phosphate solutions particularly those with Al/P ratios of 1:2 to 1:3. Referring to Figure 6.1, aluminum orthophosphates occur only in the area bounded by the lines Al/P = 3:1 and 1:3 and the $Al(OH)_3$ to $AlPO_4$ and $AlPO_4$ to $Al_2O_3 \cdot 3P_2O_5 \cdot 6H_2O$ line. Aluminum phosphates made with an Al/P ratio of 1:1 are classified as neutral, those greater than 1:1 as basic, and those less than 1:1 as acid. The studies have shown the existence of at least 18 aluminum salt solid phases. These crystalline aluminum phosphates are as follows:

Basic salt

$Al_4(OH)_3(PO_4)_3$	$Al_2(OH)_3(PO_4)$	$Al_3(OH)_3(PO_4) \cdot H_2O$
$Al_3(OH)_3(PO_4)_3 \cdot 5H_2O$	$Al_4(OH)_3(PO_4)_3 \cdot 11H_2O$	$Al_3(OH)_3(PO_4)_2 \cdot 9H_2O$
$Al_2(OH)_3(PO_4) \cdot 4.75H_2O$	$Al_3(OH)_6(PO_4) \cdot 6H_2O$	

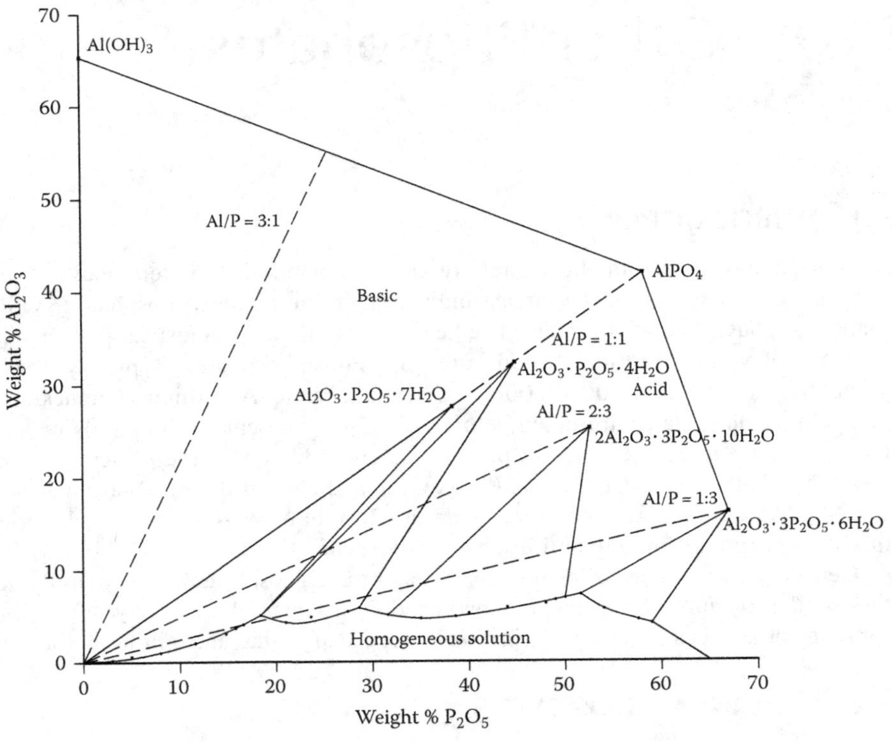

FIGURE 6.1 Al_2O_3–P_2O_5–H_2O system at 25°C.

Neutral salt

$$AlPO_4 \cdot 2H_2O \text{ (three forms)}$$

Acid salt

$Al(H_2PO_4)_3$ (three forms)	$Al(H_2PO_4)(HPO_4) \cdot H_2O$	$H_3O[Al_3(H_2PO_4)_6(HPO_4)_2] \cdot 4H_2O$
$Al_2(HPO_4)_3 \cdot 3.5H_2O$	$Al_2(HPO_4)_3 \cdot 4H_2O$	$Al(H_2PO_4)(HPO_4) \cdot 2.5H_2O$
$Al_2(H_{2-x}PO_4)(H_{3x}PO_4) \cdot 6H_2O$ $(0 \leq x \leq 1)$	$Al_2(HPO_4)_3 \cdot 6.5H_2O$	$Al_2(HPO_4)_3 \cdot 8H_2O$

It is worth noting that these compounds are often written in the form $xAl_2O_3 \cdot yP_2O_5 \cdot zH_2O$. It is well established that heating solutions or solids with an Al/P = 1:3 at 200°C give an amorphous phase that partially crystallizes to aluminum pyrophosphate, $Al_8H_{12}(P_2O_7)_8$, at 250°C. Further heating to 300°C–500°C gives aluminum tripolyphosphate, $AlH_2P_3O_{10} \cdot xH_2O$, and subsequently monoclinic and cubic aluminum metaphosphate, $Al(PO_3)_3$, at 500°C–600°C. Anhydrous $AlH_2P_3O_{10}$ hydrates rapidly to form $AlH_2P_3O_{10} \cdot 2H_2O$. Cubic $Al(PO_3)_3$ is stable until melting at 1400°C. At higher temperatures, the glass decomposes, releasing P_2O_5 and depositing $AlPO_4$; up to 1800°C further P_2O_5 is released, and the $AlPO_4$ is converted to α-Al_2O_3.

The most common application of industrial aluminum phosphates is as a binder for linings used in high-temperature furnaces, chimneys, and reactors. $AlPO_4$ is insoluble and chemically inert and shows negligible decomposition up to 1100°C; in addition to using $AlPO_4$ as a binder for high-temperature furnace linings, it is used for polymer coating on metals, for coloring quartz sand with inorganic pigments by adhering the pigment to the grain surface, as a coating for silicon steel to provide interlaminar electrical resistance for transformer cores [3], as a reactive component for bonded aggregate structures [4] and inorganic resin systems [5,6], as a reactive component of a foamed ceramic [7], and as a pharmaceutical aluminum adjuvant that enhances the immunogenicity of vaccine antigens [8].

The starting point for most commercial aluminum phosphate products is a simple reaction between aluminum hydroxide and phosphoric acid, which results in a monoaluminum phosphate solution given by the following equation:

$$Al(OH)_3 + 3H_3PO_4 + 3H_2O \rightarrow Al(H_2PO_4)_3 + 6H_2O \quad (6.1)$$

This reaction is exothermic. On an industrial scale, the production volumes are such that the reaction is usually carried out batchwise. An aliquot of solid aluminum hydroxide is added to phosphoric acid and allowed to dissolve. The precise amount added is governed by a target reaction temperature of 90°C–100°C and the cooling capacity of the reactor system. Once the reactor temperature is again stable, further hydroxide is added *quantum satis*. Figure 6.2 shows a flow sheet of a monoaluminum

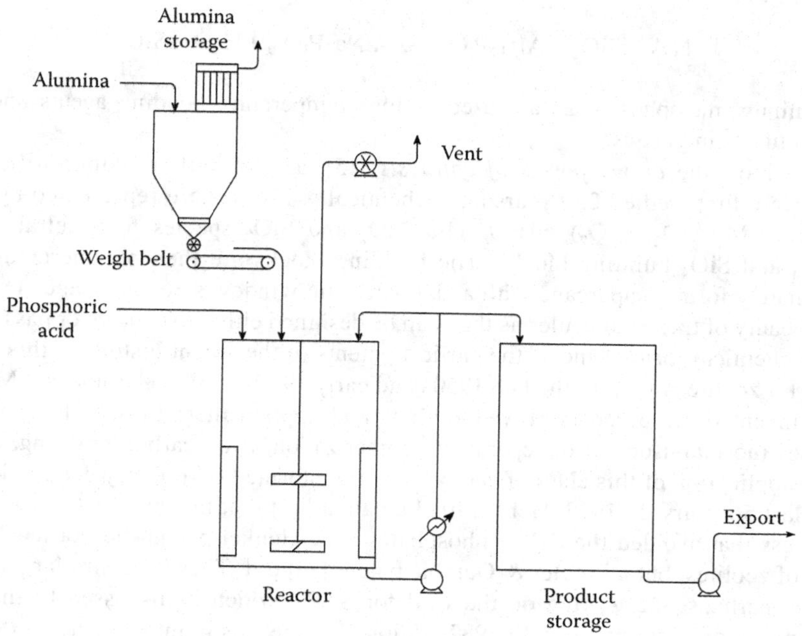

FIGURE 6.2 Monoaluminum phosphate plant flow sheet.

phosphate plant. In this variation reaction, heating or cooling is provided by an external heat exchanger supplied with cooling water and hot water as required.

Two commercial solution products are marketed at 50% and 70% concentration. With an Al/P ratio of 1:3, the 50% solution has a shelf life of over 2 years. With an Al/P ratio of 1:4, the 70% solution has a shelf life of about 1 year. The aluminum phosphate solution is reacted with aluminum hydroxide or other basic oxides and minor ingredients in a setting reaction and formed into the desired shape. This reaction is shown in the following equation:

$$Al(H_2PO_4)_3 + 2Al(OH)_3 \rightarrow 3AlPO_4 + 6H_2O \qquad (6.2)$$

A powdered monoaluminum phosphate binder is prepared by evaporating the 50% solution and then drying the product to 1% moisture. Another process fed phosphoric acid and aluminum hydroxide into a heated, kneader reactor forming a dry, weakly hygroscopic product suitable for grinding, which could be used directly for binding refractory ceramics [9]. Ramming mixtures are used for lining furnaces operating above 1100°C. Ramming mixtures comprise magnesium oxide, aluminum phosphate, and insulating material such as vermiculite [10]. Foamed ceramic materials are used for the insulation of structural steel and the repair of existing refractories and include aluminum phosphate in their composition [7,11].

Important applications for aluminum tripolyphosphate include: anticorrosive pigments [12,13]; adsorbents for basic malodorous substances such as ammonia and amines [14]; and water glass hardeners in the reaction shown in the following equation:

$$Na_2 \cdot xSiO_2 + AlH_2P_3O_{10} \rightarrow AlNa_2P_3O_{10} + H_2O \cdot xSiO_2 \qquad (6.3)$$

Aluminum metaphosphates are used as high-temperature bonding agents and as constituents in glasses.

Zeolites, the *crown jewels of catalysis* [15], are crystalline aluminosilicates and were first studied 250 years ago. Chemically, zeolites are represented by the formula $M_v(AlO_2)_x(SiO_2)_y \cdot zH_2O$. The AlO_2 and SiO_2 species form tetrahedral AlO_4 and SiO_4 building blocks. The building blocks are then polymerized and ultimately form a supercage with a characteristic window size and shape. This is the beauty of these molecules as they can be designed either to capture or pass specific chemical forms. One of the earliest patents in the recent history of this field was for zeolite A [16]. In the late 1950s and early 1960s, Union Carbide and Mobil Oil invented more and improved zeolites and applied them to oil refinery processes (normal–isoparaffin separation, isomerization, hydrocarbon cracking, etc.). The application of this class of catalyst has transformed oil refinery economics in the last 50 years. In 1973, Henkel filed a patent [17] that taught a textile washing process that avoided the use of phosphates. The Henkel patent did not teach the use of zeolites, but a Procter & Gamble patent [18] did. These and similar patents were, perhaps, the writing on the wall for STPP, which is discussed further in Chapter 7 on sustainability. In 1982, Union Carbide was granted a patent [19] for a new class of aluminophosphate molecular sieve. One advantage of the $AlPO_4$

molecular sieves is the greater range of pore size that is possible compared to the zeolites. Subsequently, the incorporation of silicon has led to another class of molecular sieve, known as *SAPO*.

6.3 AMMONIUM PHOSPHATES

The chemistry of the ammonium orthophosphates is best described with reference to the $(NH_4)_2O$–P_2O_5–H_2O phase diagram (see Figure 6.3) [20]. The following ammonium phosphates are present in the solid phases in the temperature range 0°C–75°C:

$NH_4H_5(PO_4)_2 \cdot H_2O$	$(NH_4)_2HPO_4$ (DAP)	$(NH_4)_3PO_4 \cdot 3H_2O$ (TAP)
$NH_4H_5(PO_4)_2$ (HAP)	$2(NH_4)_2HPO_4 \cdot (NH_4)_3PO_4$	
$NH_4H_2PO_4$ (MAP)		

The ammonium phosphate phase diagram is similar to other phosphate phase diagrams although more simple: the solubility line rises with increasing temperature, but there are fewer solid phases (than, for example the sodium phosphate system). Both $(NH_4)_3PO_4 \cdot 3H_2O$ and $2(NH_4)_2HPO_4 \cdot (NH_4)_3PO_4$ are unstable at room temperature and lose ammonia to form diammonium phosphate (DAP). The two products of commerce are monoammonium phosphate (MAP) and DAP—$NH_4H_2PO_4$ and $(NH_4)_2HPO_4$.

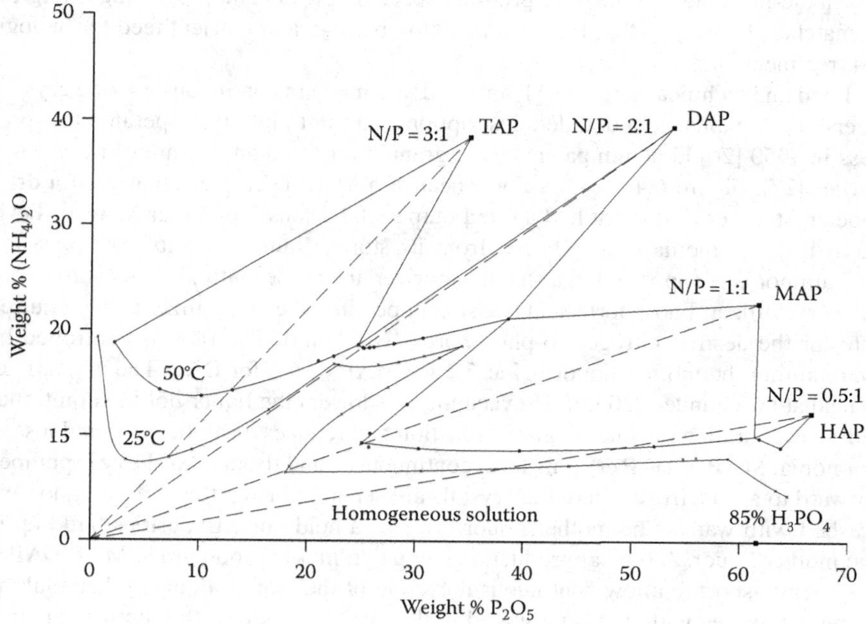

FIGURE 6.3 $(NH_3)_2O$–P_2O_5–H_2O system. (Adapted from Wendrow, B. and Kobe, K.A., The alkali orthophosphates. Phase equilibria in aqueous solution, *Chem. Rev.*, 54(6), 891–924, 1954. Copyright 1954 American Chemical Society.)

Both MAP and DAP are produced in very large quantities for fertilizer applications with wet process phosphoric acid (WPA) and ammonia as feed materials. Technical and food grade ammonium phosphates are made on a much smaller scale with either PWA or thermal acid and ammonia as feed materials. Typically, these smaller plants produce a combined MAP/DAP output of 10,000–40,000 tons per year.

Food grade MAP and DAP are used as a nutrient in the manufacture of yeast [21], in wine-making processes, and as an ingredient in compound bread improvers [22,23].

Food grade DAP is also used in cigarettes. Both MAP and DAP are used as flame retardants. In the presence of a flame, the ammonium phosphate decomposes to ammonia and phosphoric acid; the acid catalyzes the decomposition of cellulose to char, making it more resistant to ignition. Therefore, adding DAP to both the paper wrapper and the tobacco of a cigarette helps control the burn rate as the smoker sucks air through the cigarette. In this respect, DAP is used as advertised as a flame retardant for cigarettes. The tobacco industry states that ammonia compounds enhance the flavor, body, and taste of the cigarette [24]. The industry also uses ammonium phosphates to transfer nicotine out of tobacco either to improve the tobacco or to provide a nicotine source for different applications unrelated to tobacco products [25]. Ammonia modifies the nicotine content of tobacco smoke. At pH values greater than 5, nicotine is deprotonated to its volatile nonprotonated form. In the 2005 review, Willems et al. [24] concluded that ammonia increases the bioavailability of nicotine via multiple mechanisms that are based on weak bases deprotonating at high pH.

Technical grade MAP and DAP are used in specialty fertilizers, building materials, flame-proofing wood, flame-proofing specialty papers and preventing afterglow in matches, flame-proofing fabrics and cotton batting, and nutrient feed for biological treatment plants.

Food and technical grade MAP and DAP are made in continuous cooling crystallizers. TVA published a detailed description of a pilot plant that operated this process in 1950 [26] although patents were granted for the manufacture of DAP much earlier [27]. Figure 6.4 shows a flow sheet of a MAP/DAP plant that used a draft tube crystallizer. The plant had a rated output of 1.7 tons/h of either MAP or DAP. Anhydrous ammonia is transferred from its storage tank under its own pressure. The ammonia is vaporized in a steam vaporizer and passes into the base of the draft tube crystallizer. Phosphoric acid is also pumped into the crystallizer at the required rate for the desired nitrogen-to-phosphorus (N/P) ratio. The ratio is controlled by maintaining the mother liquor pH at 5.2 for MAP or 6.2 for DAP. The crystallizer is held at 60°C under 120 mm Hg vacuum; this lowers the liquor boiling point, thus aiding evaporation of water from the reaction and reduces disassociation and loss of ammonia. MAP or DAP crystals form continuously, and the crystal slurry is pumped forward to a centrifuge where the crystals are separated from the mother liquor and washed with water. The mother liquor flows to a hold tank. Impurities build up in the mother liquor. A typical production campaign making food grade MAP/DAP is sufficiently short to allow continuous recycling of the mother liquor. When making technical grades with a less pure acid, only partial recycle of the mother liquor is possible. The rest is sold as fertilizer. The crystals fall by gravity into a rotary drier. The drier operates with an air inlet temperature of 140°C–150°C and is designed to reduce crystal moisture content from 5% to 0.1%. Crystal product leaves the drier at

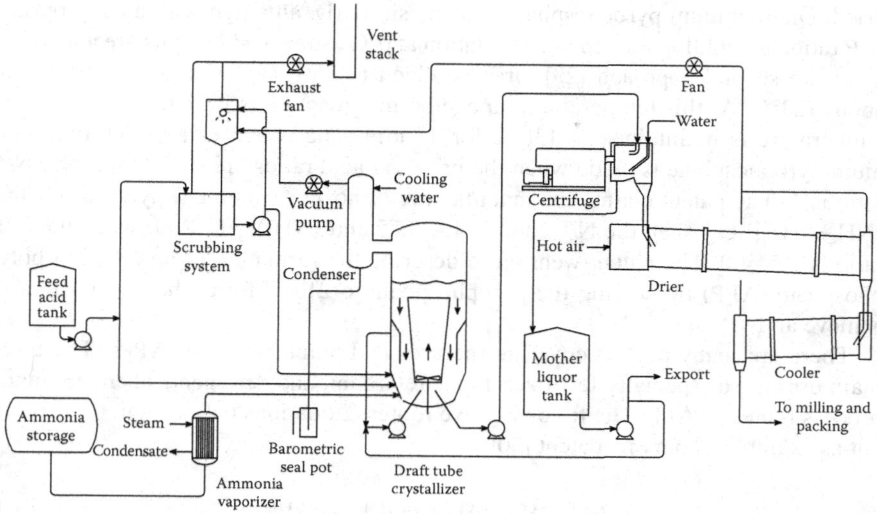

FIGURE 6.4 MAP/DAP flow sheet.

65°C and enters a rotary cooler where the product temperature is reduced to 35°C. The crystals are either packed immediately or milled to powder and packed.

Vapor from the crystallizer is condensed against cooling water and then scrubbed in a phosphoric acid scrubber. The scrubber liquor is then bled into the crystallizer. A little care is required both during design and operation to get the water balance right across the whole plant.

These plants are a little more hazardous than other phosphate plants because of the ammonia. For most plants of this scale, ammonia consumption is sufficiently large to warrant a large storage tank, either as a day tank supplied by pipeline on a larger fertilizer site or sufficient to hold a road tanker delivery. Anhydrous ammonia is a liquefied gas and is stored at 7.6 bar at 21°C; therefore, the storage vessel is a pressure vessel, and it and its ancillary equipment and piping must be designed and maintained accordingly. Ammonia is flammable; therefore, the control, instrumentation, and electrical design must account for this. It is also toxic, which adds to ventilation and operational considerations. Ammonium phosphates are prone to crystallizing out where there is a cold spot in piping or around pumps and heat exchangers; this phenomenon can make operation in colder climates challenging.

Ammonium pyrophosphates play an important role in fertilizers but are not items of commerce in the industrial phosphate sector and have been used to make the polyphosphate. Six forms are described [28] and include tetra-, tri-, and diammonium pyrophosphates, although Sears claimed novel forms [29]. Two approaches to manufacture are described.

In the first approach, pyrophosphoric acid crystals of 79.5%–80% P_2O_5 are made by adding solid phosphorus pentoxide to 90% H_3PO_4 at up to 180°C and then allowing the solution to cool. The pyrophosphoric acid crystals are then dissolved in 20%–25% ammonia solution. Tetra-ammonium phosphate crystals form and are separated and dried. Diammonium pyrophosphate is made similarly, although with an appropriate N/P ratio and holding the ammonia solution at 0°C as the acid crystals are added.

In the second approach [29], urea is added to 85% H_3PO_4 and then heated to about 123°C. At this temperature, an exothermic reaction is initiated. The reaction temperature is maintained at 130°C for 45 min. The patent claims that diammonium pyrophosphate is made when the urea and acid ratios are such that molar N/P ratio is 1. The patent claims the manufacture of novel ammonium pyrophosphates: $(NH_4)_{2.7}H_{1.3}P_2O_7$ when the N/P ratio is 1.4–1.75 and $(NH_4)_{3.3}H_{0.7}P_2O_7$ when the N/P ratio is 2.5–4.0. The patent went on to describe the formation of ammonium polyphosphate (APP) by heating the pyrophosphate to 210°C for an hour with urea to achieve an N/P ratio of 1.

There are many patents covering the uses and manufacture of APPs. The three main uses are as specialty fertilizer, flame retardant, and detergent. There are many process routes to APP. The following five routes (Equations 6.4 through 6.8) are set out as examples from one patent [30]:

$$NH_4OH + NH_3 + POCl_3 \rightarrow APP \tag{6.4}$$

$$NH_3 + H_3PO_4 + CO(NH_2)_2 \rightarrow APP \tag{6.5}$$

$$NH_3 + H_3PO_4 + P_2O_5 \rightarrow APP \tag{6.6}$$

$$H_3PO_4 + CO(NH_2)_2 \rightarrow APP \tag{6.7}$$

$$(NH_4)_4P_2O_7 + P_2O_5 \rightarrow APP \tag{6.8}$$

This patent claimed several processes for preparing substantially water-insoluble crystalline APPs having the general formula $H_{(n-m)+2}(NH_4)_mP_nO_{3n+1}$ where n is an integer greater than 10, m/n is 0.7–1.1, and $m \leq (n + 2)$. The processes were variations on the theme of thermally condensing urea (as combined ammoniating and condensing agent), phosphoric acid (ortho or condensed), and an ammonium phosphate together with a small part of APP. Thermal condensation takes place at 100°C–350°C. A later patent [31] teaches an activated APP and a process to make it with either melamine or dicyandiamide. A more recent patent [32] describes a two-step process: In the first step, ammonium orthophosphate and phosphorus pentoxide in the presence of gaseous ammonia react in a vertical reactor with kneading, mixing, and comminution tools; in the second step, the reactor discharges to a disk drier with heating disks at 240°C–300°C and a lower cooling disk at 45°C–50°C.

6.4 POTASSIUM PHOSPHATES

The chemistry of the potassium orthophosphates is best described with reference to the $K_2O–P_2O_5–H_2O$ phase diagram (see Figure 6.5) [20]. In general, the potassium system resembles the sodium system more close than it does the ammonia

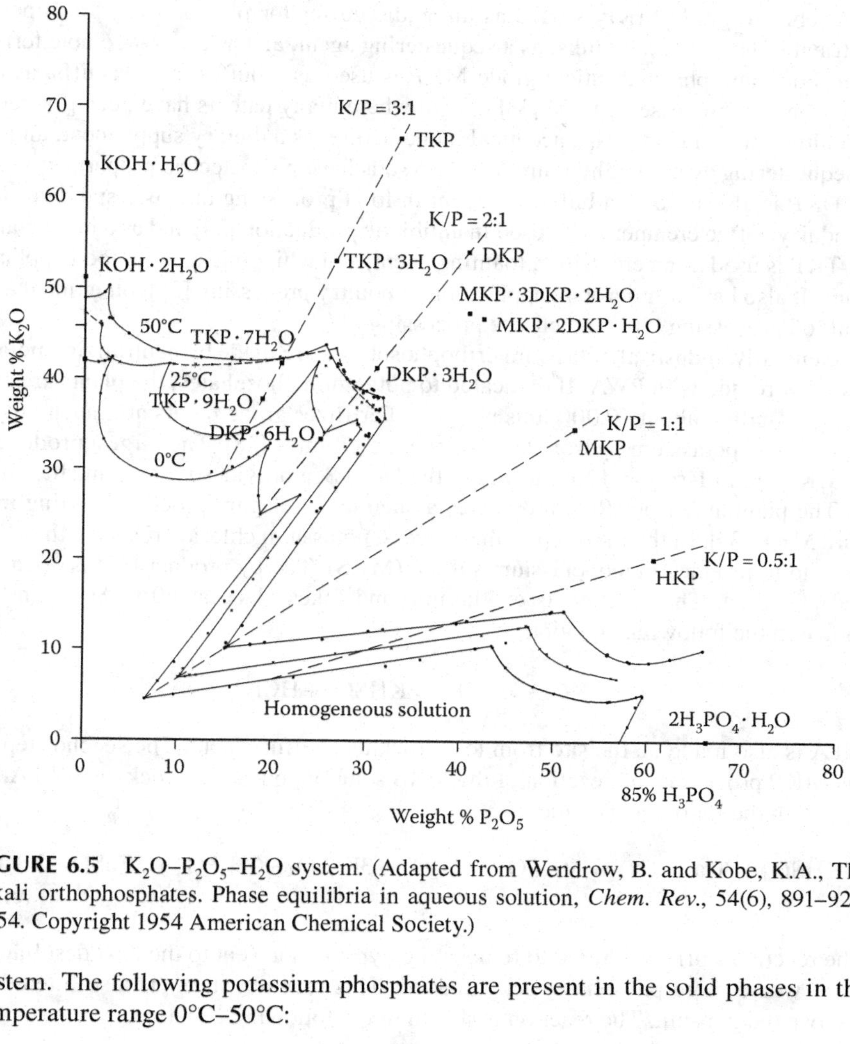

FIGURE 6.5 K_2O–P_2O_5–H_2O system. (Adapted from Wendrow, B. and Kobe, K.A., The alkali orthophosphates. Phase equilibria in aqueous solution, *Chem. Rev.*, 54(6), 891–924, 1954. Copyright 1954 American Chemical Society.)

system. The following potassium phosphates are present in the solid phases in the temperature range 0°C–50°C:

$KH_5(PO_4)_2$ (HKP)	K_2HPO_4 (DKP)	K_3PO_4 (TKP)
KH_2PO_4 (MKP)	$K_2HPO_4 \cdot 3H_2O$	$K_3PO_4 \cdot 3H_2O$
$KH_2PO_4 \cdot 2K_2HPO_4 \cdot H_2O$	$K_2HPO_4 \cdot 6H_2O$	$K_3PO_4 \cdot 7H_2O$
$KH_2PO_4 \cdot 3K_2HPO_4 \cdot 2H_2O$		$K_3PO_4 \cdot 9H_2O$

The most striking feature of the phase diagram is the high solubility of the alkaline potassium phosphates.

The potassium phosphates of commerce are monopotassium phosphate (MKP), dipotassium phosphate (DKP), and tripotassium phosphate (TKP) and are available in both anhydrous and hydrated forms. A fertilizer grade MKP is made at large scale and is a component of NPK fertilizers.

Technical grade MKP is used as an acidic buffer for pH control, as a mineral nutrient in fermentation tanks, as a sequestering agent, and as a water-soluble fertilizer. Food and pharmaceutical grade MKP is used as a buffer in foodstuffs, as an ingredient in processed cheese [33] (a field where many patents have been granted), as culture nutrients in pharmaceutical manufacture, as a dietary supplement, and as a sequestering agent. Highly pure MKP crystals have piezoelectric properties.

DKP is also used as a buffering agent in food processing and as a stabilizer for nondairy coffee creamers. It is used in antibiotic production [34] and as a humectant.

TKP is used as an emulsifier, foaming agent, and whipping agent in food applications. It also has antimicrobial application in poultry processing [35], other meat and seafood processing [36], and cheese processing [37].

Generally, industrial potassium orthophosphates are made by neutralizing potassium hydroxide with PWA. If dedicated to potassium phosphates, the plants are relatively small at about 10,000 tons per year. Fertilizer grade plants are much larger and eschew potassium hydroxide on economic grounds [38]. The largest producers of MKP are in Israel and Jordan where the Dead Sea provides potassium chloride.

The plant at Mishor Rotem operates a chemically elegant process resulting in a pure MKP [39]. In the first step of the process, potassium chloride reacts with sulfuric acid to produce monopotassium sulfate (MKS). The by-product HCl is released and absorbed. The reaction is exothermic and takes place at 60°C–80°C and is shown in the following equation:

$$KCl + H_2SO_4 \rightarrow KHSO_4 + HCl \tag{6.9}$$

WPA is also made on the site from local rock and sulfuric acid. The second step of the MKP process is the reaction of the MKS solution, phosphate rock, and WPA, as shown in the following equation:

$$6KHSO_4 + 3Ca_3(PO_4)_2 + 5H_3PO_4 + 6H_2O \rightarrow 3K_2SO_4 + 7H_3PO_4 + 3CaSO_4 \cdot 2H_2O \tag{6.10}$$

The reaction slurry is filtered to remove the gypsum and sent to the next desulfation stage where more phosphate rock is added. This reaction slurry is also filtered to remove the gypsum. The reaction is shown in the following equation:

$$3K_2SO_4 + Ca_3(PO_4)_2 + 7H_3PO_4 + 6H_2O \rightarrow 6KH_2PO_4 + 3H_3PO_4 + 3CaSO_4 \cdot 2H_2O \tag{6.11}$$

The filtrate is then neutralized with lime or calcium carbonate to a pH in the range 3.5–5 as shown in the following equation:

$$6KH_2PO_4 + 3H_3PO_4 + 3CaO \rightarrow 6KH_2PO_4 + 3CaHPO_4 + 2H_2O \tag{6.12}$$

Above pH 4.6, the lime starts to neutralize the MKP to DKP as shown in the following equation:

$$2KH_2PO_4 + CaO \rightarrow K_2HPO_4 + CaHPO_4 + H_2O \tag{6.13}$$

The calcium phosphate (DCP) is separated by filtration. Depending on the degree of neutralization, the K/P ratio of the resulting solution is variable. This solution can be used in NPK fertilizers. Equally if pure MKP is required, the solution resulting from Equation 6.13 is pumped to a cooling crystallizer. The patent claims a high purity with heavy metals, including cadmium and lead less than 1 ppm. The only obvious exception to food grade purity was the fluoride level:

$$K_2SO_4 + CaHPO_4 + H_3PO_4 + 2H_2O \rightarrow 2KH_2PO_4 + CaSO_4 \cdot 2H_2O \quad (6.14)$$

The industrial potassium phosphate plants that neutralize potassium hydroxide are very similar in configuration to sodium phosphate plants, that is, a neutralization vessel, followed by a spray drier and kiln for higher phosphates and milling and sizing equipment. Indeed, there are examples where multiproduct sodium phosphate plants replace sodium with potassium hydroxide and adjust the operating conditions slightly and make potassium phosphates. In practice, simply switching over from sodium to potassium hydroxide is not easy and requires careful adjustment to achieve smooth operation. MKP in particular is difficult to process without stoppages. One cause of stoppage is the ease with which MKP solution will crystallize and cause blockages. The inspection of the phase diagram shows that this is not surprising. One patent [40] teaches that after the neutralization, the MKP solution should be cooled below 50°C to precipitate MKP crystals, which are then homogenized in a high shear mill. The crystal slurry is then pumped to a spray drier and dried with inlet air temperatures in the range 350°C–400°C. The patent teaches that the product is a free-flowing powder with a particle size range of 75–100 μm.

The mono- and trihydrate potassium pyrophosphates are known, but only the anhydrous salt, $K_4P_2O_7$, is of commercial value. On the other hand, 60%–65% solutions of tetrapotassium pyrophosphate (TKPP) are sold commercially. There are a vast range of applications for TKPP, which include use as a sequestering, buffering, and emulsifying agent [41] and as a texturizer in processed meat, fish, and cheese and as a gelling agent in instant puddings [42]. Other examples of its use include as part of a pharmaceutical composition for the therapeutic administration of pyrophosphate for dialysis patients [43], as an aid to the purification of allyl alcohol [44], as a pet food palatability enhancer [45], and as part of an anticalculus composition [46]. In the 1950s, it was introduced as a builder for light duty liquid detergents.

TKPP is made by dehydrating a DKP solution at about 430°C. On some plants, a 50% KOH solution is co-fed to a neutralization vessel with 85% H_3PO_4 with both feeds under mass flow control and at flow ratio to achieve a K/P of 2:1, which equates to a reaction pH of 9.

The DKP solution is then pumped through spray nozzles into a small 316 stainless steel cocurrent rotary kiln. The kiln is fitted with chains in the drying section to guard against the formation of potassium phosphate boulders and baffles and lifters in the calcining and cooling sections. The product emerges and is milled and sized before packing.

Potassium tripolyphosphate (KTPP) may be made on the same equipment. The main difference is that the neutralization is carried out to give a K/P ratio of about 1.70, a little above the theoretical 1.67 ratio to avoid the formation of insoluble metaphosphates, which form all too easily as compared to the sodium system.

Calcination then results in KTPP. The chemistry is analogous to STPP and is shown in the following equations:

$$3H_3PO_4 + 5KOH \rightarrow 2K_2HPO_4 + KH_2PO_4 + 5H_2O \quad (6.15)$$

$$2K_2HPO_4 + KH_2PO_4 \rightarrow K_5P_3O_{10} + 2H_2O \quad (6.16)$$

KTPP applications include detergents [47], as an additive to extreme pressure lubricants [48], and as a cleaning component [49].

6.5 LITHIUM PHOSPHATES

Until recently, lithium phosphates were of relatively little commercial interest. It appears that the only published study of the Li_2O–P_2O_5–H_2O system was that carried out by Rollet and Lauffenburger in 1934 [20]. Figure 6.6 depicts this system and is drawn from the data in Wendrow and Kobe [20]. $Li_3PO_4 \cdot 12H_2O$ is insoluble and Li_3PO_4 is only slightly soluble; Li_2HPO_4 does not appear in the phase diagram; LiH_2PO_4 is known but does not appear on the phase diagram [50]. Lithium phosphates can be made through the reaction of lithium carbonate with phosphoric acid [51].

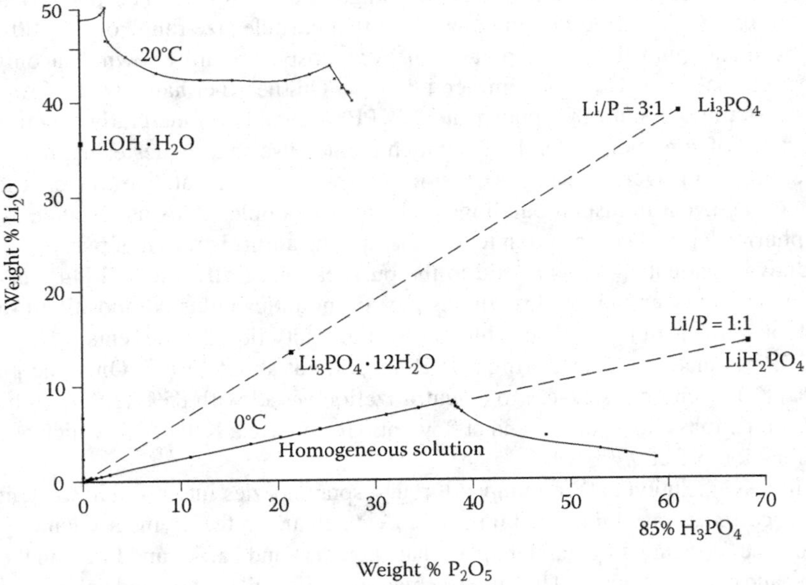

FIGURE 6.6 Li_2O–P_2O_5–H_2O system. (Adapted from Wendrow, B. and Kobe, K.A., The alkali orthophosphates. Phase equilibria in aqueous solution, *Chem. Rev.*, 54(6), 891–924, 1954. Copyright 1954 American Chemical Society.)

The sleepy backwater of lithium phosphates was transformed in the late 1990s with the invention of lithium iron phosphate (LFP) batteries. LFP, $LiFePO_4$, has several advantages over the first lithium ion batteries that use cobalt: cobalt is toxic and a rare and expensive metal, and lithium cobalt batteries are chemically and thermally unstable above 180°C. The whole field of lithium-based energy storage has been the subject of intense and competitive research and development in the last 20 years. A number of different manufacturing routes to LFP are taught in the patents. Broadly speaking, a lithium source, an iron source, and phosphoric acid undergo an aqueous reaction to form the LFP precursor. This precursor is then processed further to form the cathode of LFP batteries.

From an industrial phosphate's standpoint, the production of LFP is a matter for the battery manufacturers who are customers for phosphoric acid and in some cases iron phosphate.

6.6 MAGNESIUM PHOSPHATES

The chemistry of magnesium phosphates is similar to that of calcium phosphates. The phase diagram of the $MgO–P_2O_5–H_2O$ orthophosphate system is shown in Figure 6.7. At 25°C, the point of highest solubility is at 8.3% MgO and 33.1% P_2O_5 compared to 5.79% CaO and 24.1% P_2O_5 in the $CaO–P_2O_5–H_2O$ system showing the

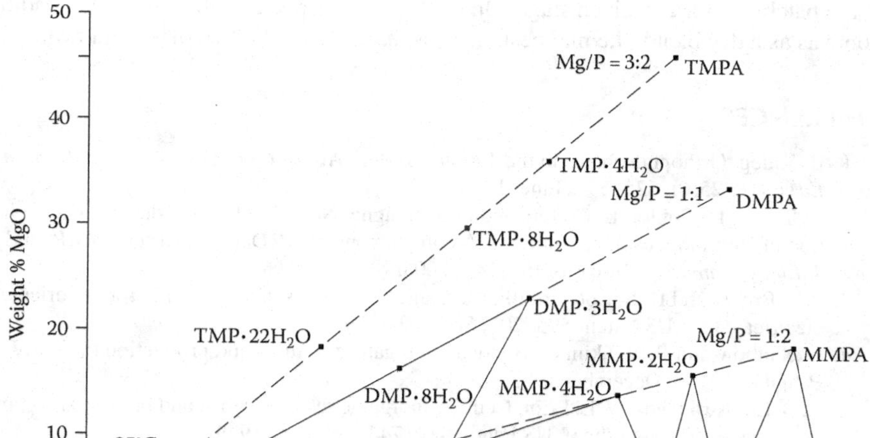

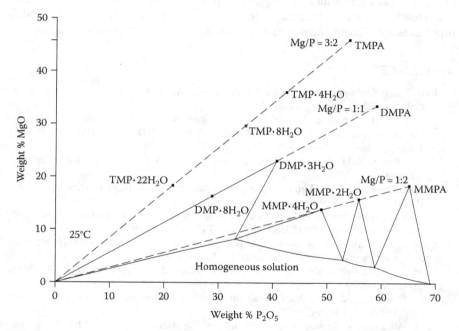

FIGURE 6.7 $MgO–P_2O_5–H_2O$ system. (Adapted from Van Wazer, J.R., *Phosphorus and Its Compounds*, vol. 1, Interscience, New York, 1958, figure 9.25, p. 539. This material is reproduced with permission of John Wiley & Sons, Inc.)

greater solubility of the magnesium system. The following magnesium phosphates are present in the solid phases in the temperature range 0°C–50°C:

$Mg(H_2PO_4)_2$ (MMPA)	$MgHPO_4$ (DMPA)	$Mg_3(PO_4)_2$ (TMPA)
$Mg(H_2PO_4)_2 \cdot 2H_2O$	$MgHPO_4 \cdot 3H_2O$	$Mg_3(PO_4)_2 \cdot 4H_2O$
$Mg(H_2PO_4)_2 \cdot 4H_2O$	$MgHPO_4 \cdot 7H_2O$	$Mg_3(PO_4)_2 \cdot 8H_2O$
		$Mg_3(PO_4)_2 \cdot 22H_2O$

Magnesium phosphates are used as castable refractories and rapid setting cements. In the manufacture of refractories, the magnesium orthophosphates are converted to metaphosphates. The orthophosphates are also used in dietary supplements. The most common use of both dimagnesium (DMP) and trimagnesium phosphates (TMP) until recently has been as an additive to dental grade DCP, which is discussed in Chapter 5. The demand for DMP and TMP has fallen in line with the demand for dental grade DCP due to the rise of silica gel–based toothpaste.

The production of both DMP and TMP is straightforward and only carried out at modest scale even when supporting historical dental grade DCP manufacture. Either magnesium oxide or hydroxide is reacted with phosphoric acid at an appropriate Mg/P ratio. The reaction slurry is then filtered. On some DCP plants where the magnesium phosphate was added at an early stage, an aliquot of magnesium filter cake was placed in the batch reactor at a given stage. On other plants, especially those where the addition was as a dry blend, the magnesium phosphate filter cake was dried and milled.

REFERENCES

1. R. Kniep, Orthophosphates in the Ternary System Al_2O_3–P_2O_5–H_2O, *Angew. Chem. Int. Ed. Engl.*, 25(6), 525–534, June, 1986.
2. G. Tricot, D. Coillot, E. Creton, and L. Montagne, New insights into the thermal evolution of aluminophosphate solutions: A complementary XRD and solid state NMR study, *J. Eur. Ceram Soc.*, 28(6), 1135–1141, 2008.
3. J. L. Brown and D. S. Loudermilk, Inorganic/organic insulating coating for nonoriented electrical steel, US Patent 5955201, September 1999.
4. C. Shubow and R. V. Skinner, Bonded aggregate structures and production thereof, US Patent 4419133, December 1983.
5. R. S. Lindstrom and W. L. Prior, Cellular inorganic resin cements, and process and compositions for forming them, US Patent 4141744, February 1979.
6. W. L. Prior and W. C. Sargeant, Composition and process for forming inorganic resins and resulting product, US Patent 4504555, March 1985.
7. M. S. Vukasovich and M. S. Vukasovich, Foamed ceramic, US Patent 3148996, September 1964.
8. L. Khandke and J. Perez, Process for producing aluminum phosphate, US Patent 2009/0016946, January 2009.
9. R. Adrian and R. von Schenk, Verfahren Zur Herstellung Von Monoaluminium-orthophosphat, German Patent 26227980, December 1977.
10. P. V. Salazar, Refractory compositions and method, US Patent 4432799, February 1984.
11. I. Iwami and A. Yoshino, Inorganic foam and preparation thereof, US Patent 4207113, June 1980.

12. H. Kondo and Y. Taketani, Anticorrosive pigment composition and coating compositions containing the same, US Patent 6010563, January 2000.
13. M. Okuda, An anticorrosive pigment composition and an anticorrosive coating composition containing the same, European Patent 0389653, October 1990.
14. K. Kamiya, M. Suzuki, Y. Nagai, M. Tsuhako, and M. Kobayashi, Adsorbent composition for malodorous gases, US Patent 5135904, August 1992.
15. C. H. Bartholomew and R. J. Farrauto, *Fundamentals of Industrial Catalytic Processes*, 2nd edn., John Wiley, New York, 2010.
16. R. M. Milton, Molecular sieve adsorbents, US Patent 2882243, April 1959.
17. G. Jakobi, C. H. Krauch, A. Sanner, and E. Schmadel, Washing process with acid monomer grafted cellulose fabric in bath to absorb metal cations hardening water, US Patent 3955920, May 1976.
18. T. W. Gault and H. K. Krummel, Detergent compositions, US Patent 3985669, October 1976.
19. S. T. Wilson, B. M. Lok, and E. M. Flanigen, Crystalline metallophosphate compositions, US Patent 4310440, January 1982.
20. B. Wendrow and K. A. Kobe, The alkali orthophosphates. Phase equilibria in aqueous solution, *Chem. Rev.*, 54(6), 891–924, December 1954.
21. I. Shibuya, H. Okano, Y. Kanaoka, and N. Takesue, Method for producing yeast with high glutamic acid content, US Patent 0223287, September 2011.
22. J. F. Conn, Yeast leavened bread improver agent and compositions and processes employing the same, US Patent 3404983, October 1968.
23. W. W. Hodgson, Method for incorporating bread improvers in continuous breadmaking, US Patent 3666486, May 1972.
24. E. W. Willems, B. Rambali, W. Vleeming, A. Opperhuizen, and J. G. C. van Amsterdam, Significance of ammonium compounds on nicotine exposure to cigarette smokers, *Food Chem. Toxicol.*, 44(5), 678–688, May, 2006.
25. T. M. Larson, T. B. Moring, and M. S. Ireland, Nicotine transfer process, US Patent 4215706, August 1980.
26. H. L. Thompson, P. Miller, P. M. Johnson, I. W. McCamy, and G. Hoffmeister, PILOT PLANTS. Diammonium phosphate, *Ind. Eng. Chem.*, 42(10), 2176–2182, October 1950.
27. B. G. Klugh and W. B. Seyfried, Process for the manufacture of diammonium phosphate, US Patent 1822040, September 1931.
28. A. W. Frazier, J. P. Smith, and J. R. Lehr, Fertilizer materials, characterization of some ammonium polyphosphates, *J. Agric. Food Chem.*, 13(4), 316–322, July, 1965.
29. P. G. Sears, Ammonium pyrophosphates, processes for preparing, and uses, US Patent 3645675, February 1972.
30. C. Y. Shen, Ammonium polyphosphates, US Patent 3397035, August 1968.
31. A. Maurer and H. Staendeke, Activated ammonium polyphosphate, a process for making it, and its use, US Patent 4515632, May 1985.
32. W. Becker, H. Neumann, and T. Staffel, Process for the preparation of ammonium polyphosphate, US Patent 5277887, January 1994.
33. N. L. Lacourse, J. M. Lenchin, and G. A. Zwiercan, Imitation cheese products containing high amylose starch as total caseinate replacement, US Patent 4695475, September 1987.
34. J. A. Marquez, M. G. Patel, D. Taplin, R. T. Testa, and M. J. Weinstein, Antibiotic W-10 complex, antibiotic 20561 and antibiotic 20562 as antifungal agents, US Patent 4232006, November 1980.
35. C. R. Mostoller, Antibacterial composition and methods thereof comprising a ternary builder mixture, US Patent 7354888, April 2008.
36. F. G. Bender, E. M. Frankovich, and C. Mostoller, Process for treating red meat, poultry and seafood to control bacterial contamination and/or growth, US Patent 5700507, December 1997.

37. S. J. H. Ernster, Method for manufacturing a whole milk cheese, US Patent 0004644, January 2013.
38. N. Friedman, A. Langham, S. Manor, G. Pipko, and A. Steiner, Process for the manufacture of monopotassium phosphate, US Patent 4836995, June 1989.
39. I. Alexander and Bar-on Menachem, Process for the manufacture of monopotassium phosphate, US Patent 4678649, July 1987.
40. J. Iannicelli and J. Pechtin, Process for the manufacture of monobasic potassium phosphate, US Patent 7601319, October 2009.
41. P. A. Deuritz, A. K. Legg, and D. C. W. Reid, Dairy product and process, US Patent 0213906, August 2012.
42. R. S. Kadan and J. G. M. Ziegler, Flan-type pudding, US Patent 4919958, April 1990.
43. A. Gupta, Parenteral administration of pyrophosphate for prevention or treatment of phosphate or pyrophosphate depletion, US Patent 8187467, May 2012.
44. S. Matsuhira, Process for purifying allyl alcohol, US Patent 4743707, May 1988.
45. C. Shao and Y. Stammer, Potassium pyrophosphate pet food palatability enhancers, US Patent 0170067, August 2005.
46. D. Anderson, Calculus dissolving dental composition and methods for using same, US Patent 8298516, October 2012.
47. B. Song and F. DeNome, Automatic dishwashing detergent compositions containing potassium tripolyphosphate formed by in-situ hydrolysis, US Patent 0069003, March 2006.
48. N. Shissler, D. Athans, D. Hunsicker, and J. S. Aubin, Extreme pressure additives and lubricants containing them, US Patent 0329687, December 2012.
49. U. Pegelow and P. Schmiedel, Cleaning agent components, US Patent 0045441, February 2008.
50. A. D. F. Toy, *The Chemistry of Phosphorus*, Pergamon Press, New York, 1975.
51. C. Keffer, A. D. Mighell, F. Mauer, H. E. Swanson, and S. Block, Crystal structure of twinned low-temperature lithium phosphate, *Inorg. Chem.*, 6(1), 119–125, January, 1967.
52. J. R. Van Wazer, *Phosphorus and Its Compounds*, vol. 1, Interscience, New York, 1958.

7 Sustainability, Safety, Health, and the Environment

7.1 INTRODUCTION

Question marks have continued to be raised over most aspects of phosphates including the sustainability of phosphate resources, the environmental impact of processing of phosphates, and the health issues associated with the use of phosphate products. Many of the greater concerns are with issues that do not directly involve the purification of phosphoric acid and the production and use of its derivative products; however, the whole phosphate industry is inextricably intertwined.

In this chapter, we will consider the sustainability of the raw material, the manufacturing processes, and the final product applications. The chapter concludes with a short section on recycling within the industrial phosphate industry.

7.2 PHOSPHATIC RESOURCES

Our survival depends on our food supply. Food supply is governed by the productivity of farming land. Crops are grown on farming land and processed for our consumption directly or indirectly as animal feed, which becomes our meat supply. Farming land productivity depends on ambient warmth, light, moisture, and soil nutrients. If soil nutrients are not replenished, crop yields reduce and ultimately fail and lead to famine. The return of animal and human manure to the land helps replenish soil nutrients, but even with this replenishment, crop failure and consequent famine were common in the middle ages. The first significant development in agriculture in the last 1000 years was crop rotation first implemented widely about 500 years ago. Different crops take up different proportions of nutrients from the soil, and by rotating them through different fields, including a fallow year when no crop was sown, nutrient levels could recover. As populations grew and became urbanized, a greater productivity was needed from the land. The second major development of agriculture was the discovery of artificial *manures*, phosphatic fertilizer made from animal bone and sulfuric acid, in the first half of the nineteenth century. Bone was rapidly consumed and briefly guano was used. Finally, in the second half of the nineteenth century, phosphate rock was used to make fertilizers. The usage grew, and processing technology became more and more industrialized up to the late 1940s. The 1950s saw the beginnings of even greater growth of fertilizer use going hand in hand with increasing economic prosperity. The 1960s saw the continued growth and the discovery and development of new phosphate mines. In the same period and downstream of fertilizers, sodium tripolyphosphate (STPP) market growth continued skyward, yet at the same time alarm bells were beginning to ring concerning excessive levels of nutrients in water courses. The causes of excess nutrient were, in part, laid at the door of STPP. The 1970s saw the growth of phosphate rock and fertilizer production to true global scale with prices strongly influenced by world economics like other commodities such as oil. In the 1980s, larger plants were built, smaller plants closed, companies merged, and eutrophication gained political traction.

Eutrophication is taken to mean a process where a body of water acquires an excessive amount of nutrients, by implication nitrates and phosphates, and responds by hosting a harmful level of algae growth. When the algae dies, the subsequent decomposition processes reduce oxygen levels to an extent that other organisms including fish die. In some cases, the water may become hazardous to humans.

The use of STPP as a detergent undoubtedly made a small contribution to the phosphate load in wastewater. The contribution from human and animal waste and from runoff (where rainfall falls on fields and carries away waste and fertilizer into water courses) from fields far outweighed the STPP contribution. Nevertheless, STPP made an easy political scapegoat, and legislators started to talk about bans. The European phosphate industry responded with the creation of Centre Europeén d'Etudes sur les Polyphosphates (CEEP) in 1971 as a joint research fund for the polyphosphate industry providing a forum for scientific research and the circulation of information concerning the impact of phosphates on the environment. CEEP is still active and produces bimonthly newsletters (http://www.ceep-phosphates.org). During the 1990s, various studies were commissioned including demonstration projects to take phosphate out of wastewater at sewage treatment plants [1]. In general, these schemes worked adequately but were not a technical slam dunk and required the municipal owners of sewage treatment plants to make substantial capital investments. It was, however, a step toward closing the phosphorus cycle. Legislators carried on banning STPP, and something was seen to be done.

In Europe, the bovine spongiform encephalopathy (BSE) crisis of the 1990s had generated thousands of tons of phosphatic ash, and trials were carried out to use this material to make both phosphorus and phosphoric acid. Both P_4 and wet process phosphoric acid (WPA) plants require a relatively consistent raw material supply, and the trials did throw up some challenges. As a result, this small closure of the phosphorus cycle has not taken off.

In 2007, Déry and Anderson [2] invigorated the topic by coining the phase *peak phosphorus* applying Hubbert's concept of *peak oil* to the phosphate world. Other academics took up the baton [3] and the hypothesis has proved so attractive that it has been taken up in the media and led to the foundation of GPRI, the Global Phosphates Research Initiative (www.phosphatefutures.net), a collaboration between independent research institutes in Europe, Australia, and North America and is based in the Institute for Sustainable Futures at the University of Technology, Sydney, Australia.

In 1949, M. King Hubbert, a geophysicist with the Shell oil company, proposed the concept of *peak oil* [4]. In 1956, he presented a paper to a meeting of the American Petroleum Institute predicting that petroleum production would peak in the United States about 1965 and gas production about 1970 [5]. His hypothesis behind this prediction was that the rate of production of oil and any other finite resource is exponential and follows a logistic distribution curve often referred to as a bell curve, which is given in the following:

$$y = \frac{e^{-t}}{(1+e^{-t})^2} \tag{7.1}$$

The y-axis represents production rate and the t-axis time. The area under the curve is the ultimately recoverable resource (URR) and the mean of the curve the date of the peak. Hubbert also applied this thinking to nuclear fuel, and it has been applied to many other resources. The hypothesis works in very limited cases where the resource reserves are known and are treated as equally readily extracted. In real life, if production rates start to drop and the resource is still in demand, the price goes up, which permits the exploration for more resource and the development and operation of technology to produce less readily extracted material.

Déry plotted the United States Geological Survey (USGS) data of the production of phosphate rock on the island of Nauru. This was an ideal data set as Nauru is mined out and its peak production was in 1973. Déry found that the data fitted the Hubbert model and then went on to attempt to fit the US production data and apply it to world phosphate rock production. Cordell et al. [3] reinforced the earlier paper. The year of peak phosphate production was declared as 1989. The USGS production data for 2004–2006 undermined this claim as each exceeded the 1989 figure (the kickup in production is clear in Figure 1.15 in Chapter 1). Other groups including Fertecon Research Centre Limited strongly contested the idea [6] suggesting that rather than a few decades of phosphate rock remaining, the resource may be good for 1000 years. Subsequently, Vaccari and Strigul examined the Hubbert curve extrapolation using USGS data to 2009 and concluded that "the 'bell-shaped curve' is not robust towards predicting the peak of phosphate rock production when the URR is not known *a priori*" [7]. Other workers have studied the Hubbert methodology itself, noting that Hubbert himself was not wedded to his methodology and suggesting that a *broad* methodology might be more fruitful than a *narrow* one [8].

There is a consensus that in practical terms phosphate is a finite resource but widely varying views on just how finite. The phosphate peak is moved further into the future in three ways:

1. New resources discovered and mined
   a. New resources continue to be discovered. In recent years, work has commenced on new resources in Canada, Chile, Kazakhstan, Peru, and Saudi Arabia.
   b. Other resources: as yet unrecovered are known but await financing.
2. More efficient use is made of the recovered resource
   a. Modern phosphate rock mining techniques are more efficient and waste less P_2O_5 than in previous times.
   b. Phosphate rock consumers, the vast majority being WPA producers, are always looking to maximize the efficiency of their production processes

and minimize P_2O_5 losses. Given that the producers are commercial entities, the extent of P_2O_5 efficiency is governed by its economics. As the price of phosphate rock rises so it becomes worthwhile spending more on its efficient use. In general, P_2O_5 efficiencies on WPA plants are around 95%.

c. Much of the agricultural land in developed economies is saturated with nutrients applied liberally over many years (which is a factor in eutrophication). These economies have environment agencies which enforce legislation that limits the amount of nutrients that may be added to the soil (to combat the effects of runoff that contributes to eutrophication). At the same time, farmers constantly look to make savings and improve yields—fertilizer is expensive so only productive quantities are bought. Meanwhile, fertilizer companies are constantly developing the composition and delivery mechanisms of fertilizers as they respond to the demands of the farmer, the environment (reinforced by the agency), and competitive pressure. Fertilizer companies have offered a wide range of formulations for decades; now, these formulations are tailored for the crop type and growth stage, the soil, the rainfall, and the weather and are designed with coatings that permit a variable release of nutrient over time. This modern design of fertilizer is known as *precision fertilizer technology* and is a growing area of study and application. Therefore, we can say that in general legislators, fertilizer companies and farmers are aligned in wanting to maximize the efficiency of phosphate usage.

d. Once harvested crops are directed immediately either for processing and human consumption or for animal feed that in turn becomes meat and poultry. The ratio of cost in the supermarket of meat and vegetables is about 10/1 and reflects closely the proportion of crops directed to animal feed for meat production. The efficiency of conversion of animal feed to animal weight is vital for the efficient use of phosphates. In the last 20 years the development and introduction of phytase into animal feeds have led to significant efficiency gains. For example, adding phytase to pig feed has increased take-up of phosphate by 50% and obviously reduced the phosphate lost in manure by the same percentage. Unfortunately, the loss of crops and spoilage of food remain a huge challenge globally.

e. Only about 8% of the world's phosphate resources end up as industrial phosphate products. Compared to the fertilizer stream, the industrial phosphate stream is efficient. In part, this is because industrial products often go through more production stages, which add greater cost that must be recovered in sales. The imperative for production and delivery efficiency is greater, and more effective means of loss reduction are financially justifiable.

3. Demand on the resource is reduced by closing the phosphorus cycle

a. Many workers have carried out substance flow analysis (SFA) to attempt to compile a mass balance for phosphorus at city, national, and global scale in recent years [9]. The purpose of this work is to assess and direct efforts at the most critical and potentially fruitful opportunities for closing the phosphorus cycle. Figure 7.1 shows a Sankey diagram of global

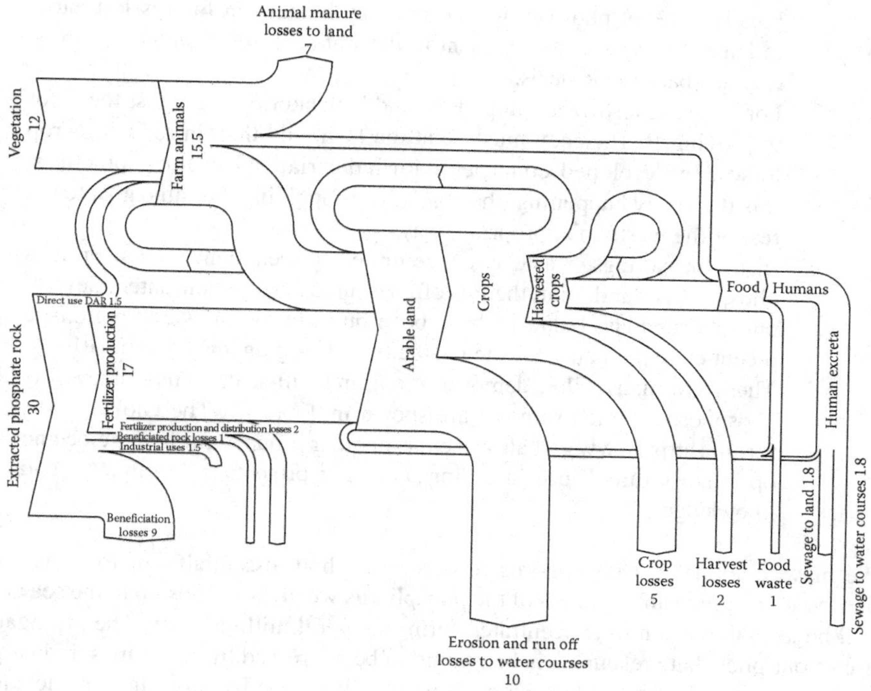

FIGURE 7.1 Global phosphorous of Sankey diagram.

phosphorus flows in millions of tons based on data from a number of sources. At best, the numbers are indicative; for some, there is very a wide range either side. Nevertheless, the amount of phosphorus in mined phosphate rock compared to that which ends up consumed by humans is striking. Also striking is that the amount of *natural P* entering the system through animal grazing as vegetation; of course, prior to manufactured fertilizers, this was the predominant source.

Mine losses in beneficiating the ore coming out of the ground are huge; 30% losses are used here, which may be considered an average but depends greatly on the quality of rock that is available. Losses in the manufacture of fertilizers and industrial phosphates are relatively low. The biggest losses are at the farm: runoff from the soil, crop loss, harvest loss, and poor recycle rates of animal manure account for 50% of the total *P* losses. It is said that globally enough food is grown for the world, yet the richer countries consume more per capita than poorer countries and together with large losses in storage and distribution many still go hungry. Losses in the food commodities—food ready for consumption—are about one-third: in developed markets, losses due to produce going past *sell by* date is a real issue. Pressure on landfill is leading to growth in anaerobic digesters, which allow food waste to rot down, recovering methane, and produce compost that recycles the *P* value.

Finally, 90% of phosphorus consumed by humans in food is lost either to landfill or via waste treatment to the watercourse. A small amount is recycled back to the fields.

b. For farmers, fertilizer companies, and legislators, the biggest focus for improving *P* efficiency must be at the farm and then in reducing crop losses. In developed countries with industrialized farming practices, this is already happening; the challenge is applying the same goal to the rest of the world in appropriate ways.
c. In urban settings, waste is concentrated at treatment plants. From a phosphorus standpoint, the *P* is often diluted and contaminated. Several phosphorus mass balances have been built up for cities and regions, a recent example being for the megacities of Beijing and Tianjin [10].
d. There are many pilot, demonstration, and full-scale plants recovering phosphorus; some examples are shown in Table 7.1. The Global TraPs forum (http://www.globaltraps.ch) is another group seeking to guide and optimize future *P* use including assessing projects, performance, and knowledge gaps.

We may conclude that phosphorus resources are both essential and, for practical purposes, nonrenewable—much of the phosphorus we process ends up in the sea and will be available again in concentrated form in several million years. The profligate use of our phosphate resources is beginning to be addressed by academics, industry, and legislators. The urgency of the problem remains the subject of vigorous debate, the outcome of which is critical. If the problem is overstated, unnecessary and wasteful resources will be applied to conserving this resource; if understated, global food supply will be compromised.

It is likely that industrial phosphate producers will continue to make or purchase WPA made from phosphate rock and sulfuric acid for the foreseeable future because the demand is so small compared to the fertilizer demand. However, several phosphate salts plants import PWA over considerable distances, and there may be cases where it would make both technical and financial senses to reprocess locally recovered phosphate back to sales grade products.

7.3 MANUFACTURING PROCESSES

Processes to make PWA and phosphate derivatives have an environmental impact but one far less significant than WPA plants. The three main environmental issues for WPA plants are the gypsum stack, which is the safe storage (or sometimes disposal to sea) of the calcium sulfate by-product of the WPA process and the metal impurities therein, which include uranium and other radionuclides and heavy metals; fluoride emissions; and water management. There is a large body of literature on these subjects of which Becker is a good start [11].

Most plants require an environmental permit to operate. Typically during the design and planning phases of a project to build a new plant, the chemical engineers and site managers will prepare an application for a permit and present this to the local environment agency. The key components of the application are the location

TABLE 7.1
Examples of *P*-Recovery from Waste

Country	Location	Type	Description
Japan	Senboku	MSW	*P*-recovery from wastewater by calcium phosphate precipitation since 2009
Japan	Fukuoka	MSW	*P*-recovery from sludge liquor by struvite precipitation since 1999
Japan	Gifu	MSW	*P*-recovery from sewage sludge incineration ash forming calcium phosphates since 2010
Japan	Tokyo	Ind	*P*-recycling from industrial wastewater by calcium phosphate precipitation 1998
Japan	Hofu	Ind	*P*-recycling from industrial wastewater by calcium phosphate precipitation 2007
United Kingdom	Slough	MSW	*P*-recovery from sludge liquor by struvite precipitation—Ostara Pearl process
United Kingdom	Various	Ind	Chicken litter incinerator—ash sold as fertilizer
Germany	Various	MSW	*P*-recovery from sludge liquor by struvite precipitation—PCS AirPrex process, LysoGest
Germany	Nürnberg	MSW	*P*-recovery from sewage sludge incineration ash—Mephrec SUN Ignitec process
Germany	Neuburg	MSW	*P*-recovery from industrial wastewater by calcium phosphate precipitation—P-RoC process (phosphorus recovery from wastewater by crystallization of calcium phosphate compounds)
Netherlands		Ind	*P*-recovery from dairy and potato processing liquor by struvite precipitation—NuReSys Biostru process
Netherlands	Wieden-Echten	MSW	*P*-recovery from sludge liquor by struvite precipitation—PCS AirPrex process
Russia	Leningrad	Ind	Chicken litter incinerator—ash sold as fertilizer
Switzerland	Zürich	MSW	*P*-recovery from sewage sludge incineration ash

Source: CEEP, Scope Newsletter—ScopeNewsletter91.pdf [Online], Available: http://www.ceep-phosphates.org/Files/Newsletter/ScopeNewsletter91.pdf, Accessed April 13, 2013.
Note: MSW, municipal sewage works; Ind, industrial.

and description of the process; its raw materials, production rate, products, and prescribed substances in these three groups; emissions to air, water, and solid wastes; and the location of release points. The application also includes a description of techniques to prevent release and monitoring provisions of these releases. The operator is then obliged to report actual emissions to the agency annually. Often, the agency will require the operator to improve the process, thereby reducing emissions. These applications become public records and as such are potentially available to industry competitors. Sometimes, by agreement with the authorities, some elements of the application are redacted for commercial confidentiality reasons although the emissions are always left visible.

7.3.1 Environmental Aspects of PWA Processes

Clearly, the environmental impact of a PWA process depends on the extent of that process. A process that imports a 28% P_2O_5 WPA that requires a full pretreatment from desulfation, fluoride reduction, heavy metal removal, and carbon treatment to concentration will have more emissions than a process importing a 59% P_2O_5 WPA ready for solvent extraction.

Taking the most comprehensive case, the emissions are as follows:

- Desulfation—sulfate levels are reduced by the addition of a calcium (or barium) source followed by solid separation. Separation is carried out either by settling and centrifuging or filtration; or directly by filtration with water washing. In the filtration step, the solids are predominantly calcium sulfate with up to 5% acid and up to 30% water depending on filter. On those PWA plants located adjacent to a WPA plant, it is sometimes feasible to return this material to the WPA reactor or transfer it to the gypsum stack. In the United States, mining activities come under the Resource Conservation and Recovery Act (RCRA) and the Bevill Amendment to this act. WPA plants process mined rock and are permitted to make a gypsum stack so that the gypsum by-product of the mining process is safely stored. The gypsum arising from desulfation is no different from that coming from the WPA plant filter; however, it would be a matter for discussion with the authorities whether those solids could be added to the stack. The quantity of solids from desulfation is readily estimated by assuming a reduction from 2% to 0.5% sulfate in the WPA, an appropriate estimate of water content and 5% by weight acid content.
- Fluoride reduction—fluoride can be reduced either during concentration or through precipitation by adding sodium as soda ash or sodium hydroxide. Fluoride removal by precipitation adds sodium to the process that may not all be removed again by precipitation; however, this route requires the Na/P ratio to be raised to 0.15 and the acid then to be filtered of solid sodium fluorosilicate crystals. Potentially, the fluorosilicate is a saleable product rather than an effluent; either way, its quantity is estimated from the fluoride reduction from typically 1.5% (of the incoming 28% P_2O_5 acid) to 0.1%. If the waste from desulfation cannot be sent to the gypsum stack and if there is no market for the fluorosilicate, these processes are potentially combined, also combining the waste streams. Fluoride removed during concentration is condensed as fluorosilicic acid, also a potentially saleable product, but often neutralized and sold or disposed of as the fluorosilicate.
- Heavy metal removal—the heavy metal sulfides filtered out of this process step are toxic, hazardous, and always directed to a safe, registered hazardous waste disposal. The predominant component is arsenic sulfide; other coprecipitates are cadmium, copper, and molybdenum sulfides. Again, the quantities are estimable from the impurity levels in the WPA together with

an allowance for acid in the filtrate. The unit scrubber, intended to knock out any hydrogen sulfide produced in the reaction, is a registered source point and is commonly monitored and reported.

- Carbon treatment—the organic content of WPA is reduced by passing the acid through activated carbon. Usually the active carbon is in the form of pellets held in three columns with two on line and one in reserve or undergoing regeneration. Regeneration takes place by pumping sodium hydroxide through the column. This liquor eventually passes to drain. In the context of a plant or site, the flow is small and has the effect of raising the pH of what are often slightly acidic drain flows.
- Concentration—depending upon the configuration of the concentrator, there is always an outflow of fluoride captured either as fluorosilicic acid or effectively as a fluoride content in water used for condensation. The destination of fluoride must be clearly set out as even quite low emissions to air are damaging to trees, plants, and buildings.
- Solvent extraction—the most obvious concern of designer and operator is to keep the solvent in the plant. Solvent losses from an environmental standpoint result from solvent solubility in the acid and fugitive leaks. All solvents used in PWA processes have some solubility in acid; as the acid is removed from the solvent extraction system, it conveys some solvent away with it. Most of this solvent is stripped and returned to the solvent system; however, if the solvent stripping step is not working properly, then the solvent can be conveyed to locations where it may pass to atmosphere in an uncontrolled way. The solvent system, including the solvent storage, is usually blanketed with nitrogen not only for flammability reasons but also to suppress solvent fume diffusing to atmosphere; the blankets are sealed with seal pots and vent by way of chilled stacks (depending on solvent) to condense solvent vapor. Attention is paid to the integrity of flanged pipe joints, pump, and control valve seals to ensure that fugitive leaks are minimized.
- Concentration, defluorination, and color removal—the main arising from concentration and defluorination is slightly acidic condensate with raised fluoride content. Final color removal will generate no effluent if hydrogen peroxide is used and some from regeneration if activated carbon is used. Process effluents that are largely water pass through a holding stage to check pH and the absence of organics, may have pH adjustment, and are then released appropriately.
- Rainwater is assumed to contact the process if it falls within the battery limits of the plant and is treated like process condensate and therefore checked for pH and contamination before release.
- Spillages are contained by sealed flooring and bunding and where possible returned to the appropriate part of the plant.
- Maintenance and cleaning—the pretreatment steps, solvent extraction, and solvent stripping do require periodic cleaning to remove scale. Generally, this is collected and disposed of as waste and an allowance made as part of the environmental permit.

7.3.2 Environmental Aspects of Phosphate Salts Plants

Most phosphate salts plants comprise a reaction/neutralization section, a drying section, for higher sodium and potassium salts a calcining section, and a milling and sizing section. There are three areas of concern: fluorides, dust, and oxides of nitrogen. Fluoride release is not usually a problem for food grade products as the fluoride content of the feed acid is low. For technical products, especially those like STPP that require high-temperature calcining, hydrogen fluoride is released (unless it was defluorinated before use) and must be controlled by the plant scrubbing system. The drying, calcining, milling, sizing, and transfer of powders inevitably generate dust that must be controlled by cyclones, filters, and scrubbers. Each vent is registered, monitored, and reported. Typically, the authorities will require a site to meet an aggregate standard for dust emission. Oxides of nitrogen potential arise from air heaters. Often air heaters are fired by natural gas, although oil is sometimes used and with indirect heaters, coal. At very high temperatures in the air heaters can give rise to nitrogen oxides and should be avoided. The drainage on a phosphate salts plant is treated in the same way as a PWA plant in that washings from the neutralization section and plant clean down flow to drain and must therefore be controlled and treated appropriately. Washing losses are minimized by careful housekeeping on salts plants and sweeping and vacuuming powder as much as possible before washing.

7.3.3 Safety and Health Aspects of PWA and Phosphate Salts Plants

Beyond the normal safety and health hazards of most chemical manufacturing plants (temperature, pressure, nitrogen, natural gas, acids, alkalis, etc.), both PWA and phosphate salts plants are mostly relatively low-hazard plants. Of specific safety concern is solvent management as most solvents used on PWA plants are at least to some degree flammable, and one PWA plant has burnt down. The chemical engineering design is aimed at minimizing solvent inventory and where necessary applying nitrogen blankets. Layout and structural design must account for the risk of fire, and instrument and electrical designers must specify equipment to an appropriate electrical classification.

From both a health and safety hazard standpoint, the heavy metal removal section is most hazardous. Great care is taken with both the control of hydrogen sulfide gas arising from the sulfide reactions and the handling of the arsenic and other sulfide filter cakes. Not only is hydrogen sulfide toxic, but it is potentially explosive. Safe working procedures are essential for both production and maintenance activities.

7.4 PHOSPHATE PRODUCT SAFETY

7.4.1 Food Phosphates

PWA and aluminum, ammonium, calcium, lithium, magnesium, potassium, and sodium phosphates are commercially available as food grade products. Under the US Code of Federal Regulations (CFR), these ortho- and higher phosphates have generally recognized as safe (GRAS) status when used in accordance with good manufacturing practice and at levels not exceeding that required for the intended, functional effect. The recommended daily allowance of phosphates as phosphorus in the human diet varies with age and for adults is 700 mg/day [12] and up to 4000 mg.

All the previously mentioned products are used as food additives. In recent years, the use of sodium phosphates in baking and other applications has been falling because of a need to reduce sodium levels in the human diet. Both calcium and potassium phosphates are in use as alternatives.

Concern has been expressed over the last 40 years that excess dietary phosphorus exacerbates the onset and severity of osteoporosis. The acid-ash hypothesis posits that increased excretion of *acidic* ions derived from the diet, such as phosphate, contributes to net acidic ion excretion, urine calcium excretion, demineralization of the bone, and osteoporosis [13]. The conclusion of the Fenton et al. meta-analysis was that there is no evidence that higher phosphate intakes are detrimental to bone health and that the dietary advice that dairy, meats, and grains are detrimental to bone health due to *acidic* phosphate content needs reassessment.

Concern has also been expressed that food additives might be linked to hyperkinetic behavior in children. The main concern in the last 40 years has been the effect of food colorants such as tartrazine (E102) [14,15] and others collectively known as the *Southampton six*. Investigators [16] have concluded that there is no proof that phosphate ingested by mouth triggered behavioral disturbance or caused it to persist in children.

7.4.2 Detergents

> It is a received truth that the use of phosphates as a builder in detergent formulations is damaging to the environment and that the damage caused by nutrient enrichment of the water environment—the phenomenon known as eutrophication—can be avoided by replacing phosphates with other compounds, notably zeolite A and its co-builder, polycarboxylic acid or PCA [17].

CEEP commissioned two studies that were published in 1994. The first study was carried out by the Environmental Engineering section of Imperial College of Science, Technology and Medicine and considered "the Environmental and Economic Impact of Key Detergent Builder Systems in the European Union" [18]. The second was a life cycle study and was the second in a series of three studies to evaluate the environmental impact of phosphates and zeolite A/PCA as alternative builders in UK laundry detergent formulations and was carried out by Landbank Environmental Research and Consulting [17].

The Imperial study made a number of conclusions:

- Any detergent builder system must be recognized as making a major positive contribution to the environment by enabling textiles to be recycled.
- The market recognized that zeolite-only formulations have inferior performance compared to either STPP or zeolite/PCA detergents.
- The environmental impact of detergent builders in a global context is very small.
- The issue of eutrophication is independent of phosphorus arising from STPP.

The 1994 Landbank study established an inventory for the STPP and zeolite A/PCA systems that identified the inputs and outputs associated with all stages of the life cycles of the two systems. The study goes right back to the phosphate and bauxite mines and forward to wastewater treatment. Emissions arising from processing, electricity, heat, and transport are considered and tabulated. The study section headings include the following:

- STPP production—from phosphate rock extraction and beneficiation through WPA and PWA production to STPP
- Zeolite A production—including hydrated alumina production considering bauxite mining and the Bayer process and subsequent zeolite process
- PCA production—commencing with the manufacture of acrylic acid, maleic acid, and finally PCA production
- Inorganic chemical production—including sulfuric acid, sodium hydroxide, soda ash, and lime production
- Raw material extraction and processing—including coal, oil, natural gas, and sulfur
- Energy conversion—both to electricity and heat
- Transport—considering both land- and water-based transport and the transport distances of the different builders
- The washing process and disposal as it relates to phosphates, zeolites, and PCA
- Washing performance

The study made a number of recommendations, which included the following:

- Phosphate recovery from municipal sewage works.
- The industry should formulate and offer compact phosphate-based detergents as a way of reducing packaging and transport costs. This was done with detergent tablets made from a specially processed STPP grade.
- That those responsible for promoting *green* claims for phosphate-free detergents reconsider because it took away the public pressure for phosphate recovery plants, which represent the proper response to eutrophication.

In 1995, the third part of the study was published as "The Swedish Phosphate Report – a Delphi study to compare the environmental impacts of detergent builders in Nordic countries and the implications for the sustainable management of freshwater resources in Europe" [19]. The report concluded that eutrophication could only be controlled when the nutrients from all sources in the wastewaters are removed by chemical or biological means and that phosphate bans had no influence on the course of freshwater eutrophication. The study went on to state that in Sweden, Switzerland, and the Great Lakes region of North America, eutrophication had been controlled in many cases and the lakes and rivers progressively restored to health. The study was positive about the incineration of wastewater sludge and the use of pelletized ash as a fertilizer. In one example, incineration reduced 250,000 tons of liquid digested sewage to 10,000 tons of dry pellets. Incineration deals with concerns about the bioactivity of sewage sludge and the potential for harm when it is used untreated on land for crops. Incineration does not deal with heavy metals in the sewage; the study notes that with a systematic approach to sources large and small, the problem can be stopped at source. In the context of detergents, it was noted that zeolites offer no potential for recycle and the fate of polycarboxylates is uncertain.

Over the life cycle of STPP and zeolite A/PCA, the environmental impact is very similar. Comparing the washing performance, studies show that STPP is better and that zeolite/PCA also requires soda ash to achieve similar results (when the phosphate ban was implemented in Switzerland, the consumption of detergents rose). The potential for phosphate recovery and recycling is high. On these bases, it would seem that the removal of phosphate bans, together with phosphate recycling at wastewater treatment works, would lead to an overall environmental improvement and contribute to closing a small part of the phosphorus cycle.

7.5 RECYCLING WITH THE INDUSTRIAL PHOSPHATE INDUSTRY

It is hopefully clear in previous chapters that industrial phosphate plants, whether PWA or phosphates salts, take care to maximize P_2O_5 efficiency and incorporate internal means of recycling. This is done for financial reasons; the more purchased P_2O_5 turned into the saleable product and the less P_2O_5 wasted that must be treated, the better the financial return. Happily, the environment is a co-beneficiary, along with stock holders, of this approach. In general, once the P_2O_5 product is sold, recycling and reuse is somewhat patchy. In some specific cases, for example, the sale of etchant formulations to the semiconductor industry, the sales offer includes technical backup and used product return as well as the product itself. In this case, the supplier may clean the used etchant and resell it or pass the used etchant on to other markets. In other cases, for example, the metal dipping market, the customer seeks outlets for used product, which often means a fertilizer producer incorporating the stream into the feed acid. In terms of the global phosphorus cycle, the recycling of acid used for etching and metal treatment in general is tiny, yet it is still large enough to be a very worthwhile business and is not to date fully exploited.

P_2O_5 processed into detergents, toothpastes, and food additives and most phosphate salts end up in sewage treatment plants; there is little opportunity for reuse or recycling upstream of this stage but as discussed earlier every opportunity once there.

REFERENCES

1. S. Brett, J. Guy, G. K. Morse, and J. N. Lester, *Phosphorus Removal and Recovery Technologies*, Selper Ltd., London, U.K., 1997.
2. P. Déry and B. Anderson, The Oil Drum Peak Phosphorus, August 2007. [Online]. Available: http://www.theoildrum.com/node/2882 (accessed April 13, 2013).
3. D. Cordell, J.-O. Drangert, and S. White, The story of phosphorus: Global food security and food for thought, *Glob. Environ. Change*, 19(2), 292–305, May 2009.
4. M. K. Hubbert, Energy from fossil fuels, *Science*, 109(2823), 103–109, February 1949.
5. M. K. Hubbert, Nuclear energy and the fossil fuels, Presented at the *Spring Meeting of the Southern District Division of Petroleum American Petroleum Institute*, San Antonio, TX, 1956.
6. M. Mew, Future phosphate rock production—Peak or plateau? March-04-2011. [Online]. Available: http://www.fertecon-frc.info/page15.htm (accessed March 15, 2013).

7 D. A. Vaccari and N. Strigul, Extrapolating phosphorus production to estimate resource reserves, *Chemosphere*, 84(6), 792–797, August 2011.
8. A. R. Brandt, Testing hubbert, *Energ. Policy*, 35(5), 3074–3088, May 2007.
9. D. Cordell, T.-S. S. Neset, and T. Prior, The phosphorus mass balance: Identifying "hot-spots" in the food system as a roadmap to phosphorus security, *Curr. Opin. Biotechnol.*, 23(6), 839–845, December 2012.
10. M. Qiao, Y.-M. Zheng, and Y.-G. Zhu, Material flow analysis of phosphorus through food consumption in two megacities in northern China, *Chemosphere*, 84(6), 773–778, August 2011.
11. P. Becker, *Phosphates and Phosphoric Acid: Raw Materials, Technology, and Economics of the Wet Process*, 2nd edn. rev. and expanded, Marcel Dekker, New York, 1989.
12. Dietary Guidelines for Americans, 2010. [Online]. Available: http://www.cnpp.usda.gov/dgas2010-policydocument.htm (accessed April 13, 2013).
13. T. R. Fenton, A. W. Lyon, M. Eliasziw, S. C. Tough, and D. A. Hanley, Phosphate decreases urine calcium and increases calcium balance: A meta-analysis of the osteoporosis acid-ash diet hypothesis, *Nutr. J.*, 8(1), 41, 2009.
14. D. McCann, A. Barrett, A. Cooper, D. Crumpler, L. Dalen, K. Grimshaw, E. Kitchin et al., Food additives and hyperactive behaviour in 3-year-old and 8/9-year-old children in the community: A randomised, double-blinded, placebo-controlled trial, *The Lancet*, 370(9598), 1560–1567, November 2007.
15. European Food Safety Authority annual report 2009.pdf. [Online]. Available http://www.efsa.europa.eu/fr/ar09/docs/ar09en.pdf (accessed April 13, 2013).
16. B. Walther, E. Dieterich, and J. Spranger, Does dietary phosphate change the neurophysiologic functions and behavioral signs of hyperkinetic and impulsive children? *Monatsschr. Kinderheilkd.*, 128(5), 382–385, May 1980.
17. B. Wilson and B. Jones, *The Phosphate Report—A Life Cycle Study to Evaluate the Environmental Impact of Phosphates and Zeolite A-PCA as Alternative Builders in UK Laundry Detergent Formulations*, Landbank Environmental Research and Consulting, London, U.K., January 1994.
18. G. K. Morse, J. N. Lester, and R. Perry, *Environmental and Economic Impact of Key Detergent Builder Systems in the European Union*. Selper Ltd, London, U.K., 1994.
19. B. Wilson and B. Jones, *The Swedish Phosphate Report*, Landbank Environmental Research and Consulting, London, U.K., 1995.
20. CEEP, Scope Newsletter—ScopeNewsletter91.pdf. [Online]. Available: http://www.ceep-phosphates.org/Files/Newsletter/ScopeNewsletter91.pdf (accessed April 13, 2013).

8 Commissioning

8.1 INTRODUCTION

In the world of chemical production plants, commissioning and start-up are slightly loose terms that are often used interchangeably. For the purposes of this chapter, commissioning is taken to mean bringing new plant and equipment to nameplate operation (and already an exception is made in that occasionally a plant is declared commissioned when even after the best endeavors of the commissioning team and the support of both operations and projects, the plant does not reach nameplate). Start-up is bringing a plant that is shut down back to operation. Start-up often occurs if because of a minor breakdown the operating plant is put on hold (the feeds are stopped, the intermediate storages allowed to fill, etc.), while the breakdown is fixed. This case is relatively straightforward. Start-up after a planned maintenance shutdown of 2 weeks is more challenging as intermediate storages may have been emptied. Start-up after a major maintenance shutdown of several weeks, sometimes referred to as a turnaround, where both maintenance and major modification have taken place, starts to look very like commissioning.

Commissioning is a clear step in the project life cycle as shown in Figure 8.1 and takes place after procurement and construction. At this point, the business is fully committed financially, the money is spent, and the return is awaited. The business can only begin to make a return on the investment if the plant starts producing saleable product. As well as delivering a return to stock holders, the business executive team must ensure it does so without harm to the public, employees, or the environment; all these imperatives tend to be directed toward the commissioning manager. Commissioning follows on the heels of construction, and it is normal for the commissioning team to find defects that need urgent correction. As a consequence, it is normal for a small number of constructors to be on plant as the commissioning team move into water trials and processing. Constructors construct safely but generally are not experienced in working on a live chemical plant. As well as rectifying physical defects, it is normal for the commissioning team to carry out modifications both to the plant and the controlling software. It is hopefully obvious therefore that the commissioning stage of a project is a pressurized environment with a much higher risk than normal of safety and environmental incidents. It is an excellent environment for the young engineer to gain valuable experience. A successful commissioning therefore requires a good, experienced commissioning manager supported by a good team.

The time taken for commissioning varies enormously; as a rule of thumb, one might say 3 weeks to 3 years. The shortest commissioning periods tend to happen on plants operating a familiar process, with familiar feed materials, by a dedicated and experienced commissioning team, and with a good project team and supportive site providing backup.

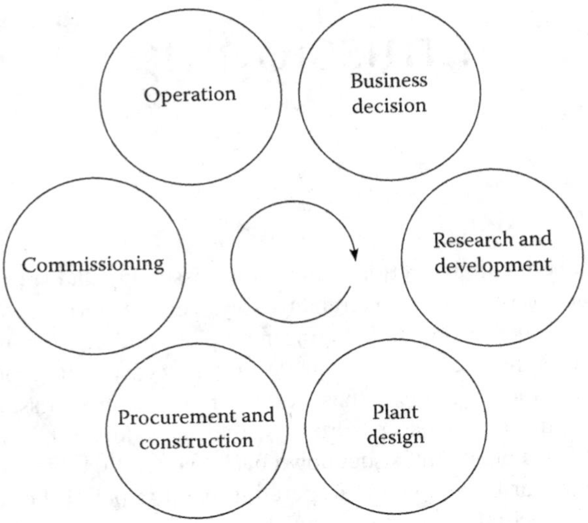

FIGURE 8.1 Commissioning in the project life cycle.

There is not much literature in the public domain on commissioning. Operating company engineering departments and production sites usually have their own commissioning procedures and occasionally produce commissioning reports. The larger contractors that carry out commissioning have their own procedures as well. Horsley [1] and Killcross [2] provide useful generic guides to this topic. So the first part of this chapter is concerned with the important general principles of commissioning a chemical plant; the second part deals more specifically with the approach to commissioning two different types of plant—the purified phosphoric acid plant and the derivative phosphate salts plant.

8.2 COMMISSIONING IN GENERAL

Commissioning follows a series of steps, and it is normal that there is clear documentation that each step is complete. In the following list in Table 8.1, both water and chemical trials are included. Water trials describe an operation where a plant item or system is operated with water rather than the intended chemical. This allows checks on the operation of most plant items and their control with a lower hazard material. A chemical trial is where the intended chemical is introduced into, for example, a storage vessel and recirculated—no reaction and therefore no processing take place. Both water and chemical trials are increasingly testing the robustness of the design and construction of the plant while minimizing the hazard in doing so. Water and chemical trials are not always carried out on every item or system, for example, using water on equipment that must subsequently be dry or water-free can add unnecessary work. Introducing a hazard or introducing a high-hazard chemical onto the plant too early might not be beneficial. Ideally, the commissioning manager is appointed very early in the project life cycle, for example, during flow sheet development, so that the design team incorporates how the plant will be commissioned into the design.

TABLE 8.1
Commissioning through the Project Stages

Project Stage	Action
Preliminary design	Appoint commissioning manager.
	Plant designated into commissioning systems.
Hazard studies	Commissioning manager or representative present.
Procurement	Commissioning representative part of team inspecting critical equipment.
Construction (early)	Commissioning team develops operating instructions from drafts written by design (process) engineers, review P+IDs, and write training courses.
Construction (late)	Deliver training courses to supervisors, operators, maintenance and laboratory personnel.
	Control software checked on simulators.
	Commissioning engineers familiarize themselves with the systems allocated to them and liaise closely with construction supervisors as these systems are completed to identify snags early.
Precommissioning (no utilities)	As construction formally offers systems for inspection, the commissioning team leads P+ID checks and punch listing (also known as snagging, where plant, equipment, piping, and instruments are checked against design—anything incorrect is written on the punch list) with operators and maintenance personnel that aids plant familiarization.
	The object is to verify that what has been designed has been installed (this cannot be taken for granted especially when the project team are in a different region or continent from the plant).
Precommissioning (with utilities)	When initial punch listing is complete and the construction team accepts the punch list, the construction manager and the commissioning manager may formally agree to the introduction of utilities, the first of which is electricity. Making the plant, or a plant system, live is essential to perform motor and instrument checks but introduces a hazard. A formal or informal permit to work system is introduced so that no one is inadvertently working on live equipment.
	The second utility is usually water as this is needed for flushing, washing, and pressure testing.
	Care is required, especially on food grade plants that the test/flushing water is clean. Using dirty water just adds to the cleaning time before chemicals are introduced.
	In hazard terms, it is easy to take water for granted; this is a mistake. Water can maim and kill and when uncontrolled be very damaging to plant and equipment.
Construction to commissioning handover	When the construction and commissioning manager mutually agree that construction is complete (with agreed exceptions), the plant is formally handed over to commissioning. Both parties usually sign a form with punch lists attached. This package also includes any pressure and other quality inspections performed on plant and piping and data books for plant and equipment. This milestone is often important contractually and may trigger various payments.

(*continued*)

TABLE 8.1 (continued)
Commissioning through the Project Stages

Project Stage	Action
	As part of the handover, some companies have a formal safety procedure whereby the commissioning manager and the site operations manager carry out a safety inspection of the plant, review the documentation, and sign off the handover.
Water trials	Water trials allow the commissioning team an opportunity to simulate plant operation without chemicals. During water trials, joint leaks may be found that were not observed during pipe pressure tests; temperature, pressure, and level instruments and control valves can be checked to a greater degree than when dry as well as pump and heat exchanger performance. During the water trials, the operators are further trained in the control of the plant. Trips and alarms and sequences are also tested as this stage of commissioning progresses. As with the punch listing stage, the plant undergoes water trials in a logical manner based on systems and usually working from front to back. Because there is no chemical contamination fixing anything that breaks and carrying out minor modifications is much less complicated than with chemicals in the plant. There is sometimes a temptation to waste time making minor improvements and put off moving to chemical trials. During water trials, any remaining utilities not already commissioned are brought on line. Three that introduce new hazards are steam, nitrogen, and compressed air. At the conclusion of water trials, the commissioning manager may formally document that they are complete and the plant is ready to receive chemicals. The site director is always informed at this stage, and the local environmental agency may also be told that the plant is about to accept chemicals (depending on local legislation and practice).
Chemical trials	Chemical trials can be as simple as importing feed materials into day tanks and circulating them. The important point is that the plant is treated as live and all the appropriate management procedures are in place for both production and maintenance. The plant is still at this stage under the control of the commissioning manager. With the raw materials in place, the plant is now ready to make the product. Two approaches are possible: either the raw materials are reacted and further processed and thus the plant gradually fills up or the plant imports product into intermediate storages. In the latter case, the plant usually attains a steady state very quickly and then allows new feed materials to move through it. Taking the former more obvious approach, it can on occasion be difficult to achieve stability.
Production	The initial rate decision is very plant-specific; one concern is that the first product may be out of specification and therefore the less made the better. Another is that too low a rate might not be controllable or supporting equipment (e.g., a steam boiler or other utility) may not yet be fully shaken down. Therefore, the start-up rate is often set at 50%–75% of nameplate.

TABLE 8.1 (continued)
Commissioning through the Project Stages

Project Stage	Action
	If the operation is stable, the rates are pushed up until a bottleneck is found. Ideally, the bottleneck is at or above nameplate, and the only remaining challenge is a qualifying run. Depending on the plant configuration and intermediate storages, it is often possible to identify more than one bottleneck and so begin to identify a development program for the plant. The final stage of commissioning is to perform the qualifying or guarantee run. This usually requires the commissioning team to demonstrate that the plant is capable of achieving design rates consistently for an agreed period of time; 72 h is typical. As well as achieving rates, the product must achieve specification and is tested. Depending on the context, the qualifying or guarantee run may be a formal, contractual matter (between client and contractor) or an informal recognition that the plant is commissioned and the team can be dispersed (often the case for a project carried out in house).
Handover to site operations	If the plant has achieved nameplate steadily, any essential modifications have been installed, all training is completed, and the documentation is in place, then the plant is formally handed over to the site. Part of this process may involve another plant inspection and review of paperwork.
Postcommissioning report	A postcommissioning report is a useful document to capture the learning during commissioning. Often, the rates achieved during commissioning are not required by the market for some time, and the operators learn to run the plant at, say, 80% rates. 80% rates may easily become nameplate so when some time later the market builds up and no one remembers what the plant achieved during commissioning, a lot of effort may be wasted reinventing a previously commissioned wheel. This problem also occurs with plants making stable long-term products with market demand that follows the economic cycle. It is too easy in a downturn to let operations and other personnel go, lose corporate memory, and not know what the plant is capable of producing. In a market with strong growth, the commissioning report is a valuable input to the next project. If market demand quickly fills plant capacity, the business has to decide whether to build new or expand the existing plant. This decision is informed by what was found during commissioning. Debottlenecking a plant is usually quicker and less expensive than building new; however, if the plant is very well balanced, a significant plant capacity increase may require so much change that the case for a new plant is justified.

8.3 COMMISSIONING A PURIFIED ACID PLANT

Commissioning a PWA plant is a highly nontrivial exercise; it can go very well and take a matter of weeks. But on the other hand, there are, sadly, examples where commissioning has gone very badly and the plants never reached their true potential. The difficulty is the nature of the process that many things can go wrong and some problems can mask others. Even when plant and equipment is correctly specified and installed, it might not perform as intended because it is being operated, inadvertently, outside its design parameters. This problem might present itself as a poor-quality item failing early, whereas the root cause might be a process control problem upstream. Getting to the bottom of these problems can be a difficult and lengthy exercise. Of the larger, complete plants—as opposed to projects where a major unit was modified on a running plant—most commissioning periods extended to several months with at least one serious technical challenge and have required a dedicated team and support from various corporate departments.

For the purposes of discussion, let us consider a fictitious PWA plant in the United States. The plant has a capacity of 75,000 tons/year P_2O_5, all of which is food grade. The PWA plant is located on a site with wet process phosphoric acid (WPA) and fertilizer plants and shares the site utilities (steam, water, and electricity); however, it has its own cooling tower for cooling water, a reverse osmosis (RO) unit to provide purified water, a liquid nitrogen tank and distribution system, and its own compressed air system. Chemical services for the plant include sodium sulfide storage, solvent storage, hydrogen peroxide storage, and a small day tank for ammonia. The sulfide, solvent, and peroxide are delivered from external suppliers by truck, whereas the ammonia is pumped in a pipeline from the site ammonia plant.

Similarly, the WPA is pumped to the plant day tank. The WPA has already undergone desulfation and crude defluorination so the only pretreatment required is heavy metal removal, organic reduction, and concentration from its import concentration of 42% P_2O_5 to the 59% determined for the feed to the solvent extraction section. Extraction, scrubbing, and stripping take place in three separate Kühni columns with a split of 50%, so 150,000 tons/year P_2O_5 of WPA is imported and 75,000 tons/year P_2O_5 exported for fertilizer production. The levels of arsenic and cadmium in the feed acid require removal through sulfiding; the cadmium level and overall split dictate that removal is carried out upstream of the extraction section. Organics are reduced to 50 ppm in activated carbon columns prior to concentration and solvent extraction. After extraction, both product and raffinate acid streams are solvent stripped. Raffinate acid is then exported for fertilizer manufacture. Product acid undergoes final concentration, defluorination to sales specification, and color treatment with hydrogen peroxide.

8.3.1 Precommissioning

The plant is broken down into the following nine systems and subsystems; in the majority of cases, each plant P+ID comprises a subsystem:

1. 100—Feed materials, chemical utilities, and heavy metal removal
   100.1 WPA day tank
   100.2 Solvent storage and treatment
   100.3 Sodium sulfide storage
   100.4 Heavy metal removal
   100.5 Hydrogen peroxide storage
   100.6 Ammonia
   100.7 Sodium hydroxide
   100.8 Activated carbon, filter aid and other dry additives

2. 200—Heavy metal removal
   200.1 Sulfide reaction and filtration
   200.2 Hydrogen sulfide scrubbing
   200.3 Carbon treatment
3. 300—Feed acid concentration
   300.1 Feed acid concentrator intermediate storage
   300.2 Feed acid concentrator
   300.3 Concentrated feed acid intermediate storage
4. 400—Solvent extraction
   400.1 Extraction column
   400.2 Scrubbing column
   400.3 Stripping column
   400.4 Product acid intermediate storage
   400.5 Raffinate acid intermediate storage
5. 500—Solvent removal
   500.1 Product acid solvent stripping
   500.2 Raffinate acid solvent stripping
   500.3 Stripped product acid intermediate storage
   500.4 Stripped raffinate acid intermediate storage
6. 600—Concentration
7. 700—Defluorination
8. 800—Color treatment
9. 900—Utilities
   900.1 Site process water
   900.2 Cooling water
      900.2.1 Cooling towers, pumps, and dosing
      900.2.2 Cooling water system on plant
   900.3 Instrument and process compressed air
   900.4 Steam (intermediate and low pressure)
   900.5 Nitrogen
   900.6 RO water

The plant is built in the most efficient way from a construction standpoint; however, as mechanical completion draws near, the construction emphasis moves toward completing systems in the order requested by the commissioning team. In this case, it is system 900, utilities, followed by system 100, chemical services. Usually, it is preferable to offer whole systems to the commissioning team; however, the relatively self-contained subsystems like intermediate storages (e.g., 400.4) may be offered and accepted because of their relative simplicity.

The plant is punch listed against a master set of P+IDs always kept in the control room. These P+IDs are labeled and color coded to identify the different systems. Each system is carefully checked on plant against copies of the final, signed design P+IDs, and notes taken. All equipment items, piping, fittings, insulation, tracing, low spots that might not drain, and valve accessibility are checked. It is a matter of judgment, but for some systems, a detailed checklist, pipeline by pipeline, fitting by fitting, is sometimes prepared for use by the commissioning engineers. Any difference between design and built is recorded and reported to the commissioning manager. On completion, the commissioning P+IDs are marked up in the control room, and progress is reviewed in the daily meeting.

Punch list items are reported both to the construction manager and project engineer. When there is an obvious difference, it is usually rectified promptly. The area of debate occurs when, for example, a valve is deemed inaccessible by the commissioning team, the construction manager demonstrates that it is installed to design, and the project engineer refers to the design approval meeting minutes that showed the commissioning manager signed off this piece of design and that both time and budget are pressing.

The second phase of precommissioning comprises the field instrument, loop, and electrical checks. These are led by the commissioning control/electrical engineer and technicians.

The commissioning mechanical engineer is responsible at this stage for ensuring that the plant and equipment data books, maintenance instructions, and any pressure vessel documentation are collated and available. The data books for vessels include the verification of the metal plate used, fabrication inspections, and pressure tests, all of which establish essential benchmarks for future maintenance. On the fictitious plant, a number of heat exchangers are steam heated and are protected from overpressure by safety relief valves, which are registered on the site system and checked for the correct setting. The ammonia line is registered on the site system because it is a long line carrying a hazardous material that could overpressure the line if locked in.

The final stage of precommissioning is water flushing. Not every pipe is water flushed; air, nitrogen, and steam lines are blown through by their respective fluid later. Water flushing is intended to ensure that lines are clear of debris and blockages. The water used for flushing is most conveniently clean; otherwise, it introduces rubbish itself. Vessels are inspected by removing manhole covers. While there are no utilities or chemicals on the plant vessel, entry is not subject to the same level of precaution and procedure as when they are. Visual inspection from the outside is usually sufficient to establish the absence of tramp material; however, on occasion, items are left by constructors that must be removed. Agitated vessels where the agitator blades are bolted to the shaft are always inspected for tightness and occasionally tack welded. The Kühni columns and defluorination column (PTFE and graphite lined with graphite packing) are checked over by both the plant commissioning team and service engineers from the manufacturers.

8.3.2 Commissioning Utilities

With water flushing complete and the plant boxed up with the majority of punch list items corrected, the plant is ready for handover for commissioning. At this point, the plant is deemed mechanically complete, and the commissioning manager formally accepts the plant.

Utilities are made live first; in general, the precise order will vary by circumstance. In this case, the first service to be made live is process water; this is done by opening valves at the high points of the system and then opening the main isolation valve to let the water in. The newly installed plant pipe work should be clean following inspection and flushing; however, interconnecting pipe work from the existing site to the new plant may have carried standing water for weeks or months and can often be discolored with rust. It is worthwhile therefore allowing the flow to run to drain until clear. As soon as water comes out of the high points, the system is free of air and all vents and drains are closed.

With water in the system, the cooling water system is commissioned. Cooling tower basins often attract rubbish and bits of wood that pass easily into cooling water pump inlets and jam them. A forced draft tower has one or more large fans at the top, which will quickly destroy themselves if there is anything that might impede the free movement of fan. Therefore, it is critical that the cooling tower basin is thoroughly cleaned and the cooling tower fans are inspected. For this plant, the cooling water pipe work is generally of a size that only permits water flushing rather than dismantling (being largely continuously welded and large bore); therefore, it is normal to insert temporary *top hat* filters in the pump inlets. This type of filter is conical in form with many holes large enough for the water to pass but small enough to catch things that might damage the pump. The conical form is intended to minimize the inlet pressure drop and avoid cavitation. Temporary inlet filters are frequently used during commissioning and are managed with a register. It is not unknown for a problem to arise during a later stage of commissioning to be caused by a forgotten filter. Once the system is checked as clean and the pumps boxed up, water is introduced to the cooling tower basin and then allowed into the pump inlets. A pump is switched on and water starts to flow. At the same time, a commissioning engineer operates the valve at the highest point to vent the system. Once circulation is established, pressure and other instrument checks are carried out. Then the fans are started and temperatures monitored. Depending on ambient temperature and humidity, some cooling should be detected even though at this stage there is no process load. The final stage of commissioning the cooling water system is to open up the process heat exchanger and check flows. Cooling water dosing is essential to avoid legionella growth that causes Legionnaires' disease, to inhibit microbial growth (algae, bacteria, fungi, and protozoa) that causes slimes to form that clog up heat transfer surfaces and can induce corrosion, and to control scale formation and corrosion due to pH changes. Typically, a site engages a specialist company to manage cooling water dosing across a site. These companies supply a dosing unit that comprises a number of dosing pumps and control system and dosing chemicals and will also control blowdown. As water is evaporated in the cooling system, freshwater is added; however, the concentration of additives changes, and occasionally some cooling water is released to drain in order that the levels can be restored to those desired—this is blowdown.

Compressed air and instrument air come from the same compressor system but are regulated to different pressures. Modern air compressor systems are oil-free and designed to produce high-quality air with far greater energy efficiency than even 20 years ago. Nevertheless, the air quality and pressure required for an air-operated maintenance tool and that needed for instruments are different, so the instrument air passes through a drier and filter. The commissioning procedure is to start up the compressor and open up valves on the system to blow it through. It is a good practice to request that the compressor manufacturer sends a service engineer to ensure that the compressor and its ancillaries are installed and commissioned correctly.

Steam systems, including the main pipelines that bring steam to the plant, must be commissioned with care. It is possible to *jump* a large bore (250 mm diameter and above) steam main off its supports if it is brought up to temperature too quickly; the consequences of doing so are at best inconvenient. Essentially, the commissioning procedure is to open vents and steam trap bypasses and slowly introduce steam into the steam piping. This allows some rust and debris to be blown clear, and when steam, rather than condensate and rust, starts to emerge from the steam trap bypasses, the valves are closed. Eventually, the system is buttoned up, and if there are no leaks, it is deemed commissioned. The flow from blowing long length of large bore main clear is considerable and hazardous. One approach is to bolt a blank flange on the venting branch but stepped off with screwed rod. So taking a 300 mm main, terminating in a 300 mm branch, the bolts would be replaced temporarily with screwed rod of the same diameter as the bolts but 400 mm long and with four nuts on each rod; the blanking flange is then bolted to one end of the rods; the other end of the rods is locked onto the flange on the pipe work. When the steam reaches full flow, rust and any other debris are pushed out of the main, hit the blank flange, and then fall to the ground at low velocity. Once the blow through is complete, the blank flange is replaced with its original bolts. During commissioning of the steam system, heat exchangers on the plant are isolated from the steam supply.

Nitrogen is commissioned in a similar way to compressed and instrument air. Nitrogen is a hazardous gas adding asphyxiation to the pressure hazard. Several workers in the chemical industry have lost their lives due to inhalation of nitrogen, usually in a closed space; consequently, once nitrogen is introduced to the plant, the formality of procedures to enter equipment is tightened up.

RO water is required for the final stripping stage of solvent extraction. The RO unit is a *packaged unit*. In general, most chemical plant projects buy several packaged units; examples include refrigeration units, air systems, and automatic bagging systems. They tend to be purchased to an input/output specification; in this case, it would be that the plant will supply x cubic meters per hour of process water of a given analysis and range of composition, and the RO unit will process this and supply y cubic meters of purified water to a stated specification back to the plant. Packaged units are usually fairly complex with several equipment items and their own control system. Part of the supply agreement usually includes a few days commissioning on site by the manufacturer service engineers.

8.3.3 Water Trials

Water trials are not carried out on every item or system. Here, the water trial strategy is to water trial most of the wet systems up to and including defluorination. The water used in water trials should not make a system dirty that has been cleaned. In this case, a temporary filter is rigged up to ensure that the site process water is reasonably clean to carry out water trials.

Initially, the main storage systems (WPA, solvent, sodium sulfide, sodium hydroxide) are filled with water, which is then circulated around the system pipe work. Pump and instrument performance is measured, and flowmeters are checked. When the water trials on these systems are completed satisfactorily, the water from the WPA system is pumped forward to the heavy metal removal system, simulating WPA. Water is circulated around the heavy metal removal reaction and scrubbing subsystems 200.1 and 200.2, and again instrument readings are noted for clean water, for example, the pressure drop on the hydrogen sulfide scrubbing column packing and pump delivery pressures. Water from the sodium sulfide storage is pumped into the system. The carbon treatment system is essentially a pump through to check for leaks. Most of the water ends up in the feed acid concentrator intermediate storage tank.

It is possible to trial the feed acid concentrator system extensively. Cooling water flow is established to the condensing exchangers and steam flow to the steam ejector system and main acid heater on the pump circuit. The boiling point of partially treated WPA is higher than water so the set operating temperature for water trials is lower; similarly, the specific gravity of water is less than WPA so the load on the circulating pump is lower; nevertheless, instruments, control loops, and equipment performance such as pump curves or heat transfer are noted. This is also an opportunity for the plant operators, especially those unfamiliar with this type of plant, to gain experience. Most defects shown up by water trials are minor and are corrected quickly. Occasionally, more significant problems become apparent, for example, in this fictitious case, the main circulating pump did not develop the head expected needed to pump WPA around the circuit. Further investigation showed that the impellor was reduced in size to meet an early design specification that was not altered when plant layout changes increased the required head. The manufacturer was contacted, and a new impellor shipped and fitted within days.

Having carried out water trials on the pretreatment section, there is comparatively little to be done with the solvent extraction columns other than pumping water forward from the concentration system and in from the solvent system. Nevertheless, circulating pumps, instruments, and some control loops can be exercised if not fully tested.

System 500, solvent removal, is, like the feed acid concentration system, amenable to more extensive trials with water. Steam is used to heat water, cooling water to condense.

System 600, from a water trial standpoint, is the same as system 300, the feed acid concentration system. System 700 is treated similarly.

System 800 is a relatively simple system and is not water trialed.

On completion of the water trials, all equipment holding water is emptied. Drain valves are closed and blanked, and equipment and systems declared ready for chemicals. In this case, the only system requiring more than a simple draining of water is the solvent system where the main storage would undergo drying, either by entry and literally with a mop and bucket or by blowing air through the vessel. In operation, the solvent absorbs a small amount of water so the drying process does not have to be perfect.

Prior to introduction of chemicals, final checks are made so that punch list items are completed, maintenance and other engineering documentation is in place, and the commissioning team and operations team have completed their training. The commissioning manager usually informs the site manager that the plant is ready to receive chemicals.

8.3.4 Chemical Trials

The conventional approach to commissioning is to introduce feed materials to the front of the plant and start producing, commissioning each unit sequentially. Another approach is to work backward by importing finished goods. In this case, the solvent extraction and downstream plant are partially commissioned first with imported purified acid. Once systems 400, 500, 600, 700, and 800 are steady and partially proven, systems 100 and 200 are commissioned.

To do this, imported purified acid is pumped into the concentrated feed acid intermediate storage (system 300.3), and the intermediate column dump tanks between the Kühni columns. Solvent is imported into the solvent storage system (system 100.2). Solvent is then pumped to the Kühni columns charging the solvent circuit; steam flow commences to ensure that the solvent is up to operating temperature. Acid is then pumped into the Kühni columns so that the normal quantities of both acid and solvent are present. The column agitators are started, and mixing occurs with acid extracted into the solvent in all three columns. At this stage, the system is full of solvent and acid but static.

The *feed* acid, solvent, and RO water feeds are now initiated at 80% rates. (In this scenario, the scrubbing stage of solvent extraction takes place with RO water, not ammonia, so both product and raffinate acid can be recycled.) The overall system should stabilize quickly and start producing *purified* and *raffinate* acid into the respective storages. As these fill the solvent, removal system is made ready and then started up.

The stripped underflow acid would normally be exported at this stage; however, it is essentially the same as the stripped product acid and is pumped forward through temporary connections into the purified stream heading for concentration.

The concentrator brings the solvent stripped acids up to the design concentration, and the acid is pumped forward for *defluorination*. Having started with PWA, it is unlikely that any further defluorination will take place. The defluorination system produces a cooled acid that is then transferred to road barrels and taken back to the concentrated feed acid intermediate storage.

The plant is now ready to move to production at low rates. Control set points are input at 80% rates. The hydrogen sulfide scrubbing column and ancillaries are set up, and filter aid is made up and pumped around the filter in system 200. The carbon treatment columns, already charged with activated carbon, have their inlet, outlet, and bypass valves set appropriately. Temporary connections in downstream systems are removed and the plant restored to its normal arrangement.

WPA is then pumped to the sulfiding reactor and is allowed to react with the sodium sulfide solution. Acid purified of heavy metals is pumped forward and through carbon treatment to the feed acid concentrator intermediate storage. The concentrator is then brought on stream at 80% rates. As this is happening, the solvent extraction system is brought back on line but now with ammonia scrubbing and creating a real product and raffinate stream.

As concentrated, pretreated WPA starts to replace the PWA used for commissioning so the plant starts to work. Acid flows downstream to concentration, defluorination, and finally color treatment, where it is contacted with hydrogen peroxide.

Samples are taken frequently, at critical locations, initially perhaps the most import being the feed acid to solvent extraction. If this acid does not meet the design specification, the solvent extraction plant will not stop working but will fairly quickly deteriorate. The prompt analysis of samples is critical to progress at this stage, and a well-organized commissioning program will have this aspect covered both at plant and site level as some analysis requires equipment that would not normally be operated in the plant laboratory.

If all is satisfactory, the rates are raised to 85%. Again, if everything is satisfactory, rates continue to be raised until nameplate rates are achieved or stable rates above or below nameplate. The timing of rate increases is a matter of judgment. Clearly, it is best done only when the plant is stable because a rate change when the plant is unstable may aggravate the instability. During a rate increase, particularly when close to a bottleneck, the whole plant should be monitored closely. Therefore, when rates are increased for the first time, it is usually done during the day shift when the full team is on the plant. If the rate increases are well managed, the practical plant response time will become apparent. At this point, the process control engineers come center stage to start tuning the plant control system.

Inevitably, problems and breakdowns arise at this early stage of operation, and usually they are analyzed and solved quickly. Problems arising during this stage can be categorized into equipment- or process-related issues and are potentially myriad. Some problems build on others; consequently, it can take time to get to the root cause and fix it, and during commissioning, there is, or should be, a sense of urgency, hence one of the challenges of working in this environment. When addressing commissioning problems, it helps to rule out the obvious first; for example:

- Was the equipment being run as intended because maloperation can lead to a breakdown?
- Was the equipment running at or near design conditions (temperature, pressure, flow)?
- Did the equipment achieve manufacturer test conditions during water or chemical trials?
- Was the equipment attempting to process out of specification material (e.g., did the WPA to section 200 contain much higher levels than design of both solids and heavy metals that overwhelmed the filter)?
- Was the process asked to achieve more than designed (e.g., were the impurity levels in the feed acid far higher than specification such that there were insufficient scrubbing stages in the second column to achieve design purification)?

Occasionally, as cited in an earlier example, something is missed during the design and procurement stage; even with the best project procedures, there is still human error. More rare is a fault within an equipment item because most equipment is not truly bespoke and has undergone design and testing at the manufacturers prior to testing on the plant. Nevertheless, if all other possible causes have been ruled out, it is worthwhile checking the fundamental design of the relevant item.

Perhaps the two most common process problems are subtle changes to the feed acid processing upstream of the plant that have an unforeseen but material effect on the downstream purification and trying to operate plant systems away from design because the preceding system is not producing product to specification. Two obvious examples of the former are a slight change in phosphate rock quality leading to a slightly different impurity profile or changes to the foam dispersant in the WPA process carrying through to solvent extraction and disrupting that process step. The latter might occur when the processing problems on one unit have not yet been solved and appear intractable, yet at the same time commissioning must press on. The judgment call is made that downstream systems will be able to handle running outside design and still produce even at below full rates. This situation may arise when a problem needs study back in the laboratory and some time to get to the root of the problem. Sometimes the call is good; sometimes it compounds the problem.

Commissioning is deemed to be complete when the plant is stable and has achieved its design rates producing products within specification. Depending on ownership and who is responsible for design and for commissioning, a test run completes commissioning. A test run is not normally attempted unless there is confidence among the commissioning team that it will pass. The criteria are usually 72 h continuous running, at design rates and agreed product quality.

8.4 COMMISSIONING A PHOSPHATE SALT PLANT

Nearly all phosphate salts plants have a similar broad configuration. The plants are split wet side and dry side. Wet side comprises a PWA day tank and feed system; calcium, sodium, potassium, or other metal compounds in either powder or liquid form; a wet phase (usually) reactor; and sometimes a filter. Dry side comprises a drier (spray, spin flash, and rotary are most common); sometimes a calciner (essential for polyphosphates and most commonly but not essentially a rotary device); a mill and sizing equipment (screens of different types); and packing equipment.

The commissioning of the wet side is straightforward. The procedures and practices are very similar to those followed in the commissioning of the PWA plant. The two main challenges with the chemical commissioning of the wet side are gaining precise and tight control over the M/P ratio (where M = Ca, Na, K, etc.) and dealing with almost no intermediate storage. It is normal for the neutralization step to have sufficient capacity to ensure a complete reaction and a stable feed to drying but little more. The drier is often connected directly to the calciner. It is only after the unsized powder product is made by the calciner that there is any intermediate storage. The ease of milling, sizing, and transport of phosphate powders varies from relatively straightforward to mildly difficult depending on the phosphate and the particle size. Packing, especially automatic packing, is fiddly to set up often requiring many small adjustments to get it just right; once set up, a high-quality, well-designed packing machine runs very well. Therefore, the chemical commissioning of a phosphate salts plant should not commence before the route through to the final sales package is clear and ready. Consequently, the first system to be commissioned should be the packing equipment. If it is properly incorporated into the plant design, it is possible to commission the plant from the unmilled storage through to final packing using a relatively small amount of imported powder (2–3 tons). With this amount of powder, commissioning is not comprehensive but sufficient to prove rotary valves, conveying equipment, the mill, and the packing system. Having proved the last section of the plant commissioning can start at the front and keep going. Given that the plants are basically continuous, the sooner they are established in continuous operation, the better. Alternatively, the plant starts up from the wet side and eventually fills the unmilled storage. If in the wet side, drying and calcining are stable, then the commissioning team can focus on the milling, sizing, and packing. If not, then attempting to commission these systems compounds the commissioning workload. Generally, the two largest pieces of equipment are the drier and the calciner. The drier is always bought as a package comprising air heater (usually gas fired), drier, cyclones, and filter or scrubbing unit. Most of the package is straightforward to commission and is usually carried out with the drier service engineers. A rotary calciner requires careful installation and initial setup, which must be carried out by the supplier's service engineers. The aim is to ensure that the calciner is running straight and true on its roller bearings and continues to do so as it warms up. Both driers and calciners are relatively trouble-free, once set up, from an equipment standpoint; the commissioning exercise is usually about optimizing the operating conditions for the desired product.

The following outlines the commissioning of a typical STTP plant.

8.4.1 Precommissioning

As in the PWA case, the plant is broken down into the following systems and subsystems; in the majority of cases, each plant P+ID comprises a subsystem:

1. 100—Feed materials and chemical utilities
   100.1 PWA day tank
   100.2 Sodium hydroxide storage
   100.3 Minor wet and dry additives
2. 200—Neutralization
   200.1 Neutralization
   200.2 Recycle tank
3. 300—Spray drying
   300.1 Spray drier feed tank
   300.2 Spray drier, air heater, and preheater
   300.3 Spray drier and cyclone bank
   300.4 Spray drier wet scrubber
4. 400—Calcining
   400.1 Calciner mixer/feeder
   400.2 Calciner air heater
   400.3 Calciner
   400.4 Calciner cyclone
   400.5 Powder cooler

5. 500—Milling and sizing
   500.1 Unmilled product storage
   500.2 Mill
   500.3 Powder screening
   500.4 Milled product storages
6. 600—Packing
7. 700—Utilities
   900.1 Site process water
   900.2 Hot water
   900.2 Cooling water
      900.2.1 Cooling towers, pumps, and dosing
      900.2.2 Cooling water system on plant
   900.3 Instrument and process compressed air
   900.4 Steam (intermediate and low pressure)
   900.5 Natural gas

Precommissioning checks are identical to those carried out for the PWA plant earlier, similarly the commissioning and water trials of the utilities and wet side (systems 100, 200, and parts of 300).

Dry side precommissioning is very similar and includes checks by manufacturer's service engineers.

8.4.2 Commissioning the Spray Drying System

The spray drying system is commissioned, and in normal operation started up and shutdown with water. In order to heat the air to evaporate the water, the natural gas system is commissioned, and the hot air heater together with its burner management system brought on line. Water is pumped to the atomizer and evaporated. Although the effectiveness of the cyclone bank is not tested, the important control loops are checked out.

8.4.3 Chemical Commissioning and Production

As described in the introduction in this example, a few tons of STPP was acquired and introduced into various locations from the unmilled storage to the packaging system to prove the equipment and increase the likelihood of smooth running.

When the proving material is clear of the plant, wet feeds are introduced at 50% rates to the neutralization section. At these rates, there is more time to ensure the correct Na/P ratio and deal with any arising problems. If all is well, the spray drier is started up on water and when up to temperature switched to ortholiquor. Solid orthobead passes out of the spray drier and into the mixer/feeder and into the calciner. After the design residence time, the unmilled STPP flows out of the calciner and into the cooler. As more STPP is produced, so too is dust that is knocked out by both the spray drier and calciner cyclones. Both of these flows feed into the mixer/feeder, and a small ortholiquor flow is added to ensure blending. The cooled, unmilled STPP is conveyed up a bucket elevator and onto a coarse screen. Initially, the unmilled product is fine powder, and most passes through the screen and onto final packing.

As time progresses and both calciner recycle and sprays are brought up to speed, the particle sizes increase and a portion is directed to milling.

Rates are raised up to nameplate so long as production remains stable. Unlike commissioning a PWA plant, there is much less opportunity to push individual systems or equipment items to establish where the bottlenecks lie because there is little intermediate storage in the plant. It is also the case that most phosphate salts plants produce several different grades—a sodium phosphate plant producing the most—therefore, once stable operation is achieved, the focus is on demonstrating production of the different grades.

Once the mechanical, in its broadest sense, problems are resolved—and they can be very few—the main challenge in commissioning this and other phosphate salts plants is finding the exact operating range for an individual product and dealing with unexpected process problems.

The commonest of the latter is the unwanted color in the product and is often due to impurity in the feedstock. Finding the precise operating range for a particular product may take some time; occasionally, plant modifications are deemed necessary. For example, when making STPP, the proportion of phase 1 and phase 2 materials is important depending on the grade. The relative quantities are linked to calciner temperature and residence time. Changing residence time in the calciner can be as simple as slowing the feed rate or as complex as designing and installing a new dam ring in the calciner.

8.5 COMMISSIONING TEAM

A plant does not commission itself nor can a fully committed operating team be expected to commission a major modification in their spare time. Therefore, it is essential that a commissioning team is put together even for the smallest and most straightforward projects. For these, the team might be one person, usually the process engineer that conceived the design.

For a major project, the commissioning team would comprise the following:

- Commissioning manager—this person must have experience in commissioning, process design, and production and be capable of managing the commissioning team and relations with the construction team, the operations team on the plant, the site management, the project manager/director, and the company management. On a major project, much time is (or should be!) spent in the early stages of the project influencing the plant design and building good working relationships with the project manager, the senior process engineer, and the project engineer. In the later stages of construction, the commissioning manager will ensure that the right commissioning procedures are in place and that the team is coming together. During commissioning itself, much time is spent either defending the team from external interference and giving them space to commission or challenging the team and making the difficult calls. The commissioning manager is also the individual held accountable by external authorities (environmental, safety).
- Commissioning process engineer—this individual leads the hands on commissioning effort, directs the commissioning engineers, and solves or coordinates the solution of all technical problems. This engineer is always experienced and for the larger projects is someone who has commissioning,

design, and production experience. Quite regularly, this individual was the lead process engineer for the design phase of the project.

- Commissioning engineers (process)—there are at least four of these individuals in order that 24 h working is covered. The skills, qualifications, and experiences of these individuals may vary considerably from the recent graduate to the old hand.
- Commissioning chemist—this individual sets up the plant laboratory and trains the plant operators in plant laboratory–based analytical practice. This person also establishes any new procedures needed in the site laboratory and throughout is required to expedite sample analysis and participate in problem solving.
- Commissioning engineer (mechanical)—is responsible for the equipment and ensures that it is ready to process chemicals. This engineer usually liaises with supplier's service engineers, ensures that documentation is in place, and runs the commissioning modification system. The mechanical commissioning engineer also ensures that the maintenance team are familiarized with new equipment and trained to maintain it. This engineer is occasionally assisted by a technician.
- Commissioning engineer (control/electrical)—is responsible for all aspects control/electrical and is often very busy tweaking control software. The control of any modification to the plant is safety critical and particularly so for software modifications because they are so easy to carry out but can have significant consequences. Good practice protects the software from major modification by anyone other than this engineer. This engineer is responsible for setting up and managing a software modification system and is often assisted by at least two technicians. In the later stages of commissioning, the role often changes to control optimization.

8.6 CONCLUSION

All stages of a project life cycle are important and interdependent; however, commissioning is easily neglected. It is often difficult to pull together a commissioning team because no company can afford to have employees hanging around waiting for a project to come along. Nevertheless, it remains a critical step, and a good commissioning team has a material effect on the company's bottom line. In extreme cases, a good commissioning team is the difference between making an inherently difficult process work well or hardly at all.

For the individual, the experience of working on a commissioning team is invaluable. It is an excellent environment for the young engineer to develop facing engineering problems at a far higher rate than seen in normal operations, if at all.

REFERENCES

1. D. M. C. Horsley, *Process Plant Commissioning: A User Guide*, IChemE, Rugby, U.K., 1998.
2. M. Killcross, *Chemical and Process Plant Commissioning Handbook: A Practical Guide to Plant System and Equipment Installation and Commissioning*, Elsevier, Oxford, U.K., 2011.

Index

D

K

L

M

N

P

Q

R

S

T

U

W

Z

Printed in the United States
by Baker & Taylor Publisher Services